Guides to
Clinical Aspiration Biopsy

# Diagnostic Immunocytochemistry and Electron Microscopy

# Guides to Clinical Aspiration Biopsy

*Series Editor:* Tilde S. Kline, M.D.

**Prostate**
Tilde S. Kline, M.D.

**Thyroid**
Sudha R. Kini, M.D.

**Retroperitoneum and Intestine**
Kenneth C. Suen, M.B., B.S., F.R.C.P.(C)

**Lung, Pleura and Mediastinum**
Liang-Che Tao, M.D., F.C.A.P., F.R.C.P.(C)

**Head and Neck**
Ali H. Qizilbash, M.B., B.S., F.R.C.P.(C)
J. Edward M. Young, M.D., F.R.C.S.(C)

**Liver and Pancreas**
Denise Frias-Hidvegi, M.D., F.I.A.C.

**Breast**
Tilde S. Kline, M.D.
Irwin K. Kline, M.D.

**Flow Cytometry**
Philippe Vielh, M.D.

**Infectious and Inflammatory Diseases and
Other Nonneoplastic Disorders**
Jan F. Silverman, M.D.

**Diagnostic Immunocytochemistry and Electron Microscopy**
Hossein M. Yazdi, M.D., F.R.C.P.C.
Irving Dardick, M.D., M.Sc., F.R.C.P.C.

Guides to Clinical Aspiration Biopsy

# Diagnostic Immunocytochemistry and Electron Microscopy

## Hossein M. Yazdi, M.D., F.R.C.P.C.
Head, Cytopathology Service
Ottawa Civic Hospital
And
Associate Professor of Pathology
University of Ottawa
Ottawa, Ontario, Canada

## Irving Dardick, M.D., M.Sc., F.R.C.P.C.
Head, Diagnostic Electron Microscopy
The Toronto Hospital
And
Professor of Pathology and Otolaryngology
University of Toronto
Toronto, Ontario, Canada

IGAKU-SHOIN  New York • Tokyo

Published and distributed by

IGAKU-SHOIN Medical Publishers, Inc.
One Madison Avenue, New York, N.Y. 10010

IGAKU-SHOIN Ltd.,
5-24-3 Hongo, Bunkyo-ku, Tokyo

Copyright © 1992 by IGAKU-SHOIN Medical Publishers, Inc.
All rights reserved. No part of this book may be translated or reproduced in any form by print, photo-print, microfilm or any other means without written permission from the publisher.

Library of Congress Cataloging-in-Publication Data

Yazdi, Hossein M.

  Diagnostic immunocytochemistry and electron microscopy / Hossein M. Yazdi, Irving Dardick.
    p.  cm.—(Guides to clinical aspiration biopsy)
Includes bibliographical references and index.
    1. Immunocytochemistry.  2. Diagnosis—Electron microscopic.
    3. Cancer—Cytodiagnosis.  4. Biopsy, Needle.  I. Dardick, Irving.
  II. Title.  III. Series.
    [DNLM: 1. Diagnosis, Laboratory—methods.
  2. Immunohistochemistry.  3. Microscopy, Electron.  QY 95 Y35d
RC270.3.I44Y39  1991
616.07'56—dc20
DNLM/DLC
for Library of Congress                                             91-20908
                                                                                  CIP

ISBN: 0-89640-214-2 (New York)
ISBN: 4-260-14214-3 (Tokyo)

Printed and bound in the U.S.A.

10 9 8 7 6 5 4 3 2 1

*Dedicated to my parents, my wife, Shayesteh, and my son, Bahram* with love.

*Hossein M. Yazdi*

*Dedicated to the two people who mean the most to me, my son, Alexander, and my wife, Grace.*

*Irving Dardick*

# Preface

The diagnostic utility of aspiration biopsy cytology has been recognized for many years. Major application to patient management in North America, however, is more recent. It is now one of the main investigative procedures for human neoplasms. Because needle aspiration biopsy is largely supplanting diagnostic open surgical biopsy, doesn't it seem reasonable to apply the full range of diagnostic tests currently available in pathology, particularly immunocytochemistry and electron microscopy, to these specimens? One of the purposes of this book is to convince pathologists that this is the case.

As treatment regimens and therapeutic methods become more precise, clinicians will demand more specific diagnoses from pathologists. In this circumstance, the limitations in the review of routinely prepared and stained smears and cell block sections becomes only too apparent. With regard to more difficult diagnostic problems, the combination of immunocytochemistry and electron microscopy provides definitive diagnoses not attainable by routine preparations. It is essential to apply such investigative tools in a variety of diagnostic situations to eventually assess how best to use them. With such experience, immunocytochemistry and electron microscopy may not be considered adjunctive procedures in the future.

This book emphasizes a practical approach to the use of ancillary techniques in cytopathology. An introductory chapter outlines the suitability, indications, and practical aspects of applying ancillary techniques to aspiration biopsy cytology specimens. Subsequent chapters deal with techniques for specimen acquisition and processing, and the principles, value, and limitations of both immunocytochemistry and electron microscopy as applied to aspiration biopsy cytology specimens, as well as the application of such ancillary studies to the diagnosis and classification of certain tumors. We trust this text will allow pathologists, pathologists-in-training, and cytotechnologists to expand their perspectives of diagnostic cytopathology and will encourage further exploration of techniques such as immunocytochemistry and electron microscopy to improve their diagnostic skills.

*Hossein M. Yazdi, M.D.*
*Irving Dardick, M.D.*

# Acknowledgments

In undertaking to author any book, medical colleagues, both in one's own institution and outside it, make important contributions in the form of technical and clinical information. We acknowledge all assistance of this type. We particularly appreciate the support, encouragement, and assistance of the pathologists in our respective hospital departments.

We would have been unable to provide diagnostic ancillary studies in cytopathology without the cooperation of the radiologists in our hospitals. Of these clinicians, special thanks are extended to Dr. Nancy M. Hickey, then working in the Department of Radiology at the Ottawa Civic Hospital. The excellent specimens provided by Drs. Hardy H. Tao, Bernard J. Lewandowski, Andrea Lum, Fred Matzinger, and Rebecca A. Peterson of the Department of Radiology, Ottawa Civic Hospital, and by Drs. Gordon L. Weisbrod and Stephen J. Herman of the Department of Radiology, Toronto General Division of The Toronto Hospital, were also appreciated.

We were fortunate to have competent technologists, in particular Carol Brosko, William Parks, Janet Stinson, and Peter Rippstein. Without their enthusiasm and effort we would not have been able to develop these techniques at the Ottawa Civic Hospital.

Others made major contributions through their superb technical skills. At the Ottawa Civic Hospital, this group includes the cytotechnologists, Angella Gilliat, Danielle Leduc, Laurie MacDonald, Tami MacDonald, Brian Merrill, Nichole O'Hara, and Julie Pohoresky; the electron microscopy technologists, Marie Boivin, Sharon Cavell, and Dianne Hoppe; and the histotechnologists. At the Toronto General Division of The Toronto Hospital, Richard Leung, Sayoko Yamada, Sheer Ramjohn, Rosaria Glinieki, and Louis Jiminez provided technical support in the Electron Microscopy Laboratory.

Valuable assistance in the provision of literature searches, references, and bibliographies was provided by Sonja Duda of the Department of Pathology Library, Banting Institute, Colleen Mulloy of the William Boyd Library, Academy of Medicine, Toronto, and the pathology assistants at the Ottawa Civic Hospital. We gratefully acknowledge the assistance of the Audiovisual Department at the Ottawa Civic

Hospital. Especially valuable was the help of Gene Jewkowicz, Gordon Wright, and James Harbison for their skill in preparing the prints and figures.

No book can develop and reach its final form without the skills and perseverance of medical secretaries. The efforts of Diane Brazeau on our behalf are gratefully acknowledged. Donna Burke, Sheila Beck and Christine Allen provided additional secretarial support.

We must also acknowledge the direction, suggestions, and encouragement of the series editor, Dr. Tilde S. Kline, and the assistance provided by the Vice President in Editorial of Igaku-Shoin, Lila G. Maron.

Last, but not least, we gratefully acknowledge the strong support and tolerance of our families, Shayesteh, Bahram, Grace, and Alexander, throughout this endeavor, which included weekends and innumerable evenings.

*Hossein M. Yazdi, M.D., F.R.C.P.C.*
*Irving Dardick, M.D., M.Sc., F.R.C.P.C.*

# Contributors

**William K. Evans, M.D., F.R.C.P.C.**
Director, Ottawa Regional Cancer Centre
Ottawa Civic Hospital
And
Professor of Medicine
University of Ottawa
Ottawa, Ontario, Canada

**Farid M. Shamji, M.B., F.R.C.S.C., F.A.C.S.**
Thoracic Surgeon
Ottawa Civic Hospital
And
Associate Professor of Surgery
University of Ottawa
Ottawa, Ontario, Canada

**Hardy H. Tao, M.D., F.R.C.P.C.**
Chief, Division of Diagnostic Radiology
Ottawa Civic Hospital
And
Clinical Associate Professor of Radiology
University of Ottawa
Ottawa, Ontario, Canada

# Contents

Preface    vii

1. **Introduction**    1
   - The Minibiopsy Nature of ABC    2
   - The Role of Ancillary Techniques    3
   - Indications for Immunocytochemistry and Electron Microscopy    4
   - Practical Application    5

2. **Techniques for Specimen Processing**    11
   - Fixation    11
   - General Specimen Processing    13
   - Processing for Light Microscopy    14
     - Processing of Smears    14
     - Processing of Needle Rinse    16
     - Processing of Cell Block    16
   - Processing for Immunocytochemistry    17
   - Processing for Electron Microscopy    18
     - General Procedures    18
     - Technique for Handling ABC Specimens    21
     - Glycogen Preservation    25

3. **Principles of Immunocytochemistry**    29
   - Antibodies    29
   - Immunocytochemical Methods    31
     - Detection Systems    31
     - Blocking Methods    33
     - Chromogens    34
     - Controls    34
     - Automation    36
     - Quality Assurance    37

| | |
|---|---|
| Choices of Cytological Preparations for Immunocytochemistry | 38 |
| Cell Blocks | 38 |
| Smears | 38 |
| Preparing Multiple Slides from a Single Smear | 39 |

## 4. Interpretation and Limitations of Immunocytochemistry  43

| | |
|---|---|
| Patterns of Staining | 45 |
| Nonneoplastic Cells versus Malignant Cells | 45 |
| Background Staining | 46 |
| Artifacts | 46 |
| Control | 51 |

## 5. Commonly Used Antibodies  53

| | |
|---|---|
| Actin | 54 |
| Alpha-Fetoprotein | 54 |
| Alpha-1-Antichymotrypsin and Alpha-1-Antitrypsin | 55 |
| B72.3 | 55 |
| Calcitonin | 57 |
| Carcinoembryonic Antigen | 58 |
| Chromogranin | 59 |
| Cytokeratin (CK) | 59 |
| Desmin | 59 |
| Epithelial Membrane Antigen | 60 |
| Estrogen Receptor (ER) | 60 |
| Glial Fibrillary Acidic Protein (GFAP) | 60 |
| Human Chorionic Gonadotropin | 60 |
| Intermediate Filaments | 61 |
|     Cytokeratins | 62 |
|     Vimentin | 65 |
|     Desmin | 65 |
|     Neurofilament Proteins | 65 |
|     Glial Fibrillary Acidic Protein | 67 |
| Leukocyte Common Antigen (LCA) | 67 |
| LEU-M1 | 67 |
| LEU-7 | 68 |
| Lymphoid Markers | 68 |
| Melanoma-Specific Antibody | 68 |
| Neurofilament Proteins (NFP) | 69 |
| Neuron-Specific Enolase | 69 |
| Placental Alkaline Phosphates | 69 |
| Progesterone Receptor (PR) | 70 |
| Prostatic Acid Phosphatase | 70 |
| Prostate-Specific Antigen | 70 |
| S-100 Protein | 71 |
| Synaptophysin | 74 |
| Thyroglobulin | 74 |
| Vimentin | 76 |

## 6. Principles of Electron Microscopy    89

| | |
|---|---:|
| Diagnostic Architectural Features | 90 |
| Diagnostic Cellular Features | 93 |
| Junctions | 93 |
| Microvilli | 101 |
|     Secretory Products | 110 |
|     Basal Lamina | 127 |
|     Cytoplasmic Filaments | 128 |
|     Intracytoplasmic Lumens | 133 |
| Benign Lesions | 135 |

## 7. Value and Limitations of Electron Microscopy    143

| | |
|---|---:|
| Quantifying the Utility | 143 |
| Specific Limitations | 146 |

## 8. Guidelines for Classification of Neoplasms    153

| | |
|---|---:|
| Undifferentiated or Poorly Differentiated (Look-alike) Neoplasms | 153 |
|   Large/Polygonal Cell Neoplasms | 155 |
|   Pleomorphic-Cell Neoplasms | 156 |
|   Spindle-Cell Neoplasms | 157 |
|   Small-Cell Neoplasms | 159 |
| Malignant Melanoma | 159 |
| Neuroendocrine Neoplasms | 164 |
| Small Cell Neoplasms | 173 |
|   Neuroblastoma | 176 |
|   Malignant Lymphoma | 177 |
|   Rhabdomyosarcoma | 177 |
|   Erving's Sarcoma | 177 |
|   Primitive Neuroectodermal Tumor | 178 |
| Mesenchymal Neoplasms | 178 |
|   Myogenic Sarcoma | 182 |
|   Neurogenic Neoplasms | 188 |
|   Malignant Fibrous Histiocytoma | 192 |
|   Liposarcoma | 192 |
|   Synovial Sarcoma | 193 |
|   Angiosarcoma | 195 |
|   Osteogenic Neoplasms | 195 |
|   Cartilaginous Neoplasms | 198 |
|   Chordoma | 199 |
|   Ewing's Sarcoma | 199 |
|   Other Sarcomas | 200 |
| Germ-Cell Neoplasms | 200 |

## 9. Malignant Lymphoma    217

| | |
|---|---:|
| Lymphoid Markers | 218 |
|   Leukocyte Common Antigen | 218 |
|   Immunoglobulins | 219 |
|   B- and T-cell Markers | 219 |
|   Ki-1 Antigen | 219 |
|   Leu-M1 | 219 |
|   Leu-7 | 221 |
| Immunocytochemistry | 221 |
|   Processing | 221 |
|   Malignant Lymphoma Versus Nonlymphoma Neoplasms | 223 |
|   Immunophenotyping | 223 |
| Electron Microscopy | 230 |

## 10. Determination of the Primary Site  235

| | |
|---|---|
| Squamous-Cell Carcinoma versus Adenocarcinoma | 236 |
| Prostatic Adenocarcinoma | 240 |
| Gastrointestinal Adenocarcinoma | 243 |
| Hepatocellular Carcinoma versus Adenocarcinoma | 247 |
| Mesothelioma versus Adenocarcinoma | 257 |
| Thyroid Neoplasm | 266 |
| Breast Carcinoma | 269 |
| Renal-Cell Carcinoma versus Adrenal Cortical Neoplasm | 274 |

## 11. The Clinicians Viewpoint  291
William K. Evans, Hardy H. Tao, and Farid M. Shamji

| | |
|---|---|
| An Oncologist's Perspective | 291 |
| Fine-Needle Aspiration Biopsy Techniques for Superficial Lesions | 292 |
| Potential Complications of NAB | 293 |
| Clinical Uses of NAB | 294 |
|     Malignant Lymphoma | 296 |
|     Breast Cancer | 296 |
|     Lung and Mediastinal Lesions | 297 |
|     Liver Cancer | 298 |
|     Thyroid Neoplasms | 299 |
|     Prostate Cancer | 300 |
| Summary | 300 |
| A Radiologist's Perspective | 301 |
| Thoracic Surgeons and Fine-Needle Aspiration Biopsy | 305 |

## 12. Developing and Potential Applications  311

## Index  317

# Key to Abbreviations

| | |
|---|---|
| ABC | Aspiration Biopsy Cytology |
| *ABC* | Avidin-Biotin-Peroxidase Complex |
| AFP | Alpha-Fetoprotein |
| AL | Alpha-Lactalbumin |
| A1-ACT | Alpha-1-Antichymotrypsin |
| A1-AT | Alpha-1-Antitrypsin |
| CEA | Carcinoembryonic Antigen |
| CG | Chromogranin |
| CK | Cytokeratins |
| EMA | Epithelial Membrane Antigen |
| ER | Estrogen Receptor |
| GFAP | Glial Fibrillary Acidic Protein |
| hCG | Human Chorionic Gonadotropin |
| IF | Intermediate Filaments |
| LAB | Peroxidase-Labeled Avidin-Biotin |
| LCA | Leukocyte Common Antigen |
| NAB | Needle Aspiration Biopsy |
| NFP | Neurofilament Proteins |
| NSE | Neuron-Specific Enolase |
| PAP | Peroxidase-Antiperoxidase |
| PLAP | Placental Alkaline Phosphatase |
| PR | Progesterone Receptor |
| PrAP | Prostatic Acid Phosphatase |
| PSA | Prostate-Specific Antigen |
| S-100 | S-100 Protein |
| SYN | Synaptophysin |
| TG | Thyroglobulin |

# Key to Illustrations

All smear preparations are Papanicolaou-stained, except when otherwise specified. All cell block sections are hematoxylin, phloxin, and saffron–stained, which is equivalent to hematoxylin and eosin staining. All electron microscopic sections are uranyl acetate and lead citrate–stained.

# 1

# Introduction

The role of the cytopathology laboratory in diagnosis by aspiration biopsy cytology (ABC) has dramatically expanded throughout the world over the last several years. Many factors have contributed to this growth.[1-3] One only has to scan the titles of the previously published *Guides to Clinical Aspiration Biopsy* series and the table of contents of current journals dealing with cytopathology to recognize the impact ABC has had on the practice of pathology. Almost every organ or body site is accessible by ABC. In fact, an inexpensive, technically unsophisticated, nearly painless diagnostic procedure with limited complications or risks has replaced the need for diagnostic surgical biopsy in many cases. Such attributes account for the wide acceptance of ABC by clinicians, pathologists, patients, hospital administrators, and health insurers.

The increasing use of this diagnostic procedure by clinicians and cytopathologists and the considerable impact it has had on patient care governs our philosophy in maximizing the diagnostic potential of ABC specimens. As in surgical pathology, despite an adequate sample, some ABC cases will prove difficult in terms of a definite diagnosis. In such circumstances, every effort should be made to apply the ancillary investigative techniques routinely used in surgical pathology, namely, immunocytochemistry and electron microscopy, in an attempt to solve diagnostic puzzles. To take full advantage of the tissue fragments obtained by needle aspiration biopsy (NAB) for ancillary studies, protocols need to be established for handling and processing of this material (see Chapter 2) and to designate the most diagnostically useful sites for biopsy and clinical indications. For example, lesions of organs such as breast, thyroid, salivary gland, prostate, and testis will infrequently benefit from ancillary studies. However, lesions of lung, pleura, liver, kidney, adrenal, and pancreas, as well as sites such as mediastinum, retroperitoneum, pelvis, abdomen, and the soft tissues, more often pose the type of diagnostic problems that are helped by additional investigative procedures.

On the basis of our experience and the results of a prospective comparative assessment of light microscopic and ultrastructural diagnoses,[7] we have been impressed by the diagnostic contribution and general value that ultrastructural studies have given to cytopathologists. Immunocytochemistry has been of equal or greater assistance. As currently used, each has its own advantages and disadvantages and they are complementary techniques. This is a point that we stress throughout this book. A further aim of this book is to demonstrate the key immunocytochemical (see Chapters 3, 4, and 5) and ultrastructural (see Chapter 6) features that are essential for interpretation and establishment of diagnoses.

## THE MINIBIOPSY NATURE OF ABC

Specimens obtained by NAB can be appropriately considered as minibiopsies. NAB frequently yields tissue fragments in addition to isolated and small groups of cells. Evaluation of the architectural patterns and cell-to-cell associations within the fragments, in addition to cytological evaluation of the individual cells significantly contribute to the diagnosis and classification of neoplasms. In our experience as well as that of others,[9,14,23] preparation of cell blocks further improves cytological diagnosis by ABC. In some aspirates, tissue fragments are visible to the eye in the needle rinse and can be embedded in paraffin in the same way as larger biopsy specimens. In many cases, however, the needle rinse should be enriched first by centrifugation and the cell button then submitted for embedding (see "Processing for Light Microscopy" section in Chapter 2). Cell block preparation has several advantages, including (1) architectural patterns can be more readily recognized as compared with smears; (2) multiple sections may be cut for light microscopy, cytochemical, and immunocytochemical studies; (3) immunocytochemical techniques used in surgical pathology can be effectively applied to cell block sections; and (4) because of the similarity of cell block sections to tissue sections, the cytological specimens may be more easily interpreted by those histopathologists with a limited training in cytopathology.[9] Some histological patterns, such as follicular and diffuse patterns of malignant lymphomas, cannot be evaluated in ABC specimens even in cell block sections.

Cinti and colleagues,[4] in reporting on the application of electron microscopy to ABC specimens in 1983, noted that these specimens "can be regarded as actual microbiopsies." This has been our experience in many cases.[7] In fact, several researchers have commented on the ready identification of small clumps or fragments of tissue in material flushed directly from the syringe and needle into glutaraldehyde for electron microscopy,[4,7,12,15,22,28,30] and the infrequent need to prepare pellets by centrifugation, a commonly used technique for this purpose.[2,5,25,28] There is a distinct advantage when an aspiration biopsy is considered a minibiopsy. In illustrating the problem of diagnosing some spindle-cell lesions in routine cytopathology preparation, Navas-Palacios and associates[21] observed that the inability to readily recognize architectural patterns was the limiting factor in determining some diagnoses. This inability can be resolved by light microscopy of sections prepared from cell and epon blocks, and also by electron microscopy.

# THE ROLE OF ANCILLARY TECHNIQUES

One distinct advantage of ABC is the sensitivity and specificity of this diagnostic procedure to distinguish between nonneoplastic/benign and neoplastic/malignant cells. Although the accurate cytological diagnosis of the cancer by an ABC specimen is relatively simple in experienced hands, classification can be more difficult. Traditionally, cytopathological diagnoses using ABC specimens are based on the application and interpretation of well-established morphological criteria. This process, however, remains subjective. Precise classification is easier when the tumor is well differentiated or when prior surgery has been performed, so that it is possible to compare the neoplastic cells in the smear or cell block sections with the previous tumor. Classification of most poorly differentiated, undifferentiated, and some moderately differentiated neoplasms is usually difficult and the diagnosis may not be very specific. This limitation, as exemplified by such general phrases as "undifferentiated neoplasm," "small-cell neoplasm," "spindle-cell neoplasm," "large-cell neoplasm," and "consistent with . . . ," governs the reporting of these ABC specimens by cytopathologists. This limitation does not diminish the value of ABC, however; in general, these types of diagnostic categories may be sufficient for the clinical assessment and management of patients. In a number of these situations, however, it may be necessary to proceed to more invasive diagnostic surgical procedures for a specific diagnosis and classification to make appropriate therapeutic decisions. Applying ancillary procedures to ABC specimens may obviate the need for surgical biopsies.

Every diagnostic method, including the interpretation of ABC, as outlined herein, has its constraints. Yet, clinicians may expect or at least anticipate that a specific diagnosis will be available following ABC. If not, they may inquire if it is possible to use other techniques to allow a more precise classification, and, if the clinical situation warrants, they may even volunteer to obtain a further ABC specimen so additional studies can be performed. To take full advantage of ABC specimens for diagnostic purposes, it is essential to expand the capabilities and resources of this procedure. Currently, this is the role of ancillary investigative techniques such as immunocytochemistry and electron microscopy. An undifferentiated neoplasm by light microscopy is usually differentiated enough to express an antigen profile or show ultrastructural features characteristic of a cell or tumor type. These techniques are complementary, and there are situations in which immunocytochemical or electron microscopic studies may contribute valuable information not obtainable by the other technique. Ancillary techniques not only facilitate further characterization of undifferentiated neoplasms, but also they help to refine the light microscopic cytological criteria. These techniques can be easily adapted to most cytological specimens, particularly ABC specimens (see Chapter 2) and in fact are becoming a routine procedure in many cytology laboratories, including ours. At the Ottawa Civic Hospital, immunocytochemical and ultrastructural studies performed on ABC specimens represent 23 and 22% of the workload of the diagnostic immunopathology and electron microscopy laboratories, respectively. Application of these techniques resolves certain diagnostic problems and improves the accuracy of diagnoses, thereby enhancing the diagnostic value of the NAB procedure.[1,3,6-8,10,11,16,19,20,24,27-30]

Like other techniques, application of immunocytochemistry and electron microscopy to ABC specimens is not free of pitfalls and limitations (see Chapters 4 and 7). Familiarity with these pitfalls and limitations, institution of measures to identify and correct them, and cautious interpretation are fundamentally important. Presentation of indications for application of these techniques, methods, and criteria for their interpretation is the prime goal of this book. Use of these techniques adds new dimensions to the practice of cytopathology, particularly ABC. Immunocytochemistry and electron microscopy, however, should be used to support the light microscopic diagnosis. Judicious selection of these techniques in conjunction with subjective morphological criteria and complete clinical data further improve the value of ABC, and hopefully decrease the number of invasive and more costly diagnostic surgical interventions.

# INDICATIONS FOR IMMUNOCYTOCHEMISTRY AND ELECTRON MICROSCOPY

With some degree of regularity, every cytopathologist encounters smears or cell block sections that are difficult to diagnose with certainty. In its simplest terms, this is the general indication for immunocytochemistry and electron microscopy (Table 1.1). The frequency of use of these techniques varies with the diagnostic skill and experience of each cytopathologist. Such ancillary techniques may provide the critical observations necessary to arrive at an accurate diagnosis. These techniques, however, have no major role in distinguishing the benign versus malignant and the neoplastic versus nonneoplastic nature of cells in ABC specimens. The only exception is the value of demonstrating light-chain restriction, a feature supportive of malignancy, in lymphoproliferative disorders using immunocytochemistry (see the Immunophenotyping section in Chapter 9).

More specific indications for ancillary studies of ABC specimens parallel those used for selection of cases for similar studies in surgical pathology (Table 1.2). The combination of immunocytochemistry and electron microscopy or either one of these techniques alone can be used in an attempt to differentiate between those neoplasms that, in their nonclassic cytological presentation, tend to look very much alike (see Chapter 8). These methods are also useful in subclassifying carcinomas and sarcomas, as well as in confirmation of neuroendocrine differentiation. Cytopa-

TABLE 1.1. General Indications for Diagnostic Ancillary Studies in ABC

> Unclassifiable neoplasms, by light microscopy.
> Supporting a diagnosis from a list of differential diagnoses.
> Supporting the light microscopic diagnosis.
> Determination of the primary site in metastatic neoplasms.
> Benign versus malignant (lymphoproliferative disorders).
> Confirmation of microorganisms.

**TABLE 1.2. Specific Indications for Diagnostic Ancillary Studies in ABC**

| |
|---|
| Clarification of undifferentiated or poorly differentiated (look-alike) neoplasms |
|     Carcinoma |
|     Malignant melanoma |
|     Malignant lymphoma |
|     Sarcoma |
|     Germ-cell neoplasm |
| Small-cell neoplasms in children |
| Neuroendocrine neoplasms (primary or metastatic) |
| Subclassification of sarcomas |
| Differential diagnosis of |
|     Squamous-cell carcinoma vs adenocarcinoma |
|     Prostatic adenocarcinoma |
|     Gastrointestinal adenocarcinoma |
|     Hepatocellular carcinoma vs adenocarcinoma |
|     Mesothelioma vs adenocarcinoma |
|     Thyroid neoplasm |
|     Breast carcinoma |
|     Adrenal cortical neoplasm vs renal-cell carcinoma |
| Immunophenotyping of malignant lymphoma |

thologists may be asked if it is possible to distinguish primary from metastatic neoplasms and, if the latter, what tissue or organ is the source. These ancillary methods are useful in defining the organ or tissue source for certain metastatic neoplasms, such as tumors of thyroid, prostate, liver, breast, kidney, adrenals, and mesothelial origin (see Chapter 10). Immunocytochemical study is extremely helpful in immunophenotyping of lymphoid disorders for differentiation of malignant lymphoma from nonneoplastic lymphoid proliferation and nonlymphoma neoplasms (see Chapter 9). All of these distinctions can have significant impact on patient management.

# PRACTICAL APPLICATION

Without doubt, immunocytochemistry or electron microscopy, or both, make important diagnostic contributions to the practice of surgical pathology and cytopathology. In fact, an increasing number of cytopathologists rely on these techniques, whether performed in their own laboratory or at a referral laboratory, to arrive at an accurate diagnosis. The decision to set up these techniques in a given laboratory depends on a sufficient volume of problem cases, the interest and commitment of the cytopathologist, and financial considerations. Current financial constraints emphasize the need for increased efficiency in diagnosis and therapy; this is an important consideration for initiating or continuing any procedure used for diagnosis and treatment of patients. Cost-effectiveness of such ancillary techniques should be evaluated in the context of total cost to patients or the health-care system. If immunocytochemistry or electron microscopy can provide a definite diagnosis from a specimen obtained by NAB, then charges for these procedures may be negligible when compared with those for more invasive diagnostic surgical procedures.

A minimum number of problem cases is necessary to acquire and maintain technical and interpretive skills. When the volume of problem cases is low, it is probably more reasonable and economical to refer the cases to a referral laboratory with demonstrated expertise in applying these techniques. In either situation it is essential for cytopathologists to become familiar with the principles, indications, and interpretation of these techniques.

Immunocytochemistry can be readily applied to routinely prepared smears, cytospin preparations, and cell block sections (see Chapter 2), and should be increasingly used for diagnostic purposes. With the advent of commercial kits for diagnostic immunocytochemical staining, this procedure has become widely available to hospital laboratories regardless of size. The technique is relatively simple and can be easily learned by cytopathologists and technologists. It does not need to be available only in university hospital–based laboratories. No expensive equipment, large laboratory space, or special preparation are required for performing immunocytochemistry. We recommend starting with a few primary antibodies and the use of a "universal" kit to provide the necessary secondary antibodies and chromogen. The estimated cost of initial setup is in the range of $900 to $1,000.[18] Once experience is gained and the requirements of each institution become clear, the number of primary antibodies can be increased.

Electron microscopy can also be readily applied to cytopathology specimens (see Chapter 2). Judging from the report of another institution[26] and our own experience,[7] the application of electron microscopy to ABC specimens is increasing. It is inexpensive to fix a small portion of an ABC specimen in glutaraldehyde, depending on the clinical indications, and store this specimen pending review of the cytology preparation. If no diagnostic problem occurs, then the glutaraldehyde-fixed sample can be discarded.[17] On the other hand, if electron microscopy might be helpful, the specimen can be processed. Many cytopathologists screen and report ABC specimens used for ultrastructural studies. Other facilities rely on electron microscopists or surgical pathologists with experience in this subspecialty. Even without such assistance, however, it is possible and practical for any cytopathologist or resident training in this field to gain experience. Screening ABC specimens that are not particular diagnostic problems, correlating the ultrastructural features with cytological details, and using the principles and guidelines outlined in subsequent chapters rapidly improve diagnostic acumen.

It is obvious that the capital costs of equipping a hospital-based electron microscopy laboratory are considerably greater than those for the equipment and inventory of antibodies for an immunocytochemical facility. For use in a hospital setting, the two major pieces of equipment, an electron microscope (with installation costs) and an ultramicrotome, will require an investment of approximately $225,000 to $250,000. The majority of cytopathology laboratories, particularly those associated with nonteaching hospitals, would gain little by purchasing an electron microscope. Nevertheless, the feasibility and practical application of electron microscopy in a community hospital is exemplified in an article by Mason and associates.[17]

For those hospitals and diagnostic facilities examining cytopathology specimens that already have immunocytochemical and electron microscopy capabilities, is the cost of doing these procedures a practical consideration? Electron microscopy is

**TABLE 1.3. Comparative Costs (Technical and Supplies) for Electron Microscopic and Immunocytochemical Diagnostic Investigations†**

|  | Electron Microscopy | Immunocytochemistry |
|---|---|---|
| US Hospitals (n = 14)* | | |
| Mean ± SD | $310 ± 170 | $240 ± 140 |
| Median | $335 | $200 |
| Mode | $350 | $125 |
| Range | $52–700 | $80–500 |
| Ontario, Canada | | |
| LMS units | 225 | 225 |
| Dollar value** | $116 | $116 |

*On the basis of processing, sectioning, and photography of one electron microscopy sample and the processing and immunostaining of a diagnostic problem using a panel of five monoclonal antibodies.

**On the basis of $0.517/LMS unit in Canadian dollars for the 1989 Ontario Hospital Insurance Plan fee schedule.

SD = Standard deviation. LMS = labor, materials, and supervision.

†(From Dardick I, et al: A quantitative comparison of light and electron microscopic diagnoses in specimens obtained by fine-needle aspiration biopsy: *Ultrastruct. Pathol.* 15:105–129, 1991.)

generally thought to be more expensive than immunocytochemistry, and this finding likely influences its application. Supporting data and comparative studies of this type, however, do not seem to be available. To assess this issue, we canvassed the Departments of Pathology of 14 hospitals in the United States doing both electron microscopy and immunocytochemistry. We inquired as to the charges for the technical component and supplies for both immunocytochemical and ultrastructural examinations, but excluded the professional fees because these are very variable for each institution. The median, mode, mean, and range of costs incurred for these procedures are provided in Table 1.3. The costs refer to processing, sectioning, and photography of one electron microscopy sample and the processing and immunostaining of a diagnostic problem using a panel of five monoclonal antibodies (e.g., cytokeratin, vimentin, leukocyte common antigen, neuron-specific enolase, and HMB-45). Although the costs associated with electron microscopy are generally higher than those for immunocytochemistry, the differences are not substantial. In Ontario, Canada, the fee schedule for each laboratory procedure is assigned a relative value that reflects "the relative weighting of labor, materials, and supervision (LMS unit) for each test." The LMS unit for performing electron microscopy and for undertaking immunocytochemistry, each on a particular specimen, and the total costs (in Canadian dollars) paid by the health insurance plan in Ontario (again excluding professional fees), are also provided in Table 1.3.

# REFERENCES

1. Akhtar M, Ali MA, Owen EW: Application of electron microscopy in the interpretation of fine-needle aspiration biopsies. *Cancer* 48:2458–2463, 1981.

2. Berkman WA, Chowdhury L, Brown NL, Padleckas R: Value of electron microscopy in cytologic diagnosis of fine-needle biopsy. *Am J Radiol* 140:1253–1258, 1983.

3. Chess Q, Hajdu SI: The role of immunoperoxidase staining in diagnostic cytology. *Acta Cytol* 30:1–7, 1986.

4. Cinti S, Ferretti M, Amati S, et al: Electron microscopy applied to fine-needle aspiration: A review of six cases from various sites. *Tumori* 69:423–435, 1983.

5. Collins VP, Ivarsson B: Tumor classification by electron microscopy applied to fine-needle aspiration cytology. *Acta Pathol Microbiol Immunol Scand {A}* 89:103–105, 1981.

6. Dabbs DJ, Silverman JF: Selective use of electron microscopy in fine needle aspiration cytology. *Acta Cytol* 32:880–884, 1988.

7. Dardick I, Yazdi HM, Brosko C, Rippstein P, Hickey NM: A quantitative comparison of light and electron microscopic diagnoses in specimens obtained by fine needle aspiration biopsy. *Ultrastruct Pathol* 15:105–129, 1991.

8. Domagala W, Lubinski J, Lasota J, et al: Decisive role of intermediate filament typing of tumor cells in the differential diagnosis of difficult fine needle aspirates. *Acta Cytol* 31:253–266, 1987.

9. Domagala WM, Markiewski M, Tuziak T, et al: Immunocytochemistry on fine needle aspirates in paraffin miniblocks. *Acta Cytol* 34:291–296, 1990.

10. Flens MJ, Van der Valk P, Tadema TM, et al: The contribution of immunocytochemistry in diagnostic cytology: comparison and evaluation with immunohistology. *Cancer* 65:2704–2711, 1990.

11. Gustafsson B, Manson J-C: Methodological aspects and application of the immunoperoxidase staining technique in diagnostic fine-needle aspiration cytology. *Diagn Cytopathol* 3:68–73, 1987.

12. Hagelqvist E: Light and electron microscopic studies on material obtained by fine needle biopsy: A methodological study on aspirates from tumours of the head and neck region with special emphasis on salivary gland tumours. *Acta Otolaryngol* 354(suppl):1–75, 1978.

13. Hajdu SI, Ehya H, Frable WJ, et al: The value and limitations of aspiration cytology in the diagnosis of primary tumors: A symposium. *Acta Cytol* 33:741–790, 1989.

14. Kern WH, Haber H: Fine needle aspiration minibiopsies. *Acta Cytol* 30:403–408, 1986.

15. Kindblom LG: Light and electron microscopic examination of embedded fine-needle aspiration biopsy specimens in preoperative diagnosis of soft tissue and bone tumors. *Cancer* 51:2264–2277, 1983.

16. Mackay B, Fanning T, Bruner JM, Steglich MC: Diagnostic electron microscopy using fine needle aspiration biopsies. *Ultrastruct Pathol* 11:659–672, 1987.

17. Mason D, Pedraza MA, Doslu FA, Marsh RA: Ultrastructural and immunological methods in diagnostic pathology in a community hospital. *Ultrastruct Pathol* 2:373–381, 1981.

18. Miller RT: Immunohistochemistry in the community practice of pathology: Part I; general considerations, technical factors, and quality assurance. *Lab Med* 22:457–464, 1991.

19. Nadji M: The potential value of immunoperoxidase techniques in diagnostic cytology. *Acta Cytol* 24:442–447, 1980.

20. Nadji M, Ganjei P: Immunocytochemistry in diagnostic cytology: a 12-year perspective. *Am J Clin Pathol* 94:470–475, 1990.

21. Navas-Palacios JJ, de Agustin de Agustin PP, De Los Heros FA, Perez-Barrios A,

Alvarez-Vicent JJ: Ultrastructural diagnosis of facial nerve schwannoma using fine needle aspiration. *Acta Cytol* 27:441–445, 1983.
22. Nordgren H, Akerman M: Electron microscopy of fine needle aspiration biopsy from soft tissue tumors. *Acta Cytol* 26:179–188, 1982.
23. Olson NJ, Gogel HK, Williams WL, Mettler FA Jr: Processing of aspiration cytology samples: an alternative method. *Acta Cytol* 30:409–412, 1986.
24. Osamura RY: Application of immunocytochemistry to diagnostic cytopathology. *Diagn Cytopathol* 5:55–63, 1989.
25. Stark P, Hildebrandt-Stark HE: Electron microscopy of cells obtained by fine needle aspiration biopsy of lung lesions. *Radiology* 22:327–328, 1982.
26. Strausbauch P, Neill J, Dabbs DJ, Silverman JF: The impact of fine needle aspiration biopsy on a diagnostic electron microscopy laboratory. *Arch Pathol Lab Med* 113:1354–1356, 1989.
27. Taccagni G, Cantaboni A, Dell'Antonio G, Vanzulli A, Del Maschio A: Electron microscopy of fine needle aspiration biopsies of mediastinal and paramediastinal lesions. *Acta Cytol* 32:868–879, 1988.
28. Wills EJ, Carr S, Philips J: Electron microscopy in the diagnosis of percutaneous fine needle aspiration specimens. *Ultrastruct Pathol* 11:361–387, 1987.
29. Yam LT: Immunocytochemistry of fine needle aspirates; a tactical approach. *Acta Cytol* 34:789–796, 1990.
30. Yazdi HM, Dardick I: What is the value of electron microscopy in fine needle aspiration biopsy? *Diagn Cytopathol* 4:177–182, 1988.

# 2

# Techniques for Specimen Processing

## FIXATION

The choice of fixative is perhaps the most critical aspect of processing aspiration biopsy cytology (ABC) specimens. Optimal fixation of cells and tissues depends on the type of fixative, concentration, pH, length of exposure, and the type of ancillary studies being sought. No single fixative is considered to be optimal in all circumstances. A fixative that is ideal for immunocytochemistry, such as alcohol-based fixatives, may be detrimental to the fine structure of cells and tissues and are therefore inappropriate for electron microscopic study. In contrast, glutaraldehyde, which is the fixative of choice for electron microscopy, cross-links proteins extensively;[3] it is therefore inappropriate for immunocytochemistry. Similarly, there is no single fixative that is optimal for all antigens. Some fixatives may denature the antigen or alter the configuration of the polypeptide chains, resulting in masking of the epitope.[17,30]

Alcohol-based fixatives are widely used for cytological specimens and are ideal for preservation of most routinely sought antigens. Some antigens, however, require special fixatives. For example, in our experience, acetone and periodate lysine paraformaldehyde (PLP) fixation produces optimal results with lymphoid markers (see Chapter 9). Similarly, formaldehyde fixation is required for optimal demonstration of estrogen receptors. Smears should be fixed immediately; 5 to 10 minutes of fixation is generally adequate. Longer fixation (e.g., several hours or days) may result in a gradual loss of antigens.[22,23] Smears can be air dried first and then fixed.[21-23] This approach is particularly appropriate for cell membrane antigens such as most lymphoid markers (see Chapter 9).

Neutral buffered formaldehyde (10%) is a widely used fixative for cytology cell blocks and pathology specimens. It is, however, one of the least desirable fixatives for immunocytochemistry. In contrast, alcohol-based fixatives are generally recognized to be better than aldehydes for immunocytochemistry; however, they are currently not widely used. The main advantages and disadvantages of alcohol and

TABLE 2.1. Comparison Between Alcohol and Formalin Fixation

| | Alcohol | Formalin |
|---|---|---|
| Penetration | Slow | Fast |
| Fixation process | Immediate | Slow |
| Mechanism of fixation | Precipitation of proteins | Molecular cross-linking |
| Preservation of antigens | Excellent (more antigens) | Reasonably good |
| Potential for masking of epitopes | No | Yes |
| Protease digestion | Not required | May be required |
| Long fixation | Less antigen loss | More antigen loss and/or masking of epitopes |
| Cellular details | Excellent | Reasonably good |
| Impurities | No | Yes (formic acid, etc.) |
| Electron microscopic study | Poor | Reasonably good |
| Cell shrinkage | Relatively more* | Less |
| Cost | Relatively high** | Low |

*Can be minimized by addition of acetic acid.
**For large specimens in surgical pathology.

formalin fixatives are summarized in Table 2.1.[4,5,10,12,19,27] We fix our cell blocks in methanol-glacial acetic acid (MAA) solution (90% methanol, 10% glacial acetic acid, vol/vol) for approximately one hour prior to processing. This is perhaps the key step for our success in performing immunocytochemical stains on ABC specimens. No special processing is required except to avoid the formalin station in the tissue processor. MAA fixative is not only ideal for preservation of most antigens, but it also results in a superb cytomorphology. The latter is very important for cytological interpretation and to decide if the cells are normal/benign or neoplastic/malignant in nature. Addition of glacial acetic acid decreases the extent of cell shrinkage, and sections are easier to cut from the cell blocks. Because of our success with MAA-fixed cell blocks, we submit a representative piece of tissue in MAA fixative when staining with immunocytochemical stains is anticipated in surgical pathology. Microwave fixation may improve antigen preservation of tissues fixed in formalin. This improvement appears to be largely due to postfixation in alcohol during subsequent tissue processing.[4]

Some fixatives, such as formalin, may mask the epitopes and cause a false-negative result. In these circumstances, using a proteolytic enzyme such as pepsin, trypsin, pronase, or ficin may enhance the sensitivity of the immunocytochemical method and reduce the background staining.[5,14,26-28] The duration of treatment is critical and should be tailored to the particular type of antigen sought and the duration of exposure to formalin. Generally the longer the exposure to this fixative, the longer the period of protease digestion. Proteolytic enzyme digestion, however, is not without potential problems. It may denature the epitopes in question or unmask a cross-reacting antigen, resulting in a false-negative or a false-positive result, respectively.[26,27] The glass slides must be treated with an adhesive to prevent detachment of cell block sections from the slide. *No digestion is required for alcohol-fixed cytological or histological specimens.* In cytology cell blocks or tissues fixed in formalin, protease

**TABLE 2.2. Different Fixatives Used For Immunocytochemistry In Our Laboratory**

| |
|---|
| A. 95% ethanol: used for most immunostains on smears |
| B. Methanol acetic acid (MAA): used for all immunostains on cell blocks |
|     Methanol:    9 volume |
|     Glacial acetic acid:    1 volume |
| C. Acetone-PLP (periodate lysine paraformaldehyde): mainly used for lymphomas |
|     PLP working solution (cool to 4°C) |
|         Stock 3% paraformaldehyde:    1 volume |
|         Stock periodate lysine solution:    2 volume |
|     Stock 3% paraformaldehyde |
|         Paraformaldehyde:    6.0 gm |
|         Distilled water:    200.0 ml |
|     Stock periodate lysine (pH 7.4) |
|         Distilled water:    500.0 ml |
|         Disodium hydrogen phosphate:    13.4 gm |
|         Lysine:    4.5 gm |
|         Sodium periodate:    0.730 gm |
| D. Formaldehyde-methanol-acetone: used for estrogen receptor study |
|     3.7% formaldehyde-phosphate buffered saline (PBS) solution |
|         37% formaldehyde:    1 volume |
|         PBS:    9 volume |

digestion is only recommended for certain antigens such as cytokeratins. For preparation of different fixatives see Table 2.2.

# GENERAL SPECIMEN PROCESSING

Immunocytochemical and electron microscopic studies are labor-intensive and costly; therefore, they should not be performed routinely in every case. In addition, in many cases they are not necessary for an accurate diagnosis. On the other hand, in most cases the necessity for ancillary studies cannot be anticipated clinically, and they are considered only after examination of routinely stained smears. Therefore, it is essential to handle ABC specimens so that immunocytochemical and electron microscopic studies are possible to perform if necessary. We have developed a triage method for processing ABC specimens to satisfy this need (Fig. 2.1). This method of processing has worked very well in our institution for the last six years.

For ABC specimens obtained from deep lesions, including intraabdominal and intrathoracic organs or sites, six to eight smears are prepared by a cytotechnologist and immediately fixed in 95% ethanol. The needle is then rinsed in a balanced electrolyte solution (BES). The smears are stained by a rapid Papanicolaou method and evaluated by the cytotechnologist and cytopathologist. In the majority of cases, the final diagnosis can be reported within 20 minutes. At this stage, if the light microscopy results are not conclusive, or if more refined classification is desired, the material in BES is processed accordingly (Fig. 2.2). In these cases, a provisional report including the differential diagnoses and the favored diagnosis, if possible, is

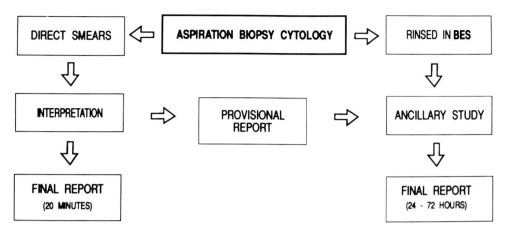

Fig. 2.1. Triage method for processing ABC specimens. BES = balanced electrolyte solution.

sent to the attending physician. The final report is usually available within 24 to 72 hours. If the aspirate is inadequate or suboptimal for cytological diagnosis or ancillary study, the aspirator is informed and a repeat needle aspiration biopsy (NAB) is usually performed. However, this repetition has not been necessary in the majority of cases. When immunophenotyping of malignant lymphoma is required, we prefer air-dried cytospin preparations (see Chapter 9).

For most ABCs of superficial organs or sites, including prostate, thyroid gland, and salivary glands, six to eight smears are prepared by a cytotechnologist and immediately fixed in 95% ethanol. The needle is then rinsed in a 50% alcohol-based fixative. After centrifugation, the specimen is processed as a cytospin preparation or a cell block depending on the size of the cell button. For some of our commonly performed aspirates, such as NAB of the breast, the aspirator is instructed to rinse the needle directly into a disposable screw cap tube containing a 50% alcohol-based fixative and to send the tube to the cytology laboratory. This approach is ideal when a cytotechnologist is not available for proper handling of ABC specimens, and can be done at any time and at any facility.[25] After centrifugation (depending on the size of the cell button), sediment smears, cytospin preparations, or cell blocks are prepared. Immunocytochemical studies can be easily performed on these specimens; however, electron microscopy is not an option for alcohol-fixed material. Due to the ease of obtaining specimens from superficial lesions, a repeat NAB is performed for electron microscopic study if necessary. This is not, however, a common occurrence.

# PROCESSING FOR LIGHT MICROSCOPY

## Processing of Smears

The number of smears prepared depends on the material in the needle. In most cases, six to eight smears are prepared and immediately fixed in 95% ethanol. The

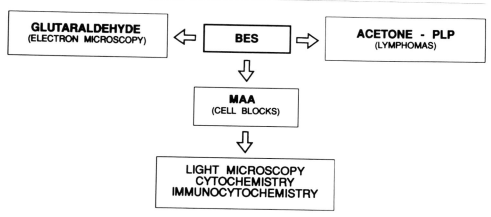

**Fig. 2.2.** Processing of ABC specimens rinsed in balanced electrolyte solution (BES). PLP = periodate lysine paraformaldehyde; MAA = methanol acetic acid.

needle is then rinsed in a conical tube containing BES (Abbott Laboratories, Chicago, IL). The material is kept temporarily (approximately 20 min) at room temperature until the smears are interpreted. The smears are stained using a modified rapid Papanicolaou method[16] (Table 2.3). This method takes approximately four minutes to perform. For interpretation and immunophenotyping of malignant lymphoma, air-dried smears or cytospin preparations for Leishman (a Romanovsky-type stain) and immunocytochemical stains are prepared (see Chapter 9).

TABLE 2.3. Rapid Papanicolaou Method

| | |
|---|---|
| 70% Ethanol | 10 dips |
| Tap water | 10 dips |
| Hematoxylin | 30–60 seconds |
| Tap water | 10 dips |
| Scott's tap water* | 10 dips |
| 50% Ethanol | 10 dips |
| 70% Ethanol | 10 dips |
| 95% Ethanol | 10 dips |
| OG-6 | 30 seconds |
| 95% Ethanol | 10 dips |
| 95% Ethanol | 10 dips |
| EA-50 | 30 seconds |
| 95% Ethanol | 10 dips |
| 95% Ethanol | 10 dips |
| Absolute ethanol | 10 dips |
| Absolute ethanol | 10 dips |
| Xylol | 10 dips |
| Xylol | 10 dips |

*Scott's tap water = magnesium sulfate anhydrous, 80 gm; sodium bicarbonate, 16 gm; tap water, 8,000 ml.

**TABLE 2.4. Processing of Needle Rinse**

1. Centrifuge at 3,000 rpm for 10 minutes.
2. If there is a visible button, proceed to step 3 or 4. If there is a minuscule button or no visible button, proceed to step 5. For processing of specimens for malignant lymphoma, see Chapter 9.
3. (a) Carefully decant all the supernatant.
   (b) Add 5–10 ml methanol-acetic acid (MAA) and keep at room temperature for approximately 20 minutes.
   (c) Remove the cell sediment, wrap in tissue paper, and place in a cassette.
   (d) Place the cassette in MAA fixative for approximately 30 minutes.
   (e) Transfer to 70% ethanol for a minimum of 30 minutes to remove acetic acid prior to further processing.*
4. If electron microscopy is required, remove a portion or entire button with a metal spatula or a disposable plastic pipette and resuspend in 2.5% glutaraldehyde for overnight fixation.
5. Pour off the supernatant, leaving 2–3 ml of balanced electrolyte solution in the tube. Then thoroughly resuspend the cells by vortexing and then proceed with cytocentrifugation.

*Acetic acid can be damaging to the fluid exchange plumbing of the processor if not removed.

## Processing of Needle Rinse

The needle rinsed in BES is processed after interpretation of the smears. The choice between different procedures depends on the indication for immunocytochemical or electron microscopic study and the size of the cell button after centrifugation (see Fig. 2.2; Table 2.4). The needle rinsed in 50% ethanol is processed similarly, except electron microscopy is not a choice because alcohol-fixed specimens are not appropriate for electron microscopy.

## Processing of Cell Block

Cell blocks are processed overnight in an automatic processor, similar to tissue specimens. *If the cell blocks are fixed in an alcohol-based fixative, however, as in our laboratory, then the formalin step should be turned off.* The tissue processor schedule can be modified to process the stat cell blocks in two to three hours (Table 2.5). This schedule is also practical for small laboratories with one automatic tissue processor. The alcohol-fixed cell or tissue blocks can be processed during the day, whereas overnight processing can be used for formalin-fixed tissues. The problem of stray (foreign) tissue that can get into the blocks on embedding, or onto the slides during cutting, drying, or staining, is much more critical with cell blocks than with tissue blocks. This complication can be misleading and can possibly lead to a false-positive diagnosis. Therefore, one should follow a protocol that focuses on using clean instruments and equipment at all stages of the procedure, from embedding to coverslipping. We stain the slides with hematoxylin, phloxine, and saffron method, which is equivalent to hematoxylin and eosin stain. If cytochemical or immunocytochemical stains are anticipated at the time of initial cutting, additional sections are cut to avoid the loss of valuable cells in the block.

**TABLE 2.5. Short Cycle Schedule for Alcohol-fixed Cell Blocks***

| Station | Reagent | Time (min) | Temperature | Vacuum |
|---|---|---|---|---|
| 01 | Formalin | Off | Off | Off |
| 02 | 50% ethanol | Off | Off | Off |
| 03 | 70% ethanol | 15 | Off | On |
| 04 | 95% ethanol | 25 | Off | On |
| 05 | Absolute ethanol | Off | Off | Off |
| 06 | Absolute ethanol | Off | Off | Off |
| 07 | Absolute ethanol | 30 | Off | On |
| 08 | Toluene | 15 | Off | On |
| 09 | Toluene | Off | Off | Off |
| 10 | Toluene | 20 | Off | On |
| 11 | Paraffin | 15 | 60°C | On |
| 12 | Paraffin | 5 | 60°C | On |
| 13 | Paraffin | 20 | 60°C | On |

*For cell blocks up to 2 mm in thickness.

# PROCESSING FOR IMMUNOCYTOCHEMISTRY

The cell block sections, direct smears, and cytospin preparations should be processed properly prior to immunocytochemical staining (Tables 2.6, 2.7). Using a diamond pen or PAP PEN (Daido Sangyo Co. Ltd. Japan) make a circle around the sections or smears (this is an optional step). There are four immunocytochemical methods used in our laboratory.

1. Alcohol-fixed smears and alcohol- or formalin-fixed cell blocks requiring staining with monoclonal antibodies are treated with primary antibody overnight and are detected by the avidin-biotin-peroxidase complex method (Table 2.8).
2. Acetone-PLP (periodate lysine paraformaldehyde)–fixed smears requiring staining with monoclonal antibodies are processed using a one-day peroxidase-labeled avidin-biotin staining method (Table 2.9).
3. All smears and cell blocks, regardless of fixative used, requiring staining with polyclonal antibodies are treated with primary antibody overnight and detected by the peroxidase antiperoxidase method (Table 2.10).

**TABLE 2.6. Preparation of Cell Block Sections For Immunostaining**

| | |
|---|---|
| 1. Cut 4-μm-thick sections and pick up on AES-treated slides. | |
| 2. Dry slides at 37°C | Overnight |
| 3. Dry slides at 45°C | 30 minutes |
| 4. Deparaffinize sections through three changes of toluene | 5 minutes each |
| 5. Hydrate sections through three changes of absolute ethanol | 1 minute each |

AES = 3-aminopropyltriethoxy-saline (Sigma Chemical Co, St. Louis, MO).

TABLE 2.7. Preparation of Smears and Cytospins For Immunostaining

A. Air-dried smears or cytospins
  1. Fix in acetone in room temperature for 2 minutes.
  2. Drain and rinse with phosphate buffered saline (PBS).
  3. Fix in freshly prepared PLP (periodate lysine paraformaldehyde) for 8 minutes.
  4. Rinse in PBS.
B. Alcohol-fixed smears or cytospins: ready for staining.
C. Papanicolaou-stained smears or cytospins
  1. Soak the slides in toluene for several hours or overnight.
  2. Remove the coverslip and immerse the slide in a change of toluene to remove residual mounting media.
  3. Rinse in several changes of absolute ethanol.
D. Preparing multiple slides from a single smear: see Table 3.3.
E. Air-dried smears for estrogen receptor study
  1. Prepare smears using Abbott adhesive-treated slides (Abbott Laboratories, Chicago, IL)
  2. Air dry the smears for less than 1 minute.
  3. Fix in fresh 3.7% formaldehyde-PBS solution for 10–15 minutes.
  4. Transfer to PBS for 4–6 minutes.
  5. Place in cold methanol ($-10°$ to $-25°C$) for 3–5 minutes.
  6. Place in cold acetone ($-10°$ to $-25°C$) for 1–3 minutes.
  7. Rinse with PBS for 4–6 minutes.
  8. Transfer to fresh PBS for 4–6 minutes.

4. Formalin-methanol-acetone–fixed smears requiring staining with monoclonal antibody to estrogen receptor are processed using a one-day PAP method (Table 2.11).

For all immunocytochemical procedures, 3,3'-diaminobenzidine tetrahydrochloride (DAB) is used as chromogen. Drying of smears and sections should be avoided in all steps.

# PROCESSING FOR ELECTRON MICROSCOPY

## General Procedures

Standard fixation, embedding, sectioning, and staining techniques are all that are required in the majority of ABC samples (Tables 2.12–2.15). This was the case for the ABCs discussed and illustrated in the following chapters. Specialized methods, however, may be needed to either harvest small fragments from the glutaraldehyde- or Karnovsky-fixed sample (see Tables 2.12–2.14 for preparation) obtained from the cytopathology laboratory or to prepare a cellular pellet that can facilitate the processing and embedding procedures, which will be detailed. Although unnecessary in the vast majority of aspiration biopsy samples, it can be helpful to be aware of other procedures and techniques employed in electron microscopy facilities for

TABLE 2.8. Avidin-Biotin-Peroxidase Complex (*ABC*) Procedure
For Monoclonal Antibodies (Alcohol-Fixed Smears, Alcohol-
or Formalin-Fixed Cell Blocks)

1. Prepare cell block sections or smears (see Tables 2.6, 2.7).
2. Incubate at room temperature with freshly prepared 3% hydrogen peroxide-methanol for 30 minutes.
3. Wash in tap water for 2 minutes.
4. Protease digestion (formalin-fixed specimens; some antigens).
5. Stain with hematoxylin for 3 minutes.
6. Wash in tap water for 3 minutes.
7. Transfer to Tris buffer until ready to proceed.
8. Incubate at room temperature with diluted normal horse serum (1:10) for 20 minutes.
9. Drain normal horse serum (do not rinse).
10. Incubate at room temperature with diluted (in 1:20 normal horse serum) primary antibody for 30 minutes (use Tris buffer for negative antibody control).
11. Incubate overnight at 4°C.
12. Wash in Tris buffer for 10 minutes.
13. Incubate at room temperature with diluted biotinylated horse antimouse immunoglobulin (Vectastain ABC Kit, Vector Laboratories, Burlingame, CA.) for 30 minutes.
14. Wash in Tris buffer for 10 minutes.
15. Incubate at room temperature with *ABC* reagents (Vectastain ABC Kit, Vector Laboratories, Burlingame, CA.) for 30–60 minutes.
16. Wash in Tris buffer for 10 minutes.
17. Incubate at room temperature with DAB solution for exactly 5 minutes.
18. Rinse with Tris buffer, then incubate at room temperature for exactly 5 minutes.
19. Wash in tap water for 5 minutes.
20. Dehydrate, clear, and mount in Eukitt.

DAB = 3,3'-diaminobenzidine tetrahydrochloride.
Tris = Tris [hydroxymethyl] aminomethane.

specific diagnostic purposes. This need can be admirably filled by consulting an excellent review article by Dvorak.[13]

In the cytopathology laboratory, following preparation of the cellular button obtained from the needle rinse (see "General Specimen Processing" section), a portion of this material is placed in 2.5% glutaraldehyde in 0.1M sodium cacodylate buffer (see Table 2.12). After a minimum of two hours or after overnight fixation in the refrigerator, the tissue is processed to resin using a routine schedule (see Table 2.15) and polymerized overnight. This process allows sectioning, screening, and reporting of the ABC specimens the following day. For comparative purposes with a semirapid (Table 2.15) and an ultrarapid (Table 2.16) schedule, the routine embedding procedure takes six to seven hours from the beginning of osmification to the final resin stage (see Table 2.15).

In certain clinical situations where a diagnosis is required quickly, the steps for fixation, processing, and embedding for electron microscopy can be considerably

TABLE 2.9. Peroxidase-Labeled Avidin-Biotin (LAB) Procedure for Monoclonal Antibodies (Acetone-periodate Lysine Paraformaldehyde–Fixed Smears)

1. Prepare smears (see Tables 2.7 and 9.3).
2. Incubate at room temperature with diluted primary antibody for 30 minutes. (Use phosphate buffered saline [PBS] for negative antibody control.)
3. Wash gently with PBS (two changes) for 2 minutes.
4. Incubate at room temperature with diluted biotinylated horse antimouse immunoglobulin (Vector Laboratories, Burlingame, CA.) for 30 minutes.
5. Wash gently with PBS (2 changes) for 2 minutes.
6. Incubate at room temperature with horseradish peroxidase avidin D (Vector Laboratories, Burlingame, CA.) for 30 minutes.
7. Wash briefly in PBS for 1–2 minutes.
8. Incubate at room temperature with DAB solution for exactly 5 minutes.
9. Rinse in PBS, then incubate at room temperature with 0.5% copper sulfate solution for exactly 5 minutes.
10. Wash with distilled water.
11. Stain with hematoxylin for 3 minutes.
12. Wash in water.
13. Dehydrate, clear, and mount in Eukitt.

DAB = 3,3'-diaminobenzidine tetrahydrochloride.

attenuated so that screening of the specimen can take place a few hours after receiving the specimen. The small size and fragmentation of the specimen means that fixation can be shortened to 15 to 30 minutes if necessary. This fixation can be followed by what might be considered a semirapid schedule (see Table 2.15), where the time between the osmium tetroxide fixation and the fresh epon stages encompasses approximately four hours. An ultrarapid technique (see Table 2.16) has been adapted by Dr. Kurtz (Veterans Administration Hospital, Dallas, TX) from a schedule developed by Bencosme and Tsutsumi.[6] Only 1.5 hours is required for the primary fixation to the fresh epoxy resin stages; this procedure is routinely used in Dr. Kurtz's and other laboratories for electron microscopy of ABC specimens. This time frame allows results to be reported the same day. Even using the routine method outlined in Table 2.15, however, ultrastructural findings can be available to cytopathologists the day following commencement of the processing schedule. Thus, a report integrating the cytological and ultrastructural aspects could be generated a maximum of 24 to 48 hours after receipt of the sample.

Regarding the ABC samples used for the series in this book, all specimens were fixed routinely in 2.5% glutaraldehyde in sodium cacodylate buffer, postfixed in 1% osmium tetroxide containing 1.5% potassium ferrocyanide (see "Glycogen Preservation" section), stained en bloc with saturated aqueous uranyl acetate, dehydrated in an ascending methanol series, and embedded in epon-araldite resin. Thin sections cut on a diamond knife were picked up on formvar-coated 100-mesh copper grids and stained with lead citrate. Formvar-coated grids were advantageous in permitting screening of a greater proportion of the section because tumor tissue fragments were often at the periphery in a small but relatively adequate specimen.

TABLE 2.10. Peroxidase Antiperoxidase (PAP) Procedure for Polyclonal Antibodies (All Smears and Cell Block Sections, Regardless of Fixative Used)

1. Prepare cell block sections or smears (see Tables 2.6 and 2.7).
2. Incubate at room temperature with freshly prepared 3% hydrogen peroxide methanol for 30 minutes.
3. Wash in tap water for 2 minutes.
4. Protease digestion (formalin-fixed specimens; some antibodies).
5. Stain with hematoxylin for 3 minutes.
6. Wash in tap water for 3 minutes.
7. Wash in distilled water (two changes).
8. Wash in Tris buffer (two changes). Slides can be held in Tris buffer until ready to proceed.
9. Incubate at room temperature with diluted normal swine serum (1:20) for 20 minutes.
10. Drain normal swine serum (do not rinse).
11. Incubate at room temperature with diluted (in 1:20 normal swine serum) primary anterisera for 30 minutes. (Use normal rabbit serum [1:500] for negative antibody control.)
12. Incubate at 4°C overnight.
13. Wash in Tris buffer (two changes) for 15 minutes.
14. Incubate at room temperature with swine antirabbit immunoglobulin for 30 minutes.
15. Wash in Tris buffer for 15 minutes.
16. Incubate at room temperature with rabbit peroxidase antiperoxidase solution for 30 minutes.
17. Wash in Tris buffer for 15 minutes.
18. Incubate at room temperature with DAB solution for exactly 15 minutes.
19. Rinse with Tris buffer, then incubate at room temperature with 0.5% copper sulfate solution for exactly 15 minutes.
20. Wash with distilled water.
21. Dehydrate, clear, and mount in Eukitt.

DAB = 3,3'-diaminobenzidine tetrahydrochloride.
Tris = Tris [hydroxymethyl] aminomethane.

## Techniques for Handling ABC Specimens

Harvesting the ABC specimen for electrion microscopy generally requires one of four procedures.

1. The technical aspects of simple centrifugation in microtubes (generally 1,500 rpm for 10 min) to harvest ABC specimens for electron microscopy are detailed by a number of authors.[1,2,9,15,18,24] Subsequent dehydration and embedding in plastic resin can be performed in these microtubes.[2] More elaborate techniques and specialized equipment for harvesting ABC specimens for electron microscopy are also available.[1,15,18,24]
2. Approximately 70 to 75% of needle rinse specimens consist of a few tissue

TABLE 2.11. Peroxidase Antiperoxidase (PAP) Procedure For Estrogen Receptor Study (Formalin-Methanol-Acetone–Fixed Smears)

1. Prepare smears (see Table 2.7).
2. Remove excess phosphate buffered saline (PBS).
3. Incubate at room temperature with normal goat serum for 15 ± 2 minutes.
4. Drain normal goat serum (do not rinse).
5. Incubate at room temperature with primary antiestrogen receptor antibody* for 30 ± 2 minutes. (Use normal rat antibody* for negative antibody control.)
6. Rinse with PBS (two changes) for 5 ± 1 minutes.
7. Incubate at room temperature with goat antirat antibody* for 30 ± 2 minutes.
8. Rinse with PBS (two changes) for 5 ± 1 minutes.
9. Incubate at room temperature with rat peroxidase antiperoxidase solution* for 30 ± 2 minutes.
10. Rinse with PBS (two changes) for 5 ± 1 minutes.
11. Incubate at room temperature with DAB solution* for 6 ± 1 minutes. (Dissolve one DAB tablet in 5 ml of substrate 10–15 minutes prior to color development.)
12. Rinse with distilled water for 5 ± 1 minutes.
13. Stain with 1% (V/V) Harris hematoxylin for 4–6 minutes.
14. Rinse gently in tap water for 5 ± 1 minutes.
15. Dehydrate, clear, and mount in Eukitt.

*Abbott ER-ICA system; Abbott Laboratories, Chicago, IL.
DAB = 3,3'-diaminobenzidine tetrahydrochloride.

TABLE 2.12. Preparation of Glutaraldehyde (2.5%) in 0.1M Phosphate Buffer

Stock solution "A"
   27.6 gm of sodium phosphate monobasic ($NaH_2PO_4 \cdot H_2O$)
   Dissolve in 800 ml of distilled water and adjust to 1,000 ml
Stock solution "B"
   53.65 gm of sodium phosphate dibasic ($Na_2HPO_4 \cdot 7H_2O$)
   Dissolve in 800 ml of distilled water and adjust to 1,000 ml
0.2M phosphate buffer stock solution
   Add 115 ml of solution "A" to 385 ml of solution "B" and adjust pH to 7.3 using either solution "A" or "B"
Phosphate buffered glutaraldehyde
   100 ml of 0.2M phosphate buffer stock solution
   Add 10 ml of 50% glutaraldehyde
   Adjust to 200 ml with distilled water

### TABLE 2.13. Preparation of Glutaraldehyde (2.5%) in 0.025M Cacodylate Buffer

Stock cacodylate buffer solution (0.025M)
    Dissolve 1.07 gm cacodylate ($C_2H_6AsNaO_2 \cdot 3H_2O$) in 150 ml of distilled water
    Adjust pH to 7.4 with 1M hydrochloric acid
    Bring final volume to 200 ml with distilled water
Cacodylate buffered glutaraldehyde
    Add 10 ml of 25% glutaraldehyde to 90 ml of stock cacodylate buffer solution
    Check osmolarity (should be between 300 and 330)

### TABLE 2.14. Preparation of Karnovsky's Fixative

Supplies:  0.2M phosphate buffer
             0.1M phosphate buffer
             Paraformaldehyde
             Glutaraldehyde (50% solution)
             1N sodium hydroxide (NaOH)

1. Add 20 grams of paraformaldehyde to 250 ml of distilled water while heating and stirring the water (do not exceed 60°C).
2. Use 20 to 40 drops of 1N NaOH to dissolve the paraformaldehyde, then allow the solution to cool.
3. Add 50 ml of 50% glutaraldehyde, followed by 200 ml of 0.2M phosphate buffer and then sufficient 0.1M phosphate buffer to make 1 liter of the solution.
4. Adjust to pH 7.4.
5. The final product is 2% paraformaldehyde and 2.5% glutaraldehyde in 0.1M buffer. If this solution is aliquoted to small bottles, it can be frozen to maintain the stability for considerable periods and then thawed prior to use.

### TABLE 2.15. Routine and Semirapid Embedding Procedures for Electron Microscopy*

| | *Routine* | *Semirapid* |
|---|---|---|
| After primary fixation | | |
|   Buffer wash | 2 × 10 min | 2 × 5 min |
|   Osmium tetroxide/buffer | 1–2 hr | 1 hr |
|   Distilled water wash | 2 × 10 min | 2 × 5 min |
|   Ethanol 50% | 1 × 15 min | 2 × 5 min |
|   Ethanol 70% | 2 × 15 min | 2 × 5 min |
|   Ethanol 90% | 2 × 15 min | 2 × 5 min |
|   Ethanol 100% | 2 × 30 min | 3 × 10 min |
|   Propylene oxide | 2 × 10 min | 2 × 5 min |
|   Propylene oxide/epon (1:1) | 30 min | 15 min |
|   Propylene oxide/epon (1:3) | 30 min | 30 min |
|   Pure epon | 60 min | 30 min |
| Total processing time: | 6¼–7¼ hr | 3¾ hr |

*Low speed centrifugation may be necessary between the solution changes through to the osmium tetroxide if the tissue fragments and cell clusters are small.

**TABLE 2.16. Ultrarapid Embedding Procedure for Electron Microscopy of ABC Specimens***

| | |
|---|---|
| Aldehyde fixation (room temperature) | 30 min |
| Rinse three times in buffer | |
| Osmium tetroxide (1% in cacodylate buffer) | 15 min |
| Rinse three times in buffer | |
| Uranyl acetate (1% aqueous) | 5 min |
| 2,2-dimethyoxypropane (DMP; add 1 drop hydrochloric acid [37%] per 100 ml) | 15 min |
| DMP | 5 min |
| DMP/epoxy resin (1:1)** | 10 min |
| Epoxy resin | 10 min |
| Total processing time = | 1.5 hr |

*All steps during processing use a rotator.
**Fresh epoxy resin in capsules. Either polymerize overnight at 80°C or 1 hour at 80°C followed by 1 hour at 100°C.

fragments, small segments of cored tissue, and portions of blood clot. These components are readily visualized by swirling the fluid in the specimen vial and pipetting them off into a separate vial of buffer for conventional processing without prior centrifugation or any of the specialized techniques given below.

3. If the specimen consists of very small fragments or dissociated cells, it is centrifuged at low speed (1,500 rpm for 10 min), the supernatant is decanted, and the fragments are resuspended in a small amount of 10 or 22% bovine serum albumin[8,18] or agar.[9,29] Following recentrifugation at 1,500 rpm for 10 minutes in a disposable 2-ml conical tube, the material is refixed by gently layering 2.5% glutaraldehyde over the portion of serum albumin (avoid mixing the glutaraldehyde and albumin or a plug will not form). The solidified plug is then removed from the tube, divided into small blocks, osmicated, and further processed for embedding in epon-araldite. After the initial centrifugation of the needle rinse specimen (if the small tissue fragments are of sufficient quantity), postfixation of the resulting pellet with osmium tetroxide may result in the pellet remaining intact or at least with a few pieces large enough to be routinely processed into the epoxy resin.[7]

4. If an ABC specimen is considerably contaminated with red blood cells, it is subjected to a filtration technique.[1,20] In this procedure, a 25-mm Gelman syringe–type filter holder (Fisher Scientific Limited, Pittsburgh, PA) is fitted with a reusable nylon filter screen (20 μm pore size) cut from a sheet of Spectra/Mesh N (Fisher Scientific Limited, Pittsburgh, PA) using a 1-inch diameter circular punch (the type used for cutting leather or cork is ideal and can be purchased in hardware stores). The filter holder is attached to a 20-ml syringe barrel, which is placed over the mouth of a graduated cylinder (Fig. 2.3A). The suspension of tissue fragments and blood cells is poured into the syringe barrel and filtration is allowed to occur by gravity. When the filtration is completed, the filter with the tissue fragments lodged on it (see Fig. 2.3B) is removed from the holder

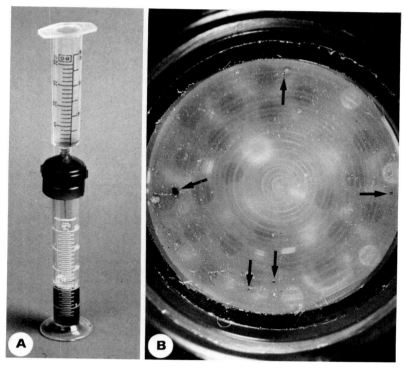

**Fig. 2.3.** A. Syringe barrel, filter holder, and graduated cylinder assembly for gravity filtration of bloody aspirations. B. Nylon mesh filter with multiple small tissue fragments (arrows).

using fine forceps. It is then shaken in buffer to dislodge adherent tissue fragments. Depending on the size of these fragments, the specimen is processed using one of the methods outlined in parts 1 to 3. The filter screen can be cleaned, dried, and reused a number of times. When needle rinse specimens that are contaminated by blood are flushed initially into a BES or glutaraldehyde, a considerable degree of clotting can be prevented if the specimen is shaken vigorously and the filtration and recovery of the tissue fragments will be expedited. This particular technique is practical and easy to employ.

Methods are available for embedding cell clusters from smears fixed in paraformaldehyde or glutaraldehyde for electron microscopy.[7,11] Some of the techniques reported by Widéhn and Kindblom[31] for visualizing small tissue specimens and embedding them in agarose could be advantageous for ultrastructural studies of ABC specimens.

## Glycogen Preservation

In certain types of tumors, such as small round-cell tumors in children or renal-cell carcinomas in adults it can be an advantage if the glycogen is not extracted or masked during the tissue processing phases, which can occur with en bloc staining

TABLE 2.17. Solution Preparation for Glycogen Stabilization

| | |
|---|---|
| Osmium tetroxide | 1.0 gm |
| Sucrose | 1.71 gm |
| Potassium ferrocyanide ($K_4Fe(CN)_6 \cdot 3H_2O$) | 1.5 gm |
| 0.1M sodium cacodylate buffer (pH 7.2) | 100 ml |

with uranyl salts. This situation is prevented by the addition of potassium ferrocyanide to the osmium solution (Table 2.17).

# REFERENCES

1. Akhtar M, Bakry M, Nash EJ: An improved technique for processing aspiration biopsy for electron microscopy. *Am J Clin Pathol* 85:57–60, 1986.
2. Akhtar M, Owen EW, Ali MA, Bakry MA: A simple method for processing fine-needle aspiration biopsies for electron microscopy. *J Clin Pathol* 33:1214–1216, 1980.
3. Angel E, Nagle RB: Magic marker: Practical problems in the use of immunoperoxidase histochemistry. *Pathologist* 39:13–18, 1985.
4. Azumi N, Joyce J, Battifora H: Does rapid microwave fixation improve immunohistochemistry? *Mod Pathol* 3:368–372, 1990.
5. Battifora H, Kopinski M: The influence of protease digestion and duration of fixation on the immunostaining of keratins. A comparison of formalin and ethanol fixation. *J Histochem Cytochem* 34:1095–1100, 1986.
6. Bencosme SA, Tsutsumi V: A fast method for processing biologic material for electron microscopy. *Lab Invest* 23:447–449, 1970.
7. Beutra C, Greer KP: Ultrastructural studies of cells in body cavity effusions. *Acta Cytol* 29:226–238, 1985.
8. Christian A: Experience with a technique for preparing cell suspensions. *Electron Microscopical Society of America Bulletin* 16:70, 1986.
9. Collins VP, Ivarsson B: Tumor classification by electron microscopy of fine needle aspiration biopsy material. *Acta Pathol Microbiol Immunol Scand {A}* 89:103–105, 1981.
10. DeLellis RA, Kwan P: Technical consideration in the immunohistochemical demonstration of intermediate filaments. *Am J Surg Pathol* 12(suppl 1):17–23, 1988.
11. di Sant'Agnese P, de Mesy Jensen K, Bonfiglio T, King D, Patten S: Plastic-embedded semi-thin sections of fine needle aspiration biopsies with dibasic staining: Diagnostic and didactic applications. *Acta Cytol* 29:477–483, 1985.
12. Domagala WM, Markiewski M, Tuziak T, et al: Immunocytochemistry on fine needle aspirates in paraffin miniblocks. *Acta Cytol* 34:291–296, 1990.
13. Dvorak AM: Procedural guide to specimen handling for the ultrastructural pathology service laboratory. *J Electron Microsc Techn* 6:255–301, 1987.
14. Gustafsson B, Månson J-C: Methodological aspects and application of the immunoperoxidase staining technique in diagnostic fine-needle aspiration cytology. *Diagn Cytopathol* 3:68–73, 1987.
15. Hultgren S, Hidvegi DF: Improved transmission electron microscopy technique for the study of cytologic material. *Acta Cytol* 29:179–183, 1985.

16. Koss LG, Woyke S, Olszewski W: *Aspiration Biopsy: Cytologic Interpretation and Histologic Bases.* New York, Igaku-Shoin, 1984.
17. Kurman RJ, Ganjei P, Nadji M: Contribution of immunocytochemistry to the diagnosis and study of ovarian neoplasms. *Int J Gynecol Pathol* 3:3–26, 1984.
18. Lazzaro AV: Technical note: Improved preparation of fine-needle aspiration biopsies for transmission electron microscopy. *Pathology* 15:399–402, 1983.
19. Leong AS-Y, Wright J: The contribution of immunohistochemical staining in tumor diagnosis. *Histopathology* 11:1295–1305, 1987.
20. Mackay B, Fanning T, Bruner JM, Steglich MC: Diagnostic electron microscopy using fine needle aspiration biopsies. *Ultrastruct Pathol* 11:659–672, 1987.
21. Miller RT, Groothuis CL: Improved avidin-biotin immunoperoxidase method for terminal deoxyribonucleotidyl transferase and immunophenotypic characterization of blood cells. *Am J Clin Pathol* 93:670–674, 1990.
22. Nadji M: The negative immunocytochemical result: What does it mean? *Diagn Cytopathol* 2:81–82, 1986.
23. Nadji M, Ganjei P: Immunocytochemistry in diagnostic cytology: A 12-year perspective. *Am J Clin Pathol* 94:470–475, 1990.
24. Odselius R, Falt K, Sandell L: A simple method for processing cytologic samples obtained from body cavity fluids and by fine needle aspiration biopsy for ultrastructural studies. *Acta Cytol* 31:194–198, 1987.
25. Olson NJ, Gogel HK, Williams WJ, Mettler FA Jr: Processing of aspiration cytology samples; an alternative method. *Acta Cytol* 30:409–412, 1986.
26. Ordóñez NG, Manning JT Jr, Brooks TE: Effect of trypsinization on the immunostaining of formalin-fixed, paraffin-embedded tissues. *Am J Surg Pathol* 12:121–129, 1988.
27. Pettigrew NM: Techniques in immunocytochemistry; application to diagnostic pathology. *Arch Pathol Lab Med* 113:641–644, 1989.
28. Pinkus GS, O'Connor EM, Etheridge CL, Corson JM: Optimal immunoreactivity of keratin proteins in formalin-fixed, paraffin-embedded tissue requires preliminary trypsinization; an immunoperoxidase study of various tumors using polyclonal and monoclonal antibodies. *J Histochem Cytochem* 33:465–473, 1985.
29. Plantholt BA: Embedding cell suspensions in agar. *Electron Microscopical Society of America Bulletin* 15:106, 1985.
30. Taylor CR, Kledzik G: Immunohistologic techniques in surgical pathology—a spectrum of "new" special stains. *Hum Pathol* 12:590–596, 1981.
31. Widéhn S, Kindblom L-G: Agarose embedding: A new method for the ultrastructural examination of the in-situ morphology of cell cultures. *Ultrastruct Pathol* 14:81–85, 1990.

# 3
# Principles of Immunocytochemistry

Immunocytochemical staining has made a substantial contribution to the field of cytopathology over the last decade, and is increasingly used as an adjunct to the traditional cytological criteria in diagnosis by aspiration biopsy cytology. These methods exploit the specific binding of an antibody to a cell or tissue antigen, followed by localization with a detection system, and subsequent visualization of the reaction product. The immunocytochemical method was first introduced by Coons and colleagues[8] in 1941, using immunofluorescent-labeled antibody. Development of enzyme-labeled antibodies, and later application of these antibodies to formalin-fixed paraffin-embedded tissue, in the late 1960s and early 1970s made immunocytochemical techniques more versatile and practical.[2,12,34,49] Introduction of more sensitive detection systems such as peroxidase-antiperoxidase (PAP) and avidin-biotin-peroxidase complex (**ABC**) methods, as well as a growing number of polyclonal and monoclonal antibodies, contributed to the growth and evolution of the technique into a valuable diagnostic aid to supplement histopathological diagnosis.[18,21,46] The first detailed paper on application and potential value of immunocytochemical staining to a variety of exfoliative and aspiration biopsy cytological specimens was published by Nadji[29] in 1980. These techniques can be easily performed on cytological specimens. This chapter briefly presents the theoretical and practical aspects of immunocytochemistry. For more comprehensive information, excellent references and workshops are available.[9,23,39,48]

## ANTIBODIES

Successful detection of antigens in cells depends on specific binding with antibodies directed to them. Although the conventional polyclonal antibodies have been useful, the introduction of monoclonal antibodies in 1975 by Köhler and Milstein[21] was a major step forward to boost the use of immunocytochemistry in diagnostic surgical pathology and cytopathology.[26,35,43] A range of monoclonal or polyclonal

TABLE 3.1. Comparison of Monoclonal and Polyclonal Antibodies*

| | Monoclonal | Polyclonal |
|---|---|---|
| Immunogen | Purity is not a major concern | Pure as possible |
| Antibody | | |
|   Composition | Homogeneous (single immunoglobulin) | Heterogeneous (immunoglobulins of different class and subtype) |
|   Chemical and immunological properties | Constant | Variable (animal to animal) |
|   Yield | Gram quantities | Submilligram to milligram quantities |
| Sensitivity | | |
|   Affinity for epitopes | Less sensitive,** often low | More sensitive, usually high (overall) |
|   Susceptibility to fixation/processing | Epitope may be masked or unavailable due to changes of the cell cycle or fixation/processing | Some epitopes resist fixation/processing because of more complex nature of antigen |
| Specificity | | |
|   Antigen recognized | A single epitope | Several epitopes on an antigen |
|   Cross-reactivity | Usually absent*** | Usually present**** |
|   Contamination by serum | Small amount | Variable amount |
|   Nonspecific irrelevant antibodies | Absent | Present |
| Reproducibility | Reproducible | May vary from batch to batch or lot to lot |
| Cost | More expensive | Less expensive |

*From Angel E, Nagle RB: Magic marker: Practical problems in the use of immunoperoxidase histochemistry. *Pathologist* 39:13–18, 1985; Edwards PAW: Some properties and applications of monoclonal antibodies. *Biochem J* 200:1–10, 1981; Saunders RL, David GS, Sevier ED: Monoclonal antibodies: Advances in in vitro and in vivo diagnostics. *Pathologist* 38:638–646, 1984; and Whiteside TL: Monoclonal antibodies in tissue section diagnosis, in Gordon DS (ed): *Monoclonal Antibodies in Clinical Diagnostic Medicine.* New York, Igaku-Shoin, 1985.

**Using monoclonal antibody "cocktail," recognizing different epitopes of the same molecules, may increase sensitivity.

***A monoclonal antibody might lack diagnostic specificity (antigen specificity) if it recognizes an epitope shared by different antigens. For example, monoclonal antibody Leu-M1 is a myelomonocytic-related marker; however, the antigen is also expressed in nonlymphoid cells, such as in a variety of carcinomas.

****Can be decreased if purified antigen is used to produce it or if the antibody itself is affinity-purified before use.

antibodies to locate corresponding antigens are available commercially. Although the use of polyclonal antibodies in immunocytochemistry is gradually decreasing in favor of monoclonal antibodies,[32] there are still some polyclonal antibodies that are preferable to monoclonal ones. A general comparison between the two types of antibodies is presented in Table 3.1. Often there are several antibodies, monoclonal and polyclonal, available for a given diagnostic problem.

Appropriate titers of every new antibody must be done on positive and negative control specimens prior to use on unknown diagnostic samples. Titers of the primary antibody or checkerboard titers of linking antibodies should be checked and documented periodically. A new batch of antibody must be compared with the old one, and retitration may be required. Optimal working dilution is a titer that gives maximum specific staining of the cells with a minimum or no background staining. If properly maintained according to the manufacturer's recommendation, most antibodies will remain stable for months and even years.[33]

# IMMUNOCYTOCHEMICAL METHODS

## Detection Systems

Success or failure of immunocytochemical methods depends on the ability of a detection system to localize the site of antibody binding (the antigen) in the cells, which can then be visualized using chromogens. Immunoperoxidase methods are particularly useful because the staining is permanent, and counterstaining of the cells allows correlation with traditional cytological criteria. More than 30 different methods have been developed since the inception of immunocytochemistry, which are well summarized in a recent publication by Myers.[28] These methods include direct and indirect methods, the latter using either labeled or unlabeled antibodies. The primary reason for development of new methods, entirely new systems, or modifications of existing methods is the increased sensitivity through detection of the smallest quantities of an antigen. This detection is done by increasing the ratio of label to antigen sought. In general, indirect methods have greater sensitivity than direct methods, and unlabeled antibodies are more desirable than chemically labeled antibodies.[16] A good detection system should adequately amplify the positive staining without contributing to the background.[38]

Despite the presence of several detection systems, the three methods illustrated in Figure 3.1 are widely used. For routine use it is ideal to utilize a single detection system for most antigens. The PAP technique employs an unconjugated primary antibody that reacts with an antigen, a secondary (link or bridge) antibody, and a soluble horseradish PAP complex. The antibody in the PAP complex and the primary antibody are raised in the same animal species; therefore, the secondary antibody can link the two and complete the reaction. Both the *ABC* and the peroxidase-labeled avidin-biotin (LAB) techniques employ a primary antibody similar to the PAP technique, followed by a secondary biotinylated antibody. In the

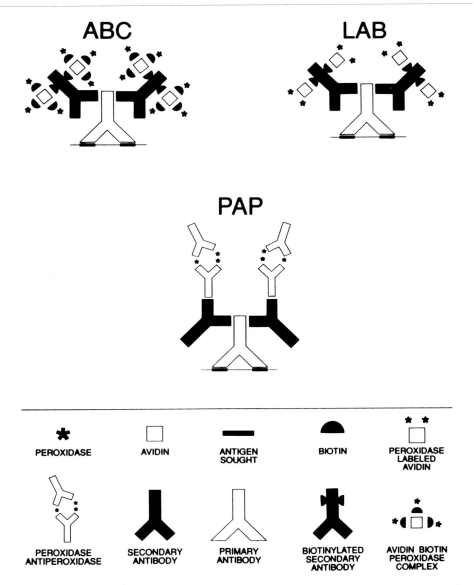

**Fig. 3.1.** The end-product of three widely used detection systems: Avidin-biotin-peroxidase complex *(ABC)*, peroxidase-labeled avidin-biotin (LAB), and peroxidase-antiperoxidase (PAP).

*ABC* method the reaction is completed by adding a preformed avidin-biotin-peroxidase complex. The last step in the LAB technique is the addition of peroxidase-labeled avidin. The *ABC* method is several times more sensitive than PAP, and the LAB method is reported to be more sensitive than *ABC*.[19,44] However, comparison between different methods is only meaningful if sensitivity, efficiency, and unique requirements of any single method are considered.[16] Swanson and colleagues[47] recently introduced a combined *ABC* and PAP method, an avidin-biotin-peroxidase-antiperoxidase complex, with significantly enhanced sensitivity. Other labels such as alkaline phosphatase and glucose oxidase can be used instead of horseradish peroxidase; however, these labels are not widely used in diagnostic pathology laboratories.

Immunocytochemical methods standardized on formalin-fixed paraffin-embedded tissues in surgical pathology can be effectively applied to cell block sections provided they are fixed in the same type of fixative. Not all immunocytochemical methods used in surgical pathology, however, can be translated directly to cytological smears or cytospin preparations. For example, an *ABC* method with aminoethylcarbazole (AEC) chromogen, which works effectively on formalin-fixed paraffin-embedded tissue, causes amplification of nonspecific or background staining on cytological smear preparations.[56] Zarbo and colleagues[56] recommended that the *ABC* immunocytochemical method (with 3,3'-diaminobenzidine tetrahydrochloride [DAB] as the chromogen), perfected on formalin-fixed paraffin-embedded tissue, can be directly applied to various cytological preparations. Our experience supports this view. The use of alcohol fixation has many advantages (see "Fixation" section in Chapter 2), and it is worthwhile standardizing the technique on alcohol-fixed paraffin-embedded tissue before applying them to cell block sections or cytological smears. Generally, no major modification is necessary. The significant advantage is that *no protease digestion is necessary for alcohol-fixed cells or tissues.*

## Blocking Methods

Background staining is perhaps the most common problem in immunocytochemistry if various blocking steps are not performed in the procedure. It is manifested by a relatively homogeneous diffuse staining of cells throughout the smear or cell block section, or nonspecific staining of the connective tissue, both of which may cause difficulty in interpretation. Although this staining is more common with polyclonal antibodies, some monoclonal antibodies may also cause background staining.[32] The staining may be due to the ionic or hydrophobic interaction between preexisting natural unrelated antibodies or serum protein in the animal used to raise the desired antibody and the tissue substrate.[33,38] Preincubation of the smear or cell block section with albumin or nonimmune/preimmune serum from the same animal species used to raise the secondary antibody can minimize undesirable staining.[28,38] This preincubation is performed before application of primary antibody. Background staining of epithelial cells may occur due to the presence of natural antikeratin antibodies. Using antisera at the highest possible dilution or incubation for short periods is usually effective.[33] Another very important blocking step is to quench endogenous enzyme activity normally present in the cells. Horseradish peroxidase is the most commonly used labeling enzyme in routine diagnostic immunocytochemistry. Endogenous peroxidase enzyme is present in relatively large amounts in polymorphonuclear leukocytes and macrophages. Red blood cells, which are commonly seen in cytological samples, also exhibit pseudoperoxidase effects.[28] The activity of the endogenous peroxidase should be suppressed or it may result in undesirable staining of the cells. This suppression can be effectively done with freshly prepared hydrogen peroxide methanol solution.[1,28,36] The traditional method of quenching endogenous peroxidase (3% hydrogen peroxide methanol solution before application of antibodies; 30-minute incubation) may destroy cell surface markers, especially lymphoid markers. For these markers it is better to use an 0.3% solution of hydrogen peroxide methanol for 10 minutes after the secondary antibody incubation. Miller and Groothuis,[27] using this method with the addition of 0.1% sodium azide, reported increased sensitivity. Quenching of endogenous alkaline phosphatase or glucose oxidase should be performed when using these labels. Similarly, endogenous activity of biotin present in liver, pancreas, or renal cells may result in nonspecific staining in detection systems using the avidin-biotin technique.[11,55] When *ABC* methods are used mast cells may react positively. This reaction is believed to be due to ionic binding of the basic molecules of avidin to the sulfate groups of heparin in mast cells.[6]

## Chromogens

Chromogens are used for visualization of the antigen-antibody binding sites. Different chromogens are available; the most commonly used one is DAB, which forms a permanent brown-colored precipitate. We prefer to use DAB, particularly when the reaction product is enhanced by copper sulfate. DAB should be handled with caution because of its possible carcinogenic potential. Prealiquoted packages, which are safer to use, are now available. There are other chromogens, such as AEC, that produce a red color. Because AEC is soluble in alcohol and other solvents, the smears or cell block sections must be coverslipped using an aqueous mounting medium.[1] The stain is therefore not permanent.

## Controls

A variety of concomitant stringent controls must be employed in immunocytochemistry to determine effective operation of the reactions and to ensure validity of the results. The control specimens should be processed and stained in exactly the same way as unknown samples.[10,22,24,28,31] This can easily be achieved with cell block sections; however, it is more difficult to provide cytology control specimens, especially positive ones, for direct smears and cytospin preparations. Because formalin-fixed paraffin-embedded tissue controls are not comparable to cytological smears, which are processed and fixed differently, an alcohol-based fixative should be used for such control tissues. For routine immunocytochemistry, two types of controls are required; tissue or cell controls and antibody control.

### Tissue or Cell Controls

Sections of tumors and normal tissues known to express the antigen under investigation are used as positive control. Negative staining of the unknown sample can be regarded as a true negative result if the known positive control demonstrates positivity. It is generally better to use tumor tissue instead of normal tissue, as well as tumors of low- or intermediate-antigen density rather than strongly immunoreactive tumors as positive control. These measures will decrease the potential for false-negative interpretation of the unknown sample. Sections of tumors and normal tissues known to be devoid of the antigen under study are used as negative controls.

Tissue control slides used for immunocytochemical study in surgical pathology can be effectively used to control unknown cell block sections. If the cell blocks are fixed in alcohol, however, as in our laboratory, the control tissue should also be fixed in an alcohol-based solution. If this is not feasible formalin-fixed tissue can be used for controls but proper caution should be exercised to avoid false interpretation. It is ideal to use cytological control for smears and cytospin preparations. Although time-consuming and perhaps not feasible for some antigens, one method is to maintain a file of imprints of tumors known to express the antigen under investigation.[29,31] It is preferable, however, to use alcohol-fixed paraffin-embedded material if tissue controls are used. The best positive control is an "internal" or "built-in" one, consisting of nonneoplastic epithelial, lymphoid, or mesenchymal cells known to express the antigen being sought. This "internal" control is particularly useful for interpretation of cell block sections.

We have employed multitissue composite (sausage) control specimens for the last few years. This method, which was first described by Battifora,[3] has several advantages.[3,42] A large number of normal and tumoral tissues with a broad spectrum of antigen densities and tissues known to be devoid of particular antigens can be mounted in a single paraffin block. This technique (Table 3.2) not only simplifies the quality control but is also economical because smaller amounts of reagents are used per case.

## Antibody Control

The primary antibody is the most critical reagent used in immunocytochemical procedures; therefore, its specificity should be checked every time an unknown

TABLE 3.2. Preparation of Composite (Sausage) Block

1. Retrieve small pieces of fresh tissue from surgical pathology and fix in 70% alcohol.*
2. Select required tissues.
3. Cut into stripes (1 × 0.2 × 0.2 cm) and store in 70% alcohol.
4. Cut off 3 cm of sausage casing (purchased from meat market) and split.
5. Pin out the casing over a cork board and moisten with 70% alcohol.
6. Stack tissue strips, all running parallel, and place on one end of the casing.
7. Roll tissues in the casing. Make sure all tissue strips remain aligned and tightly packed.
8. Staple or suture the ends of the sausage casing to prevent unravelling during processing.
9. Place in wire basket and put on tissue processor.
10. After processing, cut off ends and embed in a deep mold.
11. Run the appropriate antibodies to check for the antigens sought.

*The tissue can also be retrieved from paraffin-embedded blocks, deparaffinized in toluene, and rehydrated in alcohol.

TABLE 3.3. Preparing Multiple Slides From a Single Smear

1. Remove the coverslip by soaking the slide in toluene for several hours or overnight.
2. Immerse the slide in toluene to remove residual mounting media.
3. Apply Diatex compound (American Scientific Products, McGaw Park, IL) on the slide with a pasteur pipette to cover the entire smear and form a meniscus.
4. Incubate the slide at 37–60°C 1–2 hours or overnight to harden the compound.
5. Soak the slide in warm water for 1 hour.
6. Slowly pry the compound off at the edges with a scalpel blade. If the whole smear does not peel off easily, continue the soaking.
7. Cut the removed smear sliver with scissors or scalpel blade into desired segments.*
8. Place each segment on a separate slide (immersed in a water bath). Place the same side down as it was on the original slide.
9. Drain and incubate at 37–60°C overnight in a horizontal position.
10. Place the slide in toluene until all of the compound is dissolved.
11. Rehydrate the smears with two changes of toluene, absolute ethanol, 50% ethanol, and distilled water. Smears are ready for staining.

*The number of segments depends on the size of the smear, cellularity, and number of antibodies being performed.

sample is stained. The best method is to preabsorb the antibody with the respective purified antigen. This approach is impractical and expensive, however, and the antigen may not be available. Most laboratories substitute the primary antibody with preimmune or nonimmune serum from the same animal species, an irrelevant antibody, or buffer solution, or omit it altogether.[28,31,32] No positive staining should occur in the final preparation.

Although performing antibody control is quite feasible in surgical pathology and cytology cell block material, it may be more difficult when only cytology smears or cytospin preparations are available. The best method is to use a duplicate direct or cytospin smear. If only a few cellular smears are available, multiple slides from a single smear can be prepared (Table 3.3).[20] Another method is to etch two circles on the smear with a diamond pen or a PAP PEN (Daido Sangyo Co. Ltd. Japan) and perform the test and antibody control on separate circles.[29,31]

*Other Control*

Controls for other reagents in the immunocytochemical procedures, such as secondary and tertiary antibodies, avidin-biotin reagents, PAP complex, or other reagents, can be performed similar to the primary antibody. For example, the presence or significance of endogenous peroxidase can be checked by simply applying the substrate-chromogen solution.[28]

## Automation

Immunocytochemical procedures include multiple steps that require several applications of antibodies or reagents, incubations, and washing of the slides. Therefore, it is labor-intensive with significant potential for errors. Variables such as volume and dilution of the antibodies, precise timing, and required temperature of different steps may affect the end results. Recently, automated systems using different techniques have been introduced for immunocytochemistry.[4,45] The liquid delivery system for antibodies and reagents is based on using pipettes or capillary action. The latter system introduced by Brigati and colleagues[4] appears promising. The system automates all the steps in immunocytochemistry from dewaxing (for tissue or cell block sections) to nuclear staining. Automation will most likely minimize technical errors and improve the reproducibility of immunocytochemical results, resulting in better standardized antigen detection.[4,31,32,41,45,51] Other advantages include more comparative results between laboratories, reduction of technologists' time, and a potential for cost savings.

## Quality Assurance

Immunocytochemical stains are increasingly used in cytopathology and have become a valuable adjunct for formulation of an accurate diagnosis. Like other pathology disciplines, a well-planned quality assurance (QA) program for systematic assessment of procedures, techniques, and interpretation of results is mandatory.[3,15,41,53] The main purpose of a QA program is to detect, reduce, and correct deficiencies in the procedures or techniques, as well as maintain diagnostic excellence in the delivery of patient care.[7] This involves standardization of the methodology and its reproducibility, and checking the quality and adequacy of cytopathologist's interpretation.

Quality control of immunocytochemical procedures and techniques should focus on several areas: retrieval of cytological material, including the preparation of smears, cytospins, and cytology cell blocks; fixation; processing and sectioning of the cytology cell blocks; primary and linking antibodies; detection systems; proteolytic enzymes; blocking steps; and chromogen and other solutions and buffers. All of these steps should be carefully monitored and the solutions periodically replaced. Reproducibility of the technique can be checked from time to time by repeat staining of cytological cases (more feasible with cell block sections) with known immunophenotypes. A comparison with the previous run should be done and documented. Interlaboratory and external QA programs in immunocytochemistry are useful if available to the laboratory.[40,54]

An important component of a QA program is to assess cytopathologist's diagnostic role. Interpretation of the immunocytochemical stains should be done by the cytopathologist responsible for the case under study. The cytopathologist should have proper understanding and experience in interpretation of these stains (see Chapter 4) or consult with the pathologist in charge of the immunopathology laboratory in larger hospitals. Immunocytochemical stains are not a substitute for cytomorphological interpretation, and results should be used in conjunction with well-established cytological criteria, clinical data, and electron microscopy results when available. In our laboratory, the immunocytochemical and cytological findings are compiled into a single final report issued by the cytopathologist in charge of the case. The same reporting procedure applies to electron microscopy and other special studies. This method of reporting will significantly decrease the potential for errors. Periodic assessment and documentation of selected cytopathology reports are mandatory in a QA program. The evaluation includes typographical and transcription errors, turnaround time, adequacy, and accuracy of interpretation. The latter is checked by correlation of immunostaining results with the final diagnosis for surgically removed specimens, electron microscopic studies on cytological or histological material, or clinical follow-up.

# CHOICES OF CYTOLOGICAL PREPARATIONS FOR IMMUNOCYTOCHEMISTRY

## Cell Blocks

Immunocytochemical staining can be performed on different types of cytological preparations, including direct smears, cytospin smears, and cell block sections. Filter preparations, however, are not suitable for this type of study.[29] In our experience the best results are obtained with cell block sections because they offer maximum sensitivity. The cell blocks should be cut thin (approximately 3–4 μm) to achieve one-cell-layer-thick sections. We try to prepare cell blocks from all aspirates, especially when the necessity for immunocytochemical or cytochemical studies is anticipated. The pellet of cells and tissue fragments obtained by centrifugation of the needle rinse is fixed in methanol-acetic acid (MAA). MAA is an ideal fixative for immunocytochemical studies because it preserves the antigenicity of most routinely sought antigens (see Chapter 2). It also results in detailed nuclear and cytoplasmic morphology in hematoxylin and eosin–stained sections. The use of cell block sections for immunocytochemistry has many advantages.[13,30,31]

The most important advantages are:

1. Multiple sections can be obtained, which in most cases are adequate for multiple antibody panel staining.
2. Immunocytochemical methods standardized on tissues can be effectively applied to cell block sections.
3. Tissue control can be effectively used.
4. Parallel sections of the same cells or fragments can be cut for light microscopy as well as negative antibody control.
5. The cytoplasm and nucleus are better exposed to antibodies and reagents compared with whole cells in smears.
6. It requires less antibodies and reagents than smears; therefore, it is more cost-effective.

## Smears

All immunocytochemical methods are easily applicable to direct smears and cytospin preparations. For most antigens, with the possible exception of lymphoid cell surface markers, the best results are obtained with alcohol-fixed unstained smears.[29,37] Similarly, cytospin preparations produce good results with most antibodies,[5] provided one-cell-layer-thick preparations are used. Cytospin preparations are particularly useful for immunophenotyping of malignant lymphomas (see Chapter 9). Papanicolaou-stained smears and cytospin preparations can also be used for immunocytochemistry.[17,28,38,50] After removal of the coverslip, the smear is rehydrated and immunocytochemically stained. Decolorization is not necessary because the Papanicolaou stain fades during the immunocytochemical procedure; the decolorization procedure may also cause false-negative results. If after completing the immunocytochemical procedure the Papanicolaou stain remains strong (an infrequent result), the slide can be treated briefly with 1% acid-alcohol.

The main advantages of using smears or cytospin preparations for immunocytochemistry are that they are readily available and can be easily prepared in most laboratories. In addition, surface markers can be easily demonstrated (even better than cell or tissue block sections) in smears and cytospin preparations.[30] However, there are many disadvantages.[11,13,25,30,31,56]

The most important disadvantages are:

1. Not all immunocytochemical methods used for tissue or cell block sections can be directly translated to smears and cytospin preparations.
2. The number of smears may not be adequate for multiple antibody testing and negative antibody control.
3. Strong staining may totally obscure the nucleus, causing difficulty in interpretation (e.g., benign versus malignant cells).
4. False-negative results for cytoplasmic markers are more common with smears than with cell or tissue block sections, probably due to the intact cell membrane, which may result in inadequate penetration of antibodies and reagents into the cell. Overnight incubation with primary antibody and pretreatment of slides with Triton X-100 may enhance the detection rate. Triton X-100 will increase the permeability of the cell membrane and increases the sensitivity of the technique.[25]
5. Direct smears require more antibodies to cover all the cells and therefore increase the cost.

## Preparing Multiple Slides From A Single Smear

There are occasions where only one or two cellular smears are available and there is a need for multiple antibody testing and negative antibody control. There are a number of ways to overcome this problem. The easiest way is to encircle the areas of interest on the smear with a diamond pen or a PAP PEN and perform multiple antibody and negative antibody control stains in separate circles.[11,29,31] Another alternative is to split the glass slide with a diamond knife and perform the staining on separate pieces.[50] These techniques, however, are not practical in most cases. We have found the technique reported by Jimenez Joseph and Gangi[20] for transferring the smear to multiple glass slides very useful. We have employed this technique, with minor modifications (Table 3.3), on several occasions with successful results.

# REFERENCES

1. Angel E, Nagle RB: Magic marker: Practical problems in the use of immunoperoxidase histochemistry. *Pathologist* 39:13–18, 1985.
2. Avrameas S, Uriel J: Méthode de marquage d'antigène et d'anticorp avec des enzymes et son application en immuno diffusion. *C R Acad Sci* 262:2543–2545, 1966.
3. Battifora H: Methods in laboratory investigation; the multitumor (sausage) tissue block: Novel method for immunohistochemical antibody testing. *Lab Invest* 55:244–248, 1986.
4. Brigati DJ, Budgeon LR, Unger ER, et al: Immunocytochemistry is automated: Development of a robotic workstation based upon the capillary action principle. *J Histotechnology* 11:165–183, 1988.
5. Brown DC, Gatter KC, Dunnill MS, Mason DY: Immunocytochemical analysis of cytocentrifuged fine needle aspirates; a study based on lung tumors aspirated in vitro. *Anal Quant Cytol Histol* 11:140–145, 1989.
6. Bussolati G, Gugliotta P: Nonspecific staining of mast cells by avidin-biotin peroxidase complexes (ABC). *J Histochem Cytochem* 31:1419–1421, 1983.
7. College of American Pathologists: *Standards for Laboratory Accreditation*. Skokie, IL, College of American Pathologists, 1987.
8. Coons AH, Creech HJ, Jones RN: Immunological properties of an antibody containing a fluorescent group. *Proc Soc Exp Biol Med* 47:200–202, 1941.
9. DeLellis RA: *Advances in Immunohistochemistry*. New York, Raven Press, 1988.
10. DeLellis RA, Kwan P. Technical consideration in the immunohistochemical demonstration of intermediate filaments. *Am J Surg Pathol* 12(suppl 1):17–23, 1988.
11. Dinges H-P, Wirnsberger G, Höfler H: Immunocytochemistry in cytology; comparative evaluation of different techniques. *Anal Quant Cytol Histol* 11:22–32, 1989.
12. Di Stefano HS, Marucci AA, Dougherty RM: Immunohistochemical demonstration of avian leukosis virus antigens in paraffin embedded tissue. *Proc Soc Exp Biol Med* 142:1111–1113, 1973.
13. Domagala WM, Markiewski M, Tuziak T, et al: Immunocytochemistry on fine needle aspirates in paraffin miniblocks. *Acta Cytol* 34:291–296, 1990.
14. Edwards PAW: Some properties and applications of monoclonal antibodies. *Biochem J* 200:1–10, 1981.
15. Elias JM, Gown AM, Nakamura RM, et al: Special report: Quality control in immunohistochemistry; report of a workshop sponsored by the biological stain commission. *Am J Clin Pathol* 92:836–843, 1989.
16. Elias JM, Margiotta M, Gaborc D: Sensitivity and detection efficiency of the peroxidase antiperoxidase (PAP), avidin-biotin peroxidase complex (ABC), and peroxidase-labeled avidin-biotin (LAB) methods. *Am J Clin Pathol* 92:62–67, 1989.
17. Gal R, Aronof A, Gertzmann H, Kessler E: The potential value of the demonstration of thyroglobulin by immunoperoxidase techniques in fine needle aspiration cytology. *Acta Cytol* 31:713–716, 1987.
18. Hsu S-M, Raine L: Protein A, avidin, and biotin in immunohistochemistry. *J Histochem Cytochem* 29:1349–1353, 1981.
19. Hsu S-M, Raine L, Fanger H: Use of avidin-biotin-peroxidase complex (ABC) in immunoperoxidase techniques: A comparison between ABC and unlabeled antibody (PAP) procedures. *J Histochem Cytochem* 29:577–580, 1981.

20. Jimenez-Joseph D, Gangi MD: Application of Diatex compound in cytology: Use in preparing multiple slides from a single routine smear. *Acta Cytol* 30:446–447, 1986.
21. Köhler G, Milstein C: Continuous culture of fused cells secreting antibody of predefined specificity. *Nature* 256:495–497, 1975.
22. Kurman RJ, Ganjei P, Nadji M: Contribution of immunocytochemistry to the diagnosis and study of ovarian neoplasms. *Int J Gynecol Pathol* 3:3–26, 1984.
23. Larsson L-I: *Immunocytochemistry: Theory and Practice*. Boca Raton, CRC Press, 1988.
24. Leong AS-Y, Wright J: The contribution of immunohistochemical staining in tumor diagnosis. *Histopathology* 11:1295–1305, 1987.
25. Li C-Y, Lazcano-Villareal O, Pierre RV, Yam LT: Immunocytochemical identification of cells in serious effusions; technical considerations. *Am J Clin Pathol* 88:696–706, 1987.
26. Mason DY, Gatter KC: The role of immunocytochemistry in diagnostic pathology. *J Clin Pathol* 40:1042–1054, 1987.
27. Miller RT, Groothuis CL: Improved avidin-biotin immunoperoxidase method for terminal deoxyribonucleotidyl transferase and immunophenotypic characterization of blood cells. *Am J Clin Pathol* 93:670–674, 1990.
28. Myers JD: Development and application of immunocytochemical staining techniques: A review. *Diagn Cytopathol* 5:318–330, 1989.
29. Nadji M: The potential value of immunoperoxidase techniques in diagnostic cytology. *Acta Cytol* 24:442–447, 1980.
30. Nadji M: The negative immunocytochemical result: What does it mean? *Diagn Cytopathol* 2:81–82, 1986.
31. Nadji M, Ganjei P: Immunocytochemistry in diagnostic cytology: A 12-year perspective. *Am J Clin Pathol* 94:470–475, 1990.
32. Nadji M, Morales AR: Immunohistochemical techniques, in Silverberg SG (ed): *Principles and Practice of Surgical Pathology*, ed 2. New York, Churchill Livingstone, 1990. Pp 103–118.
33. Naish SJ: Handbook: *Immunochemical Staining Methods*. Carpinteria, CA, Dako Corporation, 1989.
34. Nakane PK, Pierce GB Jr: Enzyme-labeled antibodies: Preparation and application for the localization of antigens. *J Histochem Cytochem* 14:929–931, 1967.
35. Ordóñez NG, Manning JT Jr, Brooks TE: Effect of trypsinization on the immunostaining of formalin-fixed, paraffin-embedded tissues. *Am J Surg Pathol* 12:121–129, 1988.
36. Osamura RY: Application of immunocytochemistry to diagnostic cytopathology. *Diagn Cytopathol* 5:55–63, 1989.
37. Osamura RY, Watanabe K, Akatsuka Y: Peroxidase-labeled antibody staining for carcinoembryonic antigen of cytologic specimens for light and electron microscopy. *Acta Cytol* 29:254–256, 1985.
38. Pettigrew NM: Techniques in immunocytochemistry; application to diagnostic pathology. *Arch Pathol Lab Med* 113:641–644, 1989.
39. Polak JM, van Noorden S: *Immunocytochemistry: Modern Methods and Applications*, ed 2. Bristol, England, J Wright & Sons, 1986.
40. Reynolds GJ: External quality assurance and assessment in immunocytochemistry. *Histopathology* 15:627–633, 1989.

41. Rickert RR, Maliniak RM: Intralaboratory quality assurance of immunohistochemical procedures; recommended practices for daily application. *Arch Pathol Lab Med* 113: 673–679, 1989.
42. Rowden G, Fraser RB: Preparation of "histocomposites" for direct immunohistological screening of monoclonal antibodies. *Stain Technology* 63:49–52, 1988.
43. Saunders RL, David GS, Sevier ED: Monoclonal antibodies: Advances in in vitro and in vivo diagnostics. *Pathologist* 38:638–646, 1984.
44. Shi Z-R, Itzkowitz SH, Kim YS: A comparison of three immunoperoxidase techniques for antigen detection in colorectal carcinoma tissues. *J Histochem Cytochem* 36:317–322, 1988.
45. Stark E, Faltinat D, Von der Fecht R: An automated device for immunocytochemistry. *J Immunol Methods* 107:89–92, 1988.
46. Sternberger LA, Hardy PH Jr, Cuculis JJ, Meyer HG: The unlabeled antibody enzyme method of immunohistochemistry; preparation and properties of soluble antigen-antibody complex (horseradish peroxidase-antihorseradish peroxidase) and its use in identification of spirochetes. *J Histochem Cytochem* 18:315–333, 1970.
47. Swanson PE, Hagen KA, Wick MR: Avidin-biotin-peroxidase-antiperoxidase (ABPAP) complex; an immunocytochemical method with enhanced sensitivity. *Am J Clin Pathol* 88:162–176, 1987.
48. Taylor CR: *Immunomicroscopy: A Diagnostic Tool for the Surgical Pathologist*. Philadelphia, WB Saunders, 1986.
49. Taylor CR, Burns J: The demonstration of plasma cells and other immunoglobulin-containing cells in formalin-fixed paraffin-embedded tissues using peroxidase-labelled antibody. *J Clin Pathol* 27:14–20, 1974.
50. Travis WD, Wold LE: Immunoperoxidase staining of fine needle aspiration specimens previously stained by the Papanicolaou technique. *Acta Cytol* 31:517–520, 1987.
51. Tubbs RR, Bauer TW: Automation of immunohistology. *Arch Pathol Lab Med* 113:653–657, 1989.
52. Whiteside TL: Monoclonal antibodies in tissue section diagnosis, in Gordon DS (ed): *Monoclonal Antibodies in Clinical Diagnostic Medicine*. New York, Igaku-Shoin, 1985.
53. Wick MR: Quality assurance in diagnostic immunohistochemistry; a discipline coming of age. *Am J Clin Pathol* 92:844, 1989.
54. Wold LE, Corwin DJ, Rickert RR, Pettigrew N, Tubbs RR: Interlaboratory variability of immunohistochemical stains; results of the cell markers survey of the College of American Pathologist. *Arch Pathol Lab Med* 113:680–683, 1989.
55. Wood GS, Warnke R: Suppression of endogenous avidin-binding activity in tissues and its relevance to biotin-avidin detection systems. *J Histochem Cytochem* 29:1196–1201, 1981.
56. Zarbo RJ, Johnson TL, Kini SR: ABC-immunoperoxidase staining of cytologic preparations: Improvement of specificity. *Diagn Cytopathol* 6:134–138, 1990.

# 4

# Interpretation and Limitations of Immunocytochemistry

Interpretation of immunocytochemically stained smears and cell block sections must be done with caution, flexibility, and full understanding of the potential pitfalls. Interpretation is perhaps the most difficult aspect of immunocytochemistry. The observer must be familiar with cytopathological criteria; therefore, experience is the most important factor in diagnosis by aspiration biopsy cytology (ABC). One has to formulate a list of differential diagnoses based on light microscopic cytopathological criteria, clinical data, and statistical probabilities. If the differential diagnosis is unreasonable or if the wrong questions are being asked, the immunostaining results could be misleading. *Immunocytochemical stains are not a substitute for cytomorphological interpretation and the results should be used in conjunction with light microscopic findings, clinical data, and other ancillary studies, if available.* Using it in this context, immunocytochemical staining increases the accuracy of tumor classification, thereby preventing major errors.

The large number of antibodies available may create a problem. Proper interpretation depends on selecting an appropriate set of antibodies that give consistent results. Most available antibodies, both monoclonal and polyclonal, react with a variety of neoplasms. There are currently only a few antibodies that are single-cell lineage-specific. These more specific antibodies may prove to be nonspecific when further experience with different tumor types is gained. Therefore, one should assume that most, if not all, available antibodies (especially the newly introduced ones) are nonspecific. It is also important to become familiar with the degree of sensitivity of commonly used antibodies. Generally, more specific antibodies are not very sensitive. The immunophenotype of the neoplasm is also dependent on the degree of tumor differentiation and the type and duration of fixation and processing.

For these reasons it is mandatory to use a well-defined panel of antibodies, rather than a single antibody, to distinguish between diagnostic possibilities with confidence. It is quite useful to employ antibodies to mutually exclusive antigens, such as antibodies to cytokeratins and leukocyte common antigen, when the differential diagnoses include malignant lymphoma and undifferentiated carcinoma. The results

**TABLE 4.1. Possible Causes of Aberrant /False-Positive Reactions**

> The marker is not as specific as advertized.*
> True differentiation (not observed before).
> Necrotic and crushed cells.
> Entrapped normal cells or serum.
> Reactive cells such as myofibroblasts.
> Antigen diffusion from normal cells and subsequent absorption/phagocytosis by neoplastic cells.
> Antigens phagocytosed by macrophages or neoplastic cells.
> Antibody cross-reactivity.
> Nonspecific binding of the antibody to cells or stroma.**
> Specific binding of natural antibodies in the animal used to raise the antisera.**
> Ineffective blocking methods.**
> Artifactual staining.***
> Inappropriate fixation.****
> Drying of cells in any step of the procedure.

*Particularly applicable to new antibodies.
**See"Blocking Methods" section in Chapter 3.
***See "Artifacts" section in this chapter.
****See "Fixation" section in Chapter 2.

are reliable only when the neoplastic cells express the expected antigen or antigens and negatively stain with other antibodies. In general, a positive stain is more reliable and valuable than a negative stain. When the immunocytochemical stain is positive, the possibility of aberrant/false-positive reaction should be considered (Table 4.1), and in equivocal cases, the stains should be repeated with additional complementary antibodies.[1-3,7,10,13-18] True negative results should be differentiated from false-negative results (Table 4.2).[3-6,8-12,16,19] The latter are usually due to technical or fixation problems. False-negative results for intracytoplasmic antigens may

**TABLE 4.2. Possible Causes of False-Negative Results**

> The cytological preparation may lack positive cells (especially scanty specimens) when only a small percentage of neoplastic cells express the antigen.
> Antigen expression is below the level of the detection system (less differentiated neoplasms).
> Lack of preservation of antigens (delayed or underfixation).*
> Denaturation or masking of antigens (overfixation, especially with formalin).*
> Inappropriate, denatured, or excessively diluted antibodies.
> Antibodies or reagents are not accessible to the antigens (center of large tumor fragments or clusters in smear preparations).
> Decolorization of Papanicolaou-stained smears prior to immunocytochemical staining (this step is not necessary).
> Technical problem in any step of the procedure.
> Inappropriate enzymatic treatment (in formalin-fixed specimens).
> Inappropriate blocking methods.**
> Drying of cell in any step of the procedure.

*See "Fixation" section in Chapter 2.
**See "Blocking Methods" section in Chapter 3.

be more common in cytological preparations than histological sections if only direct smears or cytospin preparations are used. This finding may be due to the relatively intact cell membrane in smears, which impedes penetration of the reagents.[9] Cell block sections, as might be expected, produce results comparable to those obtained in histopathology.

Most antibodies and reagents used for immunocytochemistry are not approved for diagnostic use, despite the fact that many pathologists use them for this purpose. The products usually carry a label indicating that it is designed to be used for research only and not intended for in vitro diagnostic use. This is another important reason for using the results in conjunction with other findings. When the results of immunocytochemistry do not support the experienced cytological interpretation, the latter should be preferred. With the introduction of more reliable and sensitive antibodies as well as methodological improvements, the number of noncontributory results will hopefully decrease. Attention to the following guidelines, facts, and limitations will decrease the potential for error.

# PATTERNS OF STAINING

The staining intensity usually varies from cell to cell in a single tumor fragment, from fragment to fragment, and even within a single cell. The heterogeneity of antigen expression is one of the most important characteristics of a true positive (Fig. 4.1). The staining may be predominantly membranous, such as lymphoid cell surface markers, cytoplasmic (diffuse, perinuclear, dot-like), or nuclear in distribution (Fig. 4.2). An example of the latter is positive nuclear staining with antibody to estrogen receptor (see Fig. 4.2D). Some antigens, such as neuron-specific enolase and S-100 protein, are expressed both in the cytoplasm and in the nucleus (see Fig. 4.2B). Although familiarity with patterns of staining is useful for interpretation of results, with a few exceptions it is of limited value for definite differentiation of different types of neoplasms. For example, a canalicular staining with antibody to carcinoembryonic antigen is characteristic of hepatocellular carcinoma in ABC of the liver (see Fig. 10.4C). Interpretation of vacuolated cells may be difficult, because immunoreactivity may be limited to the scanty extravacuolar cytoplasm. If all the neoplastic cells stain with the same intensity (usually pale), a nonspecific reaction should be ruled out. This is particularly true in aspirates with many necrotic (Fig. 4.3) or crushed cells. When there is extensive necrosis, immunocytochemical stains should be interpreted with extreme caution.

# NONNEOPLASTIC CELLS VERSUS MALIGNANT CELLS

It is not uncommon to see nonlesional cells admixed with neoplastic cells in ABC specimens. Examples include entrapped normal cells, normal cells from other organs or sites, blood vessels in normal or tumor tissue fragments (Fig. 4.4), neuroendocrine cells, and histiocytes. It is extremely important to distinguish nonneoplastic cells from malignant cells before interpreting immunocytochemical stains. This differentiation is entirely based on the experience of the observer using well-

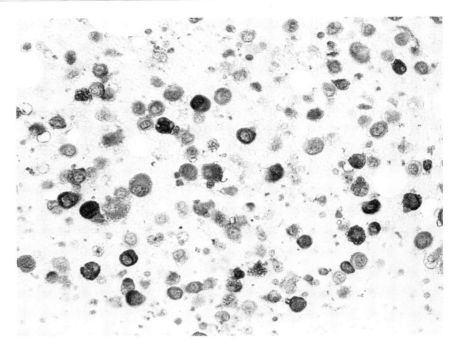

**Fig. 4.1.** ABC of malignant melanoma. Note the variable staining intensity of the melanoma cells with antibody to S-100 protein. × 260.

established cytological and architectural criteria. Currently, there are no immunocytochemical markers that can definitely distinguish nonneoplastic cells from neoplastic cells and benign neoplastic cells from malignant cells.

## BACKGROUND STAINING

Background staining is characterized by a relatively homogeneous coloration of the cells throughout the smear or cell block section, or nonspecific staining of the stroma, or both. Background staining may occur when proper blocking methods and thorough rinsing between different reagent applications are not performed (see Chapter 3). Background staining is particularly common with polyclonal antibodies. Background staining can be troublesome, and positively stained cells should be interpreted with caution. Immunostaining of neoplastic cells should be more intense than the background to be significant. Background staining decreases the sensitivity of the immunocytochemical stains because weak reactions cannot be reliably interpreted.

## ARTIFACTS

Neoplastic or nonneoplastic cells in an aspirate may contain pigments such as melanin, hemosiderin, bile, or anthracotic pigments (Fig. 4.5A). These pigments should

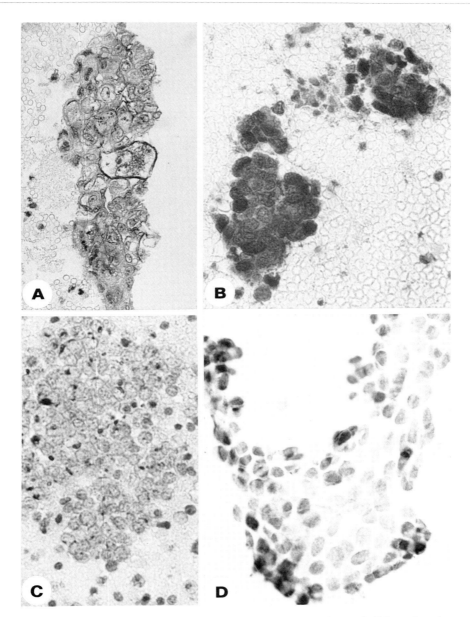

Fig. 4.2. Patterns of immunocytochemical staining, ABC. **A.** Placental alkline phosphatase. Note predominantly membranous staining. × 260. **B.** Neuron-specific enolase. Note both nuclear and cytoplasmic staining. × 416. **C.** Low-molecular-weight cytokeratin. Note dot-like cytoplasmic staining. × 416. **D.** Estrogen receptor. Note nuclear staining. × 416.

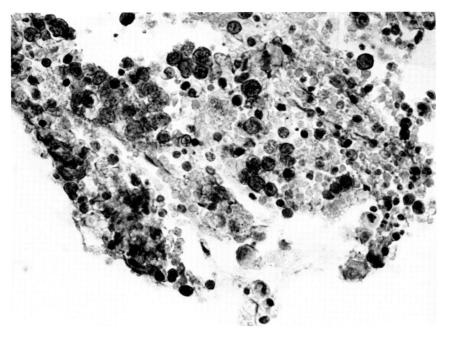

Fig. 4.3. ABC of lung. Most necrotic cells in the background stain positively with antibody to neuron-specific enolase. When there is necrosis, the immunocytochemical stains should be interpreted with extreme caution. × 416.

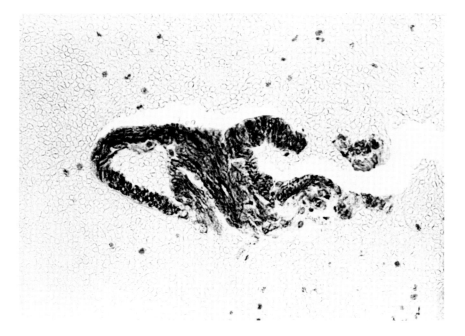

Fig. 4.4. ABC of chest wall. Blood vessels in normal or tumoral tissue are frequently present in aspirates and stain positively with antibody to vimentin. × 260.

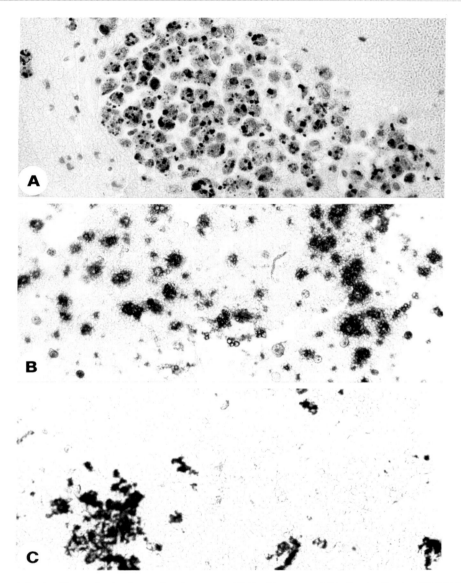

**Fig. 4.5.** Artifacts. **A.** Anthracotic pigments, ABC of lung. × 260. **B.** Antibody (UCHL1) precipitate. × 260. **C.** Partially dissolved gelatin. × 260. We currently use AES adhesive instead of gelatin.

not be mistaken as true positive staining. One can avoid the potential for misinterpretation by comparing the slide in question with the negative antibody control slide, which may also show the presence of the pigment. Stain (see Fig. 4.5B) and fixative precipitates, as well as improper deparaffinization and partially dissolved gelatin (see Fig. 4.5C), may cause difficulty in interpretation. These precipitates, however, are usually present randomly across the slide and are not confined to just the cells. Folded areas, knife nicks, and the edge of the cell block sections may artifactually stain positively; these results should be disregarded (Fig.4.6). Large

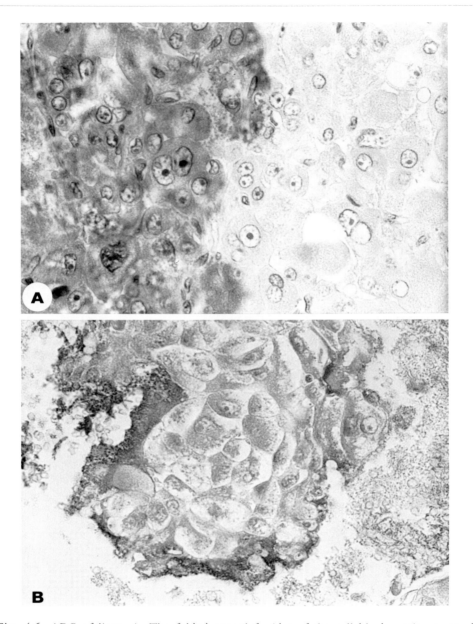

Fig. 4.6. ABC of liver. **A.** The folded area (left side) of the cell block section trapped immunocytochemical reagents and stained positive. × 416. **B.** The edge of a tissue fragment stained positively in the cell block section. × 416. These artifactual stainings should be disregarded.

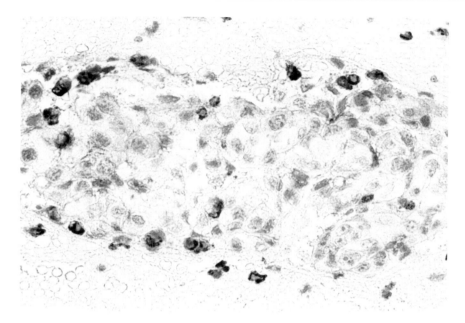

Fig. 4.7. Polymorphonuclear leukocytes express a nonspecific cross-reacting antigen with most antibodies to carcinoembryonic antigen. × 416. This positive staining can be used as a "built-in" positive control.

clusters or fragments of tumor cells in cytological smears, especially cytospin preparations, may trap different reagents used in immunocytochemistry and stain positive, usually in the center of the cluster. In these situations the antigen should also be expressed in individual cells to be of diagnostic value.

## CONTROL

Tissue or cell control and negative antibody control should be processed with each unknown sample to ensure the validity of the results (see "Control section" in Chapter 3). Tissue control derived from surgical specimens can be effectively used to control unknown cell block sections. It is more appropriate to use cytological control for smears and cytospin preparations; however, this approach is not feasible for most laboratories, and tissue controls are practical and appropriate. In these situations, because such controls are not strictly comparable, proper caution should be exercised in interpretation.

To ensure the adequacy of immunocytochemical procedures, the best positive control is a "built-in" control. This internal control may be a cross-reacting antigen, such as nonspecific cross-reacting antigen, which is present in granulocytes. Antibodies to carcinoembryonic antigen, polyclonal, and even some monoclonals stain granulocytes, which can be used as an internal positive control (Fig. 4.7). Another very useful internal control is the intermediate filament vimentin, which is expressed in mesenchymal cells, lymphoid cells, and many epithelial cells. This is one of the reasons that we include an antivimentin antibody in most of our panels. A lack of staining of cells known to express vimentin is generally indicative of improper fixation or processing.

# REFERENCES

1. Angel E, Nagle RB: Magic marker: Practical problems in the use of immunoperoxidase histochemistry. *Pathologist* 39:13–18, 1985.
2. Chess Q, Hajdu SI: The role of immunoperoxidase staining in diagnostic cytology. *Acta Cytol* 30:1–7, 1986.
3. Dinges H-P, Wirnsberger G, Höfler H: Immunocytochemistry in cytology; comparative evaluation of different techniques. *Anal Quant Cytol Histol* 11:22–32, 1989.
4. Flens MJ, van der Valk P, Tadema TM, et al: The contribution of immunocytochemistry in diagnostic cytology; comparison and evaluation with immunohistology. *Cancer* 65:2704–2711, 1990.
5. Gould VE: Synaptophysin; a new and promising pan-neuroendocrine marker. *Arch Pathol Lab Med* 111:791–794, 1987.
6. Leong AS-Y, Wright J: The contribution of immunohistochemical staining in tumor diagnosis. *Histopathology* 11:1295–1305, 1987.
7. Lloyd RV: Immunohistochemical localization of chromogranin in normal and neoplastic endocrine tissues. *Pathol Annu* 22(Part2):69–90, 1987.
8. Myers JD: Development and application of immunocytochemical staining techniques: A review. *Diagn Cytopathol* 5:318–330, 1989.
9. Nadji M: The negative immunocytochemical result: What does it mean? *Diagn Cytopathol* 2:81–82, 1986.
10. Nadji M, Ganjei P: Immunocytochemistry in diagnostic cytology: A 12-year perspective. *Am J Clin Pathol* 94:470–475, 1990.
11. Ordóñez NG: The immunohistochemical diagnosis of mesothelioma; differentiation of mesothelioma and lung adenocarcinoma. *Am J Surg Pathol* 13:276–291, 1989.
12. Papsidero LD, Croghan GA, Asirwatham J, et al: Immunohistochemical demonstration of prostate-specific antigen in metastases with the use of monoclonal antibody F5. *Am J Pathol* 121:451–454, 1985.
13. Pettigrew NM: Techniques in immunocytochemistry; application to diagnostic pathology. *Arch Pathol Lab Med* 113:641–644, 1989.
14. Reid WA, Branch T, Thompson WD, Kay J: The effect of diffusion on immunolocalization of antigen. *Histopathology* 11:1277–1284, 1987.
15. Rickert RR, Maliniak RM: Intralaboratory quality assurance of immunohistochemical procedures; recommended practices for daily application. *Arch Pathol Lab Med* 113:673–679, 1989.
16. Rosai J: *Ackerman's Surgical Pathology*, ed. 7. St. Louis, CV Mosby, 1989.
17. Shoup SA, Johnston WW, Siegler HF, et al: A panel of antibodies useful in the cytologic diagnosis of metastatic melanoma. *Acta Cytol* 34:385–391, 1990.
18. Stanta G, Carcangiu ML, Rosai J: The biochemical and immunohistochemical profile of thyroid neoplasia, in Rosen PP, Fechner RE (ed); *Pathology Annual* (Part 1). Norwalk, CT, Appleton & Lange, 1988. Pp. 129–157.
19. Taylor CR: *Immunomicroscopy: A Diagnostic Tool for the Surgical Pathologist*. Philadelphia, WB Saunders, 1986.

# 5
# Commonly Used Antibodies

To locate corresponding antigens, a wide range of monoclonal and polyclonal antibodies, as well as detection systems, are available commercially. The number of new antibodies continues to increase, which, coupled with the fact that multiple antibodies may be available for a given diagnostic problem, are sources of confusion for cytopathologists.

Only a few of the available antibodies, such as muscle-specific actin (HHF35) and melanoma-specific antibody (HMB-45), have proved specific for a single cell lineage. The specificity, however, is not absolute. For example, occasional cases of nonmelanoma tumors, such as breast carcinoma and plasmacytoma, have been reported to react with antibody HMB-45.[15,97] This possibility should be taken into consideration, especially with the new antibodies. In addition, some of the more specific antibodies, such as antibody to chromogranin and HMB-45, are not very sensitive. To avoid major mistakes in classification of neoplasms in aspiration biopsy cytology (ABC) specimens, one should assume that there is currently no absolutely sensitive and specific antibody. Furthermore, different methods of fixation and processing of ABC specimens, including the type of fixative, duration of fixation, concentration of the antibody, and the type of detection system, may significantly alter the results. Methods and antibodies suggested in this book and other publications should therefore be used only as guidelines. The results should be confirmed at each laboratory, in a number of proven cases, before application to diagnostic cytopathological specimens.

Most of the stumbling blocks, however, could be overcome by choosing an appropriate diagnostic panel of reliable antibodies to resolve differential diagnoses and to arrive at an accurate classification. This issue is addressed in this chapter, as well as in Chapters 8, 9, and 10. One should not rely on the result of one antibody study even if it is proved specific, especially when cell blocks are not available for multiple antibody studies.

A variety of factors and situations affect selection of antibodies in a diagnostic laboratory. There are no doubt some antibodies that are much more useful than others. Although many of the available antibodies are useful in a specific situation, the majority of diagnostic problems can be resolved by using a limited number of antibodies. It is also neither practical nor economical to maintain an excessive inventory of antibodies for routine use in most laboratories. The number of antibodies used in each laboratory depends on the number and type of aspirates performed and the frequency of particular diagnostic problems encountered. Familiarity with the indications, specificity, sensitivity, cross-reactivity, and aberrant reactions, and other limitations or pitfalls of each antibody should govern selection of the antibody panel. The purpose of this chapter is to review the most commonly used antibodies, in alphabetical order, that have proved useful for classification of neoplasms in ABC specimens in our laboratory (see also Table 5.6).

## ACTIN

Actin is a contractile protein and a major component of the microfilament (6 nm) that exists in virtually all eukaryotic muscle and nonmuscle cells. Often actin filaments cannot be visualized ultrastructurally. On the basis of amino acid sequences and electrophoretic mobility, there are at least six isoforms of actin that may coexist in the same tissue or even the same cell.[47,183,186] Four of these isoforms (skeletal, smooth and cardiac muscle alpha, and smooth muscle gamma) are expressed exclusively in muscle tissue. A number of polyclonal and monoclonal antibodies to actin are available commercially. Some polyclonal antibodies to actin recognize both muscle and nonmuscle isoforms and therefore are not appropriate for immunophenotyping of myogenic sarcomas.[129] We prefer muscle-specific monoclonal antibody HHF35, which recognizes an epitope common to the actin isoforms, alpha and gamma, that are specific to muscle.[182,183] In addition to normal muscle tissue, this antibody localizes to myoepithelium, myofibroblasts, and neoplasms of muscle origin (Plate 5.1A; see Figs. 8.12, 8.14, 8.15). All nonmuscle sarcomas are negative.[159,182]

## ALPHA-FETOPROTEIN

Alpha-fetoprotein (AFP) is an oncofetal glycoprotein normally synthesized in fetus, yolk sac, liver, and to some extent by the mucous glands of the gastrointestinal tract.[89,176] Elevated levels of AFP in the serum may be found in hepatocellular carcinoma, endodermal sinus tumor, embryonal carcinoma, teratoma, and a few other neoplasms such as gastric carcinoma.[89,176] The major use of AFP in ABC specimens is to support the diagnosis of hepatocellular carcinoma (see Plate 5.1B). Although many hepatocellular carcinomas are reported to express AFP, the staining may be focal and the number of positive cells may be as low as 5% of the neoplastic cells.[176] We have demonstrated a similar result in ABC specimens. In ABC of the liver, where germ-cell neoplasms are rarely a consideration, AFP expression strongly supports the diagnosis of hepatocellular carcinoma and excludes the possibility of metastatic carcinoma or cholangiocarcinoma. Negative staining, however, does not rule out the possibility of hepatocellular carcinoma. AFP is expressed in the majority of endodermal sinus tumors and embryonal carcinomas (see Plate 5.1C, Fig. 8.20) and occasionally in pure immature teratomas.[76,91]

# ALPHA-1-ANTICHYMOTRYPSIN AND ALPHA-1-ANTITRYPSIN

Alpha-1-antichymotrypsin (A1-ACT) is a single polypeptide chain with a molecular weight of 68 kd. It is mainly synthesized in the liver and has a protease inhibitory activity, a function that is shared with alpha-1-antitrypsin (A1-AT).[93,173] Indeed, there is a high level of homology between the DNA sequences of A1-ACT and A1-AT.[93] These antigens have been used as markers of histiocytes and reticulum cells. However, most sarcomas, many carcinomas, and malignant melanomas may express A1-ACT and A1-AT (Fig. 5.1).[93,96,127,173] In isolation, these polypeptides are therefore of little value in aiding the diagnosis of sarcoma or carcinoma or for confirmation of histiocytic differentiation.

# B72.3

B72.3 is a monoclonal antibody that recognizes a high-molecular–weight (> 1,000 kd) tumor-associated glycoprotein (TAG-72).[80] B72.3 is generated against a membrane-enriched fraction of metastatic breast carcinoma. Positive staining is seen around the periphery of the cells or within the cytoplasm, or both. The original studies using B72.3 demonstrated TAG-72 expression in most adenocarcinomas, particularly those arising from breast, ovary, lung, and colon, as well as some other neoplasms.[80,81,111,138] The results were similar in effusion cytology, ABC specimens, and surgically resected tumors.[80] In contrast, benign lesions from different organs (except apocrine metaplastic cells), cells in benign effusions, small-cell carcinomas of lung, malignant melanomas, malignant lymphomas, and sarcomas showed no reactivity. On the basis of these results it was concluded that B72.3 defines a tumor-associated antigen that is expressed in neoplastic cells versus benign cells and may be used as a novel adjunct for the diagnosis of neoplasms in effusions and ABC specimens.[80,81]

Recent data using B72.3 antibody in breast aspirates are controversial. In a few retrospective and prospective studies applying B72.3 to breast ABC specimens, expression of TAG-72 was demonstrated in a large percentage of breast carcinomas, and it was suggested as a potentially valuable diagnostic adjunct in ABC of the breast.[88,107,108] Bergeron and colleagues[12] evaluated the role of B72.3 antibody in histological and cytological specimens of benign and malignant breast lesions. Positive staining was found not only in a high percentage of carcinomas, but also in some atypical lobular hyperplasia, sclerosing adenosis, and scar. In a more recent study, TAG-72 antigen was demonstrated in only 10 to 27% of breast carcinomas.[175] Tubular carcinomas were negative. One of eight atypical ductal hyperplasia and most apocrine metaplastic cells and apocrine carcinomas stained positively.[175]

In a pilot study, we stained 29 negative, benign, and "suspicious" ABC smears of the breast with B72.3 monoclonal antibody.[151] Only 19% of the "suspicious" cases with proved malignancy on surgical biopsies expressed TAG-72, which is similar to the results obtained by Tavassoli and colleagues[175] in formalin-fixed tissue. Strong immunoreactivity was seen in apocrine metaplastic cells. On the basis of our experience, as well as on more recent data, we conclude that monoclonal antibody

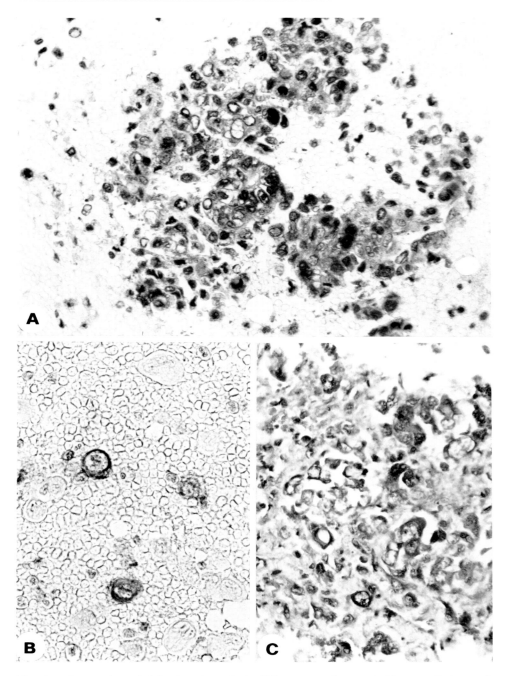

Fig. 5.1. Expression of alpha-1-antitrypsin in different neoplasms. A. Hepatocellular carcinoma, ABC of liver. × 320. B. Malignant fibrous histiocytoma, ABC of an abdominal mass. × 416. C. Metastatic adenocarcinoma, ABC of liver. × 260.

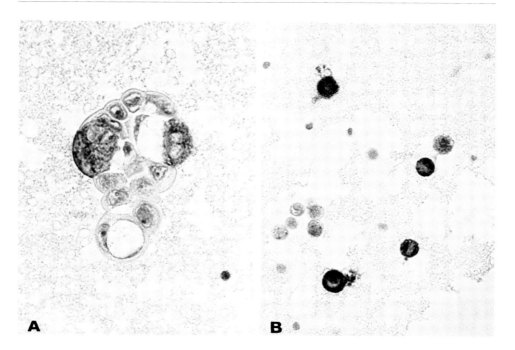

Fig. 5.2. TAG-72 expression by neoplastic cells, using B72.3 monoclonal antibody, in effusions supports the diagnosis of metastatic carcinoma. **A.** Pericardial effusion. × 416. **B.** Pleural effusion. × 416.

B72.3 is neither specific nor sensitive for malignant breast lesions and we do not anticipate that it will become a valuable adjunct for interpretation of breast ABCs.

B72.3 is also considered useful in separating malignant mesothelioma from adenocarcinoma. Szpak and colleagues[172] demonstrated positive staining in 19 of 22 lung adenocarcinomas, whereas all 20 malignant mesotheliomas were negative. We use B72.3 in a panel to distinguish mesothelial hyperplasia and malignant mesothelioma from metastatic carcinoma in effusion cytology (Fig. 5.2) and pleural or lung ABC (see Fig. 10.9). The result, positive or negative, should be interpreted in conjunction with other antibodies in the panel, light microscopic impressions, and clinical and/or radiological findings.

# CALCITONIN

Calcitonin is a polypeptide hormone with a molecular weight of 3.5 kd that is synthesized by normal and hyperplastic parafollicular C cells of the thyroid gland, medullary carcinoma of the thyroid, and a variety of nonthyroidal neoplasms such as some cases of pulmonary, pancreatic, and colonic carcinomas.[166,176] The main function of calcitonin is to suppress the release of calcium induced by resorption in the blood. C cells and corresponding medullary carcinomas may produce other substances such as somatostatin, bombesin, neuron-specific enolase, and chromogranin. Although a typical medullary carcinoma is relatively easy to diagnose in ABC specimens, the tumor is known to exhibit a variety of features that cause difficulty in interpretation.[73,114,155,166] Cytological as well as histological diagnosis, therefore, may be extremely difficult. Calcitonin is a sensitive and specific marker for confirmation of medullary carcinoma in ABC specimens (see Plate 5.1D, Fig. 10.12) because immunostaining is rarely negative or only focally positive. Other thyroid neoplasms are negative. We use calcitonin in conjunction with thyroglobulin stain in thyroid aspirates when there is cytological atypia, anaplasia, or other features suspicious for medullary carcinoma.

# CARCINOEMBRYONIC ANTIGEN

Carcinoembryonic antigen (CEA) is an oncofetal antigen present in tissues of endodermal origin and its tumor derivatives. It is composed of a large family of heterogeneous glycoproteins with an approximate molecular weight of 200 kd.[140,146,155] CEA has at least 10 to 15 epitopes, some of which are not specific for CEA and are shared by one or more CEA-related proteins such as nonspecific cross-reacting antigen (NCA) and biliary glycoprotein (BGP).[162] Antisera to CEA usually contain cross-reacting antibody to NCA, which results in staining of polymorphonuclear leukocytes.[194] Although this reaction is considered an excellent internal control for CEA in ABC specimens, there is potential for overinterpretation. This pitfall, however, can be avoided by considering the possibility of cross-reaction or by absorbing the antibody with spleen powder, which removes the antibody reactive for NCA and eliminates staining of neutrophils and histiocytes.[24] Neutrophils are also stained with some of the monoclonal antibodies, including the one we routinely use, which indicates the nonspecific nature of some of these monoclonal antibodies.[169]

Although CEA is primarily expressed by fetal tissue and malignant gastrointestinal tumors, including pancreatic carcinoma, adenocarcinomas of other organs such as lung and breast are also known to express CEA.[11,24,86,110,131,140,144,162,180,190,194,203] In contrast, CEA is not expressed in renal and prostatic adenocarcinoma as well as the majority of malignant mesotheliomas.

In our experience with monoclonal antibody, CEA expression is useful in two situations: (1) differentiation of malignant mesothelioma from adenocarcinoma in pleural or lung ABC and effusion specimens (see "Mesothelioma Versus Adenocarcinoma" section in Chapter 10), and (2) distinguishing hepatocellular carcinoma from adenocarcinoma in liver aspirates (see "Hepatocellular Carcinoma Versus Adenocarcinoma" section in Chapter 10). Most adenocarcinomas express CEA in a diffuse cytoplasmic pattern (see Plate 5.1E, Fig. 10.5). The majority of malignant mesotheliomas, however, are negative.[11,24,86,140,144,180] In a small percentage of malignant mesotheliomas that show positivity, the staining is generally weak and focal.[144,180] Recently, a peculiar staining pattern consisting of a coarse granular staining in the cytoplasm and a linear staining of the cell apices with a monoclonal antibody to CEA was reported in five of 45 consecutive cases of mesotheliomas.[169] The nature and significance of this finding remains to be determined. Most hepatocellular carcinomas express CEA in a characteristic, predominantly canalicular, pattern (see Plate 5.1F, Fig. 10.4).[203] In either situation, negative staining does not rule out the possibility of adenocarcinoma. Nonepithelial neoplasms, except for some biphasic synovial sarcomas, do not express CEA. Therefore, CEA can be used as a back-up stain to determine the epithelial lineage of an undifferentiated neoplasm. In the presence of reliable anticytokeratin antibodies, however, CEA would not be necessary.

## CHROMOGRANIN

Chromogranins (CGs) are a group of related, highly acidic proteins that are divided into three classes, namely, A, B, and C. Chromogranin A is the most widely distributed member and has a molecular weight of 68 to 75 kd.[6,101,197] CGs are found in normal tissues with neurosecretory granules, such as adrenal medulla, parathyroid gland, thyroid C cells, pancreatic islet cells, pituitary gland, and enterochromaffin cells of the gastrointestinal tract[64,103,202] and a variety of neuroendocrine neoplasms.[19,20,87,104,105,139,157] CG expression is specific for this group of neoplasms (see Plates 5.1G, 5.1H; Plate 5.2A). Several monoclonal antibodies to chromogranin A are available with comparable results. Lloyd and colleagues[102] also demonstrated the usefulness of chromogranin B expression in characterization of neuroendocrine cells in tumors.

The intensity of staining and the percentage of positive cells are highly dependent on the number of neurosecretory granules present.[202] For example, in neoplasms with many neurosecretory granules, such as carcinoid tumor (see Plate 5.1G) and pheochromocytoma, usually many cells strongly express CG. The exceptions are prolactinoma and insulinoma, which have many secretory granules but are weakly positive or negative for CG.[101] As might be expected, tumors such as small-cell carcinoma of the lung (see Plate 5.1H), Merkel-cell carcinoma, and neuroblastoma with few neurosecretory granules show focal or weak positivity. Although antibodies to CGs are very specific, they are not sensitive in poorly differentiated neuroendocrine neoplasms.

Scattered endocrine cells are usually present in lung, prostate gland, and other normal tissues, as well as in some tumors such as pancreatic and prostate adenocarcinoma.[101] These cells may be represented in ABC specimens; therefore, when the staining is focal the results should be interpreted with caution and in conjunction with other findings.

## CYTOKERATIN (CK)

See "Intermediate Filaments." Page 62.

## DESMIN

See "Intermediate Filaments." Page 65.

## EPITHELIAL MEMBRANE ANTIGEN

Monoclonal antibodies to epithelial membrane antigen (EMA), a glycoprotein isolated from human milk fat globule membrane, have been used to determine the epithelial lineage of undifferentiated neoplasms,[55,149] because this antigen is found on the surface of a variety of normal and neoplastic epithelial cells (see Plate 5.2B). Certain carcinomas, such as hepatocellular carcinoma, adrenal cortical carcinoma, and germ-cell neoplasms, do not express EMA.[149,165] In contrast, some non-Hodgkin's lymphomas, particularly Ki-1–positive large-cell lymphoma, occasional malignant melanomas, and some sarcomas such as synovial sarcoma, epithelioid sarcoma, and leiomyosarcoma, can express EMA.[32,116,149,171] EMA, therefore, is not a reliable alternative to cytokeratin stains in determining the epithelial nature of an undifferentiated neoplasm. EMA has been suggested as an effective marker to distinguish between malignant cells and reactive mesothelial cells in effusion cytology and ABC specimens,[178,195] because the latter are either negative or weakly positive. Both adenocarcinoma and malignant mesothelioma express EMA. Although our experience is similar, distinction between reactive mesothelial cells and malignant cells based on expression of EMA is not completely reliable.[187] It is also suggested that a predominantly cell-membranous EMA staining is a diagnostic clue in favor of malignant mesothelioma.[98] We have seen this phenomenon, however, in some adenocarcinomas (see Plate 5.2B). In certain circumstances, a combination of antibodies to EMA and cytokeratin may help to characterize further some carcinomas (see Fig. 8.1).

## ESTROGEN RECEPTOR (ER)

See the "Breast Carcinoma" section in Chapter 10. Page 269.

## GLIAL FIBRILLARY ACIDIC PROTEIN (GFAP)

See "Intermediate Filaments." Page 67.

## HUMAN CHORIONIC GONADOTROPIN

Human chorionic gonadotropin (hCG) is a pregnancy-associated protein that is elaborated by syncytiotrophoblastic cells of the placenta, choriocarcinoma, and embryonal carcinoma.[67,76,91] Serum hCG determination is used for patients with trophoblastic neoplasms to monitor the effect of therapy and the progress of disease. hCG is composed of two subunits. The beta-hCG subunit is used to detect tropho-

TABLE 5.1. Distribution of Different Types of Intermediate Filaments in Neoplasms

| Type | Molecular Weight (kd) | Neoplasms |
|---|---|---|
| Cytokeratins | 40–68 | Carcinoma, mesothelioma, some sarcomas |
| Vimentin | 57 | Sarcoma, malignant melanoma, malignant lymphoma, some carcinomas |
| Desmin | 55 | Neoplasms of muscle origin |
| Neurofilament proteins | 68, 150, 200 | Neoplasms of neuronal origin |
| Glial fibrillary acidic protein | 48–52 | Glioma |

blastic differentiation in germ-cell neoplasms;[155] however, it may be ectopically produced in other neoplasms.[67,155] The alpha-hCG subunit is produced by a number of nontrophoblastic neoplasms such as endocrine and exocrine tumors.[67]

# INTERMEDIATE FILAMENTS

Different cell types depend on cytoskeletal proteinaceous tubules and filaments for their internal organization, shape, and movement. These include microtubules (25nm), intermediate filaments (IFs) (8–10 nm), and microfilaments (6 nm). With some exceptions, IFs tend to be expressed in neoplasms in a pattern closely approximating that found in the tissue developing the tumor. This expression permits further characterization of undifferentiated or poorly differentiated neoplasms in cytopathology, especially in ABC specimens.[35,37,155,176] The use of antibodies to IFs can improve the accuracy of cytological diagnosis, provide information about the cell lineage or differentiation pathways of tumors, and prevent an erroneous diagnosis in cytologically difficult aspirates. Although IFs appear similar at the ultrastructural level, they are in reality five types (Table 5.1). Nuclear lamina also contains IF proteins (laminins) with molecular weights ranging from 60 to 80 kd.[47] Antibodies to IFs, especially cytokeratins, are the backbone of diagnostic immunocytochemistry, and we use them in most antibody panels. In our experience as well as others, alcohol fixation is superior to formalin fixation for preservation of filament antigenicity.[176]

Most neoplasms appear to express only a single IF type; however, coexpression of IF types not infrequently occurs. It is important to distinguish between true coexpression, pseudocoexpression, and false coexpression.[38] In true coexpression, the same tumor cell expresses more than one IF type (Table 5.2).[36,38,185] In pseudocoexpression, two or more tumor cell types each express a different type of IF (Table 5.3).[38] In many of these neoplasms, coexpression does not occur in all the cells, and tumor cells may express only one IF type. If coexpression of IF types is detected in an ABC specimen, especially from a metastatic neoplasm, the finding can be diagnostically useful because it reduces the number of differential diagnoses. False coexpression occurs when a mixture of tumor cells and normal

TABLE 5.2. True Coexpressions of Intermediate Filaments in Different Tumors*

| Combination | Tumor Types |
|---|---|
| Cytokeratin and vimentin | Carcinomas: kidney; thyroid (papillary, follicular, anaplastic); lung (adenocarcinoma and large-cell carcinoma); endometrium, ovary (serous); embryonal; and adenoid cystic carcinomas of the salivary gland |
| | Mesothelioma |
| Cytokeratin and neurofilaments | Neuroendocrine (Merkel cell) carcinoma; islet-cell tumor of pancreas; and carcinoid tumor (bronchial) and oat-cell carcinoma (rare) of lung |
| Cytokeratin, vimentin, and neurofilaments | Medullary carcinoma of thyroid |
| Cytokeratin, vimentin, and GFAP | Pleomorphic adenoma of salivary gland |
| Vimentin and desmin | Neoplasms of muscle origin |

*Coexpression does not occur in all cases.
GFAP = glial fibrillary acidic protein.

cells express different IFs. Numerous combinations may occur. For example, ABC of a lymph node with metastatic carcinoma shows positive cytokeratin staining of the neoplastic cells, and the lymphoid cells express vimentin (see Fig. 9.1). Close attention to cytomorphology to distinguish neoplastic cells from normal cells is essential to avoid misinterpretation.

## Cytokeratins

Cytokeratins (CKs) are the most complex of all IFs that are present in virtually all true epithelial cells and are therefore a sensitive marker for carcinomas. CKs are a family of water-insoluble intracellular polypeptides. On the basis of their molecular

TABLE 5.3. Pseudocoexpression of Intermediate Filaments in Different Tumors

| Combination | Tumor Types |
|---|---|
| Cytokeratin and vimentin | Synovial and epithelioid sarcomas;* phyllodes tumor; pulmonary blastoma; chordoma; embryonal carcinoma;* and malignant rhabdoid tumor |
| Cytokeratin, vimentin, neurofilaments, and desmin | Teratocarcinoma* |
| Vimentin and desmin | Triton tumor |
| Cytokeratin, vimentin, and desmin | Wilms' tumor |
| Vimentin and neurofilaments | Malignant granular cell tumor, progonoma |

*Some tumor cells may show true coexpression of vimentin and cytokeratin.

weight (40–68 kd) and isoelectric pH (5.2–7.8), Moll and colleagues[124] catalogued 19 human CKs and numbered them from 1 to 19. CKs can also be divided into acidic and neutral-basic subfamilies. Usually a subset of two to 10 CKs are expressed by a particular epithelium. CKs are always expressed as a pair (one acidic and one neutral-basic) of subunits. In normal tissues, the subset of CK subunits expressed is related to three factors: epithelial cell type (simple or stratified epithelium), state of cellular growth, and program of differentiation. Even in tumors, these factors control the expression of CK in a more or less reproducible manner compared with the tissue or organ of origin, and result in characteristic CK profiles in different epithelial neoplasms. These profiles generally remain constant in primary and secondary neoplasms. CK profiles, therefore, may help to characterize further some neoplasms for diagnostic purposes.[27,47,117,124,153,156]

For the sake of simplicity in diagnostic immunocytochemistry, CKs can be divided into two groups: lower molecular-weight CKs (LMW-CK) and higher molecular-weight CKs (HMW-CK). LMW-CKs are expressed by virtually all epithelial cells and many carcinomas; therefore, they are good wide-spectrum markers for epithelial differentiation. HMW-CKs are expressed in complex epithelia, such as squamous epithelium, and in certain neoplasms, such as squamous-cell carcinomas and epithelial mesotheliomas. Several commercially available polyclonal and monoclonal antibodies to CKs exist that may be confusing to novices. Monoclonal antibodies are more sensitive and specific. Most companies produce antibodies that are reactive to a spectrum of HMW-CKs and LMW-CKs such as AE1/AE3 or 35βH11/34βE12. The CK catalogue and patterns of reactivity of different monoclonal antibodies are summarized in Figure 5.3.[31,59] These antibodies are not necessarily comparable to each other. For example, antibodies AE1 and 35βH11 are both reactive to LMW-CKs, and antibodies AE3 and 34βE12 are reactive to HMW-CK (see Fig. 5.3). Hepatocytes and hepatocellular carcinomas are known to express LMW-CKs; however, they react with antibodies to AE3 and 35βH11. It is extremely important, therefore, to talk about CK profiles with specific commercially available antibodies rather than antibodies to LMW-CKs and HMW-CKs. Fixation is also very important. All these antibodies give better and more reliable results with alcohol-fixed cells. In this background these monoclonal antibodies may be valuable in further classification of carcinomas.[1,16,17,30,43,51,123,125,135,161,188,204] Despite the wealth of information in the literature these data should be used as guidelines only, and the sensitivity and specificity of these antibodies should be tested on proved cased in each laboratory before being used on diagnostic material. The immunocytochemical techniques employed and the fixation used have significant effects on the results.

Monoclonal antibodies to CKs are the major component of the diagnostic panels used in our laboratory to characterize further an undifferentiated neoplasm in ABC specimens. CK positivity with a broad-spectrum antibody or combination of monoclonal antibodies in the form of "cocktails"[100] supports the diagnosis of carcinoma (see Plates 5.2C, 5.2D, Figs. 8.13, and 9.1) and excludes the possibility of most sarcomas and malignant lymphomas. CK positivity is particularly useful when no evidence of epithelial differentiation is found at the ultrastructural level. Although CK positivity in tumor cells in an aspirate support epithelial origin of the neoplasms in the majority of cases, some exceptions exist. Synovial and epithelioid sarcomas, chordomas, and nerve-sheath tumors with epithelial differentiation coexpress kera-

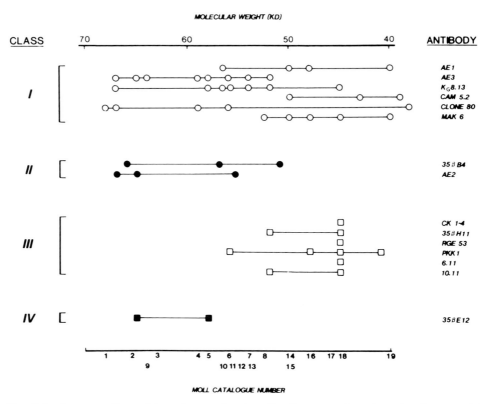

Fig. 5.3. Patterns of reactivity of monoclonal antibodies to cytokeratins: class I react with squamous, ductal, and simple epithelia (broadly reactive); class II react with higher molecular weight cytokeratins in squamous epithelium; class III react primarily with lower molecular-weight cytokeratins in simple and ductal epithelium; class IV react with cytokeratins in squamous and ductal epithelia. From DeLellis RA, Kwan P: Technical considerations in the immunohistochemical demonstration of intermediate filaments. *Am J Surg Pathol* 12 [suppl 1]:17–23, 1988. Used with permission.

tin and vimentin.[38,117] A high percentage of normal, benign, and malignant smooth muscle tissue express CK.[18,116,137] CK expression has also been demonstrated in some malignant melanomas,[118,206] rhabdomyosarcomas,[122] epithelioid vascular neoplasms,[48,61] and two cases of postirradiation malignant fibrous histiocytoma.[92,198] These observations emphasize the significance of using antibodies to CKs with other antibodies in the diagnostic panel when anaplastic tumors are evaluated.[28]

Negative CK staining does not exclude the possibility of carcinoma or malignant mesothelioma, unless it is supported by the results of other antibodies in the panel such as leukocyte common antigen or if a "cocktail" preparation of CK antibodies covering the wide range of CK molecular weights has been used. Furthermore, rare undifferentiated lung carcinomas have been shown to express only vimentin IFs.[66]

Therefore, the results should be interpreted in the background of cytological appearance and clinical findings.

## Vimentin

Vimentin is an IF with a molecular weight of 57 kd that is primarily present in neoplasms of mesenchymal derivation or differentiation.[47] Contrary to the initial belief that vimentin was a useful marker for sarcomas (see Plate 5.2E), many studies have demonstrated its expression in other neoplasms such as malignant melanomas (see Plate 5.2F) and malignant lymphoma, as well as several epithelial neoplasms (Fig. 5.4).[29,50,59,68,74,78,85,113,119,154,185,191,193] In fact, only adenocarcinomas of the colon, small intestine, gallbladder, and prostate seem not to express vimentin.[9] Epithelial cells in malignant effusions or tissue culture also switch on the expression of vimentin.[53,154] Many normal or reactive cells such as lymphocytes, macrophages, and endothelial cells express vimentin. Although this expression is useful as an internal control, extreme caution is manditory to avoid confusing them with neoplastic cells.

In cytological material, virtually all mesenchymal neoplasms, malignant melanomas, and malignant lymphomas express vimentin. Such an expression, in the absence of CK expression using a broad-spectrum or antibody "cocktail" to CKs, supports the diagnosis of these neoplasms. Further characterization can be done with other antibodies such as leukocyte common antigen, HMB-45, actin, and desmin. We use antivimentin antibody for two purposes: (1) in combination with other antibodies for possible classification of undifferentiated or poorly differentiated neoplasms, and (2) as a probe of antigen preservation. There are almost always tumor cells or normal cells in ABC specimens that express vimentin. The absence of this internal control usually indicates fixation or technical problems, or both.

## Desmin

Desmin is a relatively simple IF with a molecular weight of 50 to 55 kd that is frequently coexpressed with vimentin. Because of the close association of desmin filaments with membrane plaques and Z discs it is difficult to visualize the filaments by electron microscopy.[47] It is specifically expressed by skeletal, smooth and cardiac muscle, as well as their tumor derivatives comprising rhabdomyosarcoma and leiomyosarcoma.[3,4,10,33,47,120] Both tumors express desmin in a high percentage of cases in alcohol-fixed ABC specimens (see Plate 5.2G, Figs. 8.12, 8.14). It is, however, a more sensitive marker for rhabdomyosarcoma. Desmin is rarely expressed in myofibroblasts and nonmyogenic neoplasms such as malignant mesothelioma, malignant fibrous histiocytoma, malignant peripheral nerve sheath tumor, glial tumors, and malignant melanoma.[92,181,200] In these neoplasms the staining is usually focal in nature. We always use desmin and actin in combination when muscle differentiation is suspected.

## Neurofilament Proteins

Neurofilament proteins (NFPs) are IFs consisting of three polypeptide subunits: NF-L, NF-M, and NF-H, with molecular weights of 68, 160, and 200 kd, respec-

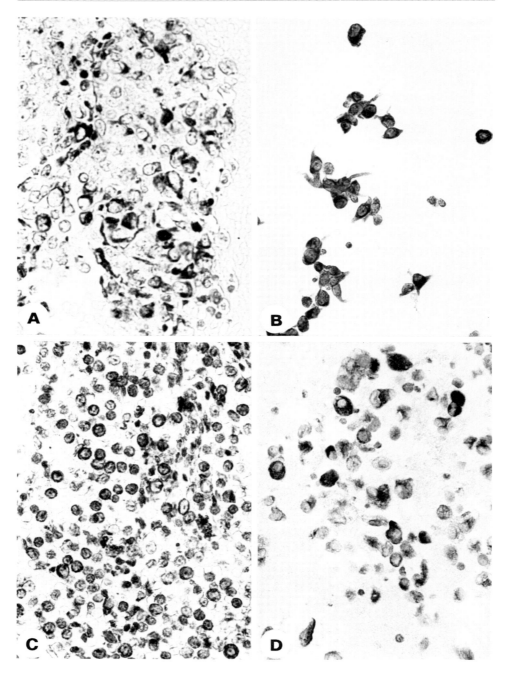

Fig. 5.4. Vimentin expression in different neoplasms. **A.** Rhabdomyosarcoma, ABC of a soft tissue mass. × 416. **B.** Metastatic malignant melanoma, ABC of an axillary lymph node. × 260. **C.** Malignant lymphoma, ABC of a pelvic mass. × 416. **D.** Adrenal cortical carcinoma, ABC. × 260.

tively.[47,70,158] Ultrastructurally they appear as randomly arranged sinuous filaments.[47] NFP expression has been shown in many neuroendocrine neoplasms.[14,94,95,121,189] Expression of NFP, in the absence of CKs, is seen in neuroendocrine neoplasms with neural differentiation such as neuroblastoma (with the exception of undifferentiated ones), pheochromocytoma, paraganglioma, medulloblastoma, and retinoblastoma.[8,99,130,142,143] Coexpression of NFP and CKs (LMW-CKs) occurs in nonneural-type neuroendocrine neoplasms such as carcinoid tumor, neuroendocrine carcinoma, and islet-cell tumor of the pancreas.[71,95,121,123,126,189] Monoclonal antibodies to NFP are available commerically and work better in alcohol-fixed specimens. NFP, as a tumor marker, is of limited value in diagnosis by ABC.

### Glial Fibrillary Acidic Protein

Glial fibrillary acidic protein (GFAP), the major protein present in glial cytoplasmic filaments, is a single peptide with a molecular weight of approximately 51 kd.[28,148] GFAP is expressed by normal, reactive, and neoplastic astrocytes; however, it is not restricted to astrocytes. Among primary brain tumors, GFAP has been demonstrated in astrocytic tumors, ependymomas, choroid plexus tumors, and some oligodendrogliomas and medulloblastomas.[5,22,26,148,179] Other nonneural intracranial neoplasms such as meningiomas, sarcomas, germinomas, malignant lymphomas, pituitary adenomas, and metastatic carcinomas are generally GFAP-negative. GFAP, has also been demonstrated in some peripheral nerve-sheath tumors, including schwannomas (mainly the deeply situated ones), neurofibromas, and, occasionally, malignant schwannomas.[62,84] The majority of pleomorphic adenomas of the salivary glands coexpress GFAP, keratin, and vimentin, which may be of value for interpretation of ABCs of head and neck tumors.[34]

In many hospitals, imprints and cytological crush preparations of brain biopsies are used as a supplement to frozen section diagnosis for intraoperative consultation. This technique is especially useful for small tissue fragments obtained by computed tomography–guided stereotatic, 14- to 22-gauge needles.[136] Immunocytochemical staining with antibodies to GFAP has been effectively applied to imprint and smear cytological preparations with satisfactory results[22,26,77] and is a useful adjunct for specimens with equivocal cytomorphology.

## LEUKOCYTE COMMON ANTIGEN (LCA)

See Chapter 9. Page 218.

## LEU-M1

See Chapter 9. Page 219.

## LEU-7

See Chapter 9. Page 221.

## LYMPHOID MARKERS

See Chapter 9. Page 218.

## MELANOMA-SPECIFIC ANTIBODY

HMB-45 is one of the few antibodies among a large number of melanoma-"specific" monoclonal antibodies that works both with alcohol-fixed and formalin-fixed paraffin-embedded tissue (see Plate 5.2H). This monoclonal antibody defines a cytoplasmic antigen. Two studies by Duray and associates[41] and Gown and colleagues[60] showed a sensitivity of over 90%. Not all malignant melanomas stain positive with HMB-45 in histological (86 to 97%)[41,60,97] and cytological (69%)[141] specimens. All eight cases of malignant melanomas in ABC studied by Pelosi and colleagues[147] were positive. The antibody is slightly more sensitive in alcohol-fixed than in formalin-fixed specimens. Staining may be focal in a number of cases and only seen in a small group of tumor cells.[60,147] Therefore, in aspirates with a limited number of cells, negative staining does not exclude the possibility of malignant melanoma.

HMB-45 has been reported to be negative in nonmelanoma tumors in surgical pathology,[41,60] and cytopathology.[141,147] Leong and Milios,[97] however, in a study of 200 nonmelanoma tumors, demonstrated positive staining in two cases of breast carcinoma and two cases of plasmacytoma. Bonetti and colleagues[15] also demonstrated positive staining in 2% of the breast carcinomas they studied. Positive staining has also been reported in a few normal tissues such as breast lobules and ducts, bronchial epithelial cells, and plasma cells.[15,97] In view of these findings and because of lesser sensitivity of this antibody as compared with S-100 protein, HMB-45 antibody should not be used alone. In summary, addition of HMB-45 to the panel of antibodies used for poorly differentiated neoplasms can enhance diagnostic accuracy (see Fig. 8.6).

Two other melanoma-associated antigens recognized by antibodies NKI/C3 and NKK/Bteb have been studied in ABC specimens and effusions.[7] Positive staining for NKI/C3 was found in 12 of 12 cases, and for NKI/Bteb in 12 of 13 malignant melanomas studied. Because NKI/C3 has also been reported in some neuroendocrine neoplasms and other cancers,[7] these antibodies may not be specific for malignant melanoma and should be used in a panel.

# NEUROFILAMENT PROTEINS (NFP)

See "Intermediate Filaments." Page 67.

# NEURON-SPECIFIC ENOLASE

Neuron-specific enolase (NSE) is an isoenzyme of the glycolytic enzyme enolase. The gamma enolase subunit is present in high concentrations in all the cells of dispersed neuroendocrine system and neuronal tissue, as well as their tumor derivatives (Plates 5.3A, B), including peripheral neuroblastoma and malignant melanoma.[65,103,176,192] A significant level of gamma enolase is present in many normal tissues, such as smooth muscle cells, myoepithelial cells, renal epithelial cells, spermatogonia, lymphocytes, plasma cells, platelets, and megakaryocytes.[65] A lesser amount is present in bronchial epithelial cells and type II pneumocytes. Furthermore, NSE may be expressed in a number of nonneuroendocrine and nonneuronal tumors, such as several types of central nervous system tumors, breast and renal-cell carcinomas, nephroblastoma, chordoma, and nonsmall cell and nonneuroendocrine lung neoplasms.[13,40,157,192,201]

Monoclonal antibody to NSE is more specific but slightly less sensitive than polyclonal antibodies. Thomas and colleagues,[177] in studying 255 nonneuroendocrine nonneuronal neoplasms, observed 2% positivity with monoclonal antibody to NSE, in comparison with 20% positivity with polyclonal antibodies. Monoclonal antibodies, therefore, appear to be more valuable. Considering these limitations, antibody to NSE may be used for screening of possible neuroendocrine differentiation (see Figs. 8.8, 8.9), but by itself is not specific and should be used in conjunction with other neuroendocrine markers such as chromogranins, bombesin, or synaptophysin.

# PLACENTAL ALKALINE PHOSPHATASE

Placental alkaline phosphatase (PLAP) isoenzyme is normally expressed by trophoblasts after the first trimester of pregnancy.[109] It is a cell surface glycoprotein consisting of two subunits, both with a molecular weight of approximately 67 kd.[45] Most seminomas, embryonal carcinomas, and endodermal sinus tumors immunoreact with polyclonal antibody to PLAP (see Plate 5.3C, Fig. 8.20). The staining is mainly on the cell membrane or in combination with cytoplasmic positivity.[109,184] Occasionally, ovarian, pulmonary, and gastrointestinal cancers may show positivity. It is suggested that monoclonal antibodies to PLAP may be more specific. Epenetos and colleagues,[45] using monoclonal antibodies, demonstrated immunoreactivity in germ-cell neoplasms; however, two cases of ovarian, one case of endometrial, and three cases of colonic carcinomas were also positive. PLAP in the serum is a useful marker for selected neoplasms.[128] PLAP is also an excellent screening marker for possible germ-cell differentiation in ABC specimens (see Fig. 8.20). Germ-cell origin of PLAP-positive cases can be further supported by antibodies to AFP and hCG.

## PROGESTERONE RECEPTOR (PR)

See "Breast Carcinoma" section in Chapter 10. Page 269.

## PROSTATIC ACID PHOSPHATASE

Prostatic acid phosphatase (PrAP) is a secretory product of prostate epithelium that is localized to lysosomal granules at the ultrastructural level, and is present in normal prostate, benign prostatic hyperplasia, and prostatic carcinoma.[83,196] PrAP is present in the semen and blood, and consists of a minimum of eight isoenzymes with multiple antigenic determinants.[163] The great majority of prostatic carcinomas are PrAP-positive (see Plate 5.3D, Fig. 10.2).[44,83,133,163,176] In one study, only 50% of prostatic carcinomas were immunoreactive when monoclonal antibody to PrAP was used.[163] PrAP is a more sensitive prostate marker than prostate-specific antigen, especially in undifferentiated carcinomas of the prostate.[49] Original studies using antisera to PrAP demonstrated no staining of primary and metastatic nonprostatic tumors in a limited number of cases.[133] It was later proved, however, that nonprostatic cells and neoplasms such as renal-cell carcinoma, islet-cell tumor of the pancreas, rectal carcinoid tumor, breast carcinoma, and adenocarcinoma of the urinary bladder may be immunoreactive.[23,25,44,46,52,83,205] PrAP is therefore not totally specific for prostatic carcinoma and should not be used in isolation. The absence of PrAP in a poorly differentiated tumor does not rule out the possibility of prostate origin.

## PROSTATE-SPECIFIC ANTIGEN

Prostate-specific antigen (PSA) is a glycoprotein with a molecular weight of 33 kd that is biochemically and immunologically distinct from PrAP.[168] It is produced and secreted almost exclusively by prostatic cells,[145] but female periurethral glandular cells also secrete PSA.[152] PSA immunoreactivity is localized over the endoplasmic reticulum, cytoplasmic vesicles and vacuoles, and within the lumina of prostatic glands at the ultrastructural level.[196] PSA is also present in benign prostatic hyperplasia and most prostatic carcinomas, both primary and metastatic (see Plate 5.3E, Fig. 10.2).[44,83,132,145] Staining is more intense and extensive than PrAP.[44,155] Most laboratories, including ours, use polyclonal antibodies to PSA; however, monoclonal antibodies to PSA give similar results and are preferable to antisera.[145] May and Perentes[112] demonstrated weak to moderately positive cytoplasmic staining of a number of adenocarcinomas, suggesting the possible nonspecific nature of this marker.

# S-100 PROTEIN

S-100 protein is an acidic calcium binding protein with a molecular weight of approximately 21 kd that was originally isolated from brain of mammals. It is called S-100 protein because of its solubility in 100% ammonium sulfate. S-100 protein is composed of at least two subunits, S-100 alpha and S-100 beta.[42,82,134,160] S-100 protein is expressed by many normal and neoplastic cells (Table 5.4, Fig. 5.5; see Plates 5.3F, 5.3G), and the list is continuing to expand.[69,72,90,106,164,167] By using monospecific antibodies, an even wider distribution of S-100 protein and the different localization of its subunits has been demonstrated.[63,174]

TABLE 5.4. Known Pattern of S-100 Protein Reactivity

| *Normal cells* | *Neoplasms* |
|---|---|
| Melanocytes | Malignant melanomas, nevi |
| Astrocytes, oligodendrocytes | Astrocytomas, oligodendrogliomas |
| Schwann cells | Nerve-sheath tumors |
| Epithelial and myoepithelial cells | |
|   Breast | Carcinomas and benign lesions |
|   Sweat gland | Eccrine carcinomas |
|   Salivary gland | Carcinomas, pleomorphic adenomas, Warthin's tumors, carcinomas of ovary, endometrium, kidney, colon, stomach, and lung |
| Neurons, neuroblasts | Neuroblastomas |
| Langerhans' cells | Histiocytosis X |
| Chondrocytes | Cartilagenous tumors |
| Adipocytes | Liposarcomas, lipomas |
| Some histiocytes | |
| Interdigitating and dendritic reticulum cells in lymph node | |
| | Granular cell tumors, chordomas |
| Muscle cells | |
| Stellate cells of the adrenal medulla | |
| Folliculostellate cells of the adenohypophysis | |

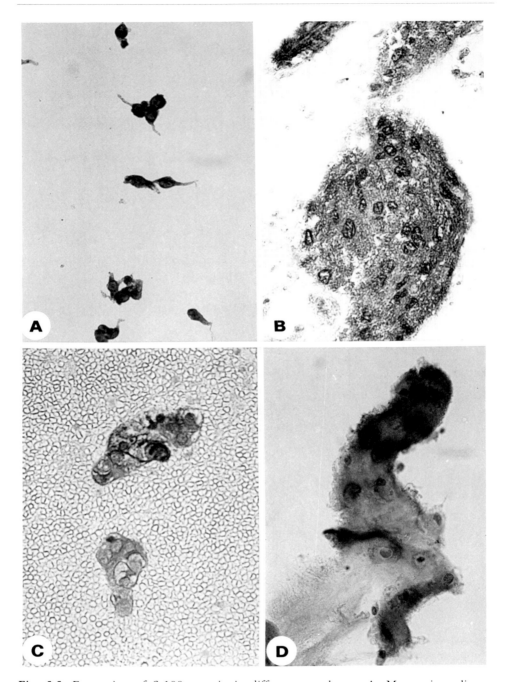

Fig. 5.5. Expression of S-100 protein in different neoplasms. **A.** Metastatic malignant melanoma, ABC of lymph node. × 260. **B.** Schwannoma, ABC of a pleural-based lesion. × 416. **C.** Renal-cell carcinoma, ABC. × 260. **D.** Chondrosarcoma, ABC of a rib mass. × 260.

TABLE 5.5 S-100 Protein Expression by Primary and Metastatic Adenocarcinomas*

| Site | Primary Neoplasms % Positive | Metastatic Neoplasms % Positive |
|---|---|---|
| Ovary | 84 | 87 |
| Salivary gland | 80 | 75 |
| Endometrium | 78 | 64 |
| Kidney | 65 | 66 |
| Breast | 60 | 62 |
| Colon | 25 | 23 |
| Stomach | 20 | 25 |
| Lung | 7 | 12 |
| Unknown primary | | 22 |

*From Herrera GA, Turbat-Herrera EA, Lott RL: S-100 protein expression by primary and metastatic adenocarcinomas. Am J Clin Pathol 89:168–176, 1988. With permission.

Polyclonal antibody to S-100 protein recognizes both alpha and beta subunits, and staining is seen in the cytoplasm as well as in the nucleus. Ordóñez and colleagues[141] demonstrated strong staining in formalin-fixed cytological preparations but a marked reduction of the number of positive cells and the intensity of the staining when acetone-fixed frozen sections or ethanol-fixed cytological material were used. In our experience, both alcohol and formalin fixation give good results.

S-100 protein is a very sensitive marker for malignant melanoma and is expressed in almost all melanomas in ABC specimens (see Plate 5.3F). Staining with antibody to S-100 protein, therefore, is potentially useful even if the aspirate is not very cellular. Despite the nonspecificity of S-100 protein and its presence in a wide variety of tumors, it continues to be a useful marker in the diagnosis of malignant melanoma in ABC specimens, provided it is used in a diagnostic panel (see Fig. 8.6). Expression of vimentin and S-100 protein in neoplastic cells and negative CK staining supports the diagnosis of malignant melanoma and excludes the possibility of carcinomas. This result, however, should be interpreted in the context of morphological and clinical findings and ideally should be confirmed with a melanoma- "specific" monoclonal antibody. S-100 protein is also a useful marker to support the diagnosis of melanoma in effusions.[150]

S-100 protein is expressed in a significant number of adenocarcinomas of different organs. Herrera and colleagues[69] found positive staining in 42% of 350 cases of primary and metastatic adenocarcinoma (Table 5.5). In a study of 400 cases of carcinomas, Drier and colleagues[39] demonstrated positivity in only 12% of cases. This lower percentage is probably due to the fact that they used only poorly differentiated carcinomas. S-100 protein expression was demonstrated in 30% of 82 carcinomas in ABC specimens and effusions.[164] Between 46 to 84% of primary or metastatic breast carcinomas, expressed S-100 protein.[39,69,170] Positive staining of adenocarcinoma cells of an unknown primary site with antibody to S-100 protein may be of some use in suggesting the possible primary site.[39,69,170] S-100 protein immunoreactivity is also a useful marker for nerve-sheath tumors (see Plate 5.3G, Fig. 8.16) and may be present in a variety of soft-tissue lesions such as liposarcoma, chondrosarcoma (see Fig. 8.19), histiocytosis X, and chordoma.[199] Other mesenchymal tumors, such as malignant fibrous histiocytomas, myogenic tumors, and hemangiopericytomas, are negative.[160,199]

When staining is focal, great care should be exercised. A significant number of normal tissues contain S-100–positive histiocytes and Langerhans' cells. Experience with cytomorphology of these cells is essential to prevent a false interpretation. These cells generally have smaller nuclei, and a stellate shape of the cytoplasm is helpful, if observed.

## SYNAPTOPHYSIN

Synaptophysin (SYN) is a glycosylated transmembrane polypeptide with a molecular weight of 38 kd, and has been postulated to have an important role in release of neurotransmitter material.[75] SYN was originally isolated from presynaptic vesicles of bovine brain neurons. SYN has been shown in neurons, paraganglion cells, pancreatic islet cells, and neuromuscular junctions.[56] Most neuroendocrine neoplasms, neural or epithelial type, such as neuroblastomas, pheochromocytomas, paragangliomas, carcinoid tumor, islet-cell tumor, medullary carcinoma of the thyroid, and many neuroendocrine carcinomas, have been shown to express SYN (see Plate 5.3H, Fig. 8.8).[21,56,58] Expression of SYN occurs in well-differentiated as well as poorly differentiated neuroendocrine neoplasms, regardless of cytoskeletal characteristics, hormonal expression, and clinical aggressiveness.[57,58]

SYN has thus far not been demonstrated in nonneuroendocrine neoplasms such as malignant melanoma, adenocarcinoma, malignant lymphoma, and sarcoma.[56,115] It has been suggested that SYN is a sensitive and specific marker for neuroendocrine differentiation.[56,57,115] In our experience, using two different commercially available antibodies, SYN is not a sensitive marker for neuroendocrine neoplasms of the lung. Only 40% of these neoplasms express SYN, and in half of the cases the staining is weak or focal in nature.

## THYROGLOBULIN

Thyroglobulin (TG) is a large glycoprotein with a molecular weight of 670 kd that is synthesized by the follicular cells of the thyroid.[54,166] Follicular cells and colloid stain homogeneously with antibody to TG. The staining, however, is less intense in colloid.[166] Most follicular neoplasms, including Hürthle-cell tumors, and papillary carcinomas produce TG; however, the staining is generally less homogeneous than normal follicular cells (Plate 5.4A).[2,79,166,176] Anaplastic carcinomas of the thyroid show variable staining or may be totally negative.[2,166] Medullary carcinomas of the thyroid generally stain negative for TG. Very occasionally, medullary carcinomas coexpress TG and calcitonin.[73]

Antibodies to TG consistently produce good results with a high degree of sensitivity and specificity. Therefore, TG is an extremely valuable marker in ascertaining the thyroid origin of most metastatic thyroid carcinomas in effusion and ABC of

TABLE 5.6. Antibodies Used in Our Laboratory

| Antibody | Monoclonal or Polyclonal | Source |
|---|---|---|
| Actin (HHF35) | M | Enzo Diagnostics, Inc (Syosset, NY) |
| Adrenocorticotropic hormone (ACTH) | P | Chemicon International, Inc (Temecula, CA) |
| Alpha-fetoprotein (AFP) | P | Dako Corporation (Carpinteria, CA) |
| Alpha-1-antitrypsin (A1-AT) | P | Dako Corporation |
| B72.3 | M | Triton Biosciences Inc. (Alameda, CA) |
| Calcitonin | P | Dako Corporation |
| Carcinoembryonic antigen (CEA) | M | Zymed Lab. Inc. (San Francisco, CA) |
| Chromogranin | M | Hybritech Inc. (San Diego, CA) |
| Cytokeratins* | | |
|   AE1* | M | Signet Laboratories, Inc. (Dedham, MA) |
|   AE3* | M | Signet Laboratories, Inc. |
|   CAM 5.2* | M | Becton-Dickinson (San Jose, CA) |
|   8.12* | M | ICN ImmunoBiologicals (Costa Mesa, CA) |
|   MA903* | M | Enzo Diagnostics, Inc. |
|   PKK1* | M | Labsystems (Raleigh, NC) |
| Desmin | M | Dako Corporation |
| Desmin* | P | Dako Corporation |
| Epithelial membrane antigen (EMA) | M | Dako Corporation |
| Estrogen receptor | M | Abbott Laboratories (Chicago, IL) |
| Factor VIII–related antigen (F VIII RAg)* | P | Dako Corporation |
| Glial fibrillary acidic protein (GFAP) | P | Dako Corporation |
| Gross cystic disease fluid protein (GCDFP-15) | M | Signet Laboratories, Inc. |
| Growth hormone (GH) | P | Dako Corporation |
| Human chorionic gonadotropin (hCG) | P | Dako Corporation |
| Keratin | P | Dako Corporation |
| Ki-67 | M | Dako Corporation |
| Leucocyte common antigen (LCA)* | M | Dako Corporation |
| Leutinizing hormone (LH) | P | Dako Corporation |
| Lymphoid markers (see Table 9.2) | | |
| Lysozyme | P | Dako Corporation |
| Melanoma-specific antibody (HMB-45) | M | Enzo Diagnostics, Inc. |

TABLE 5.6. *continued*

| Antibody | Monoclonal or Polyclonal | Source |
|---|---|---|
| Myoglobin | P | Dako Corporation |
| Myosin | P | ICN ImmunoBiologicals |
| Neurofilaments 68kd | M | Medica (Carlsbad, CA) |
| Neurofilaments 160kd | M | Amersham (Arlington Heights, IL) |
| Neurofilaments 200kd | M | Amersham |
| Neuron-specific enolase (NSE) | P | Dako Corporation |
| Placental alkaline phosphatase (PLAP) | P | Dako Corporation |
| Prolactin | P | Dako Corporation |
| Prostatic acid phosphatase (PrAP) | P | Dako Corporation |
| Prostatic specific antigen (PSA) | P | Dako Corporation |
| S-100 protein | P | Dako Corporation |
| Secretory component* | P | Dako Corporation |
| Serontonin | P | Dako Corporation |
| Somatostatin | P | Dako Corporation |
| Synaptophysin* | M | Dako Corporation |
| Thyroglobulin | P | Dako Corporation |
| Thyroid-stimulating hormone (TSH) | P | Dako Corporation |
| Tubulin (alpha) | M | Amersham |
| Ulex europaeus agglutinin-1 (UEA-1) | P | Vector Laboratories (Burlingame, CA) |
| Vasoactive intestinal peptide (VIP) | P | Serotec (Kidlington, Oxford, England) |
| Vasopressin | P | Immunocorp (Montréal, Qué) |
| Vimentin* | M | Labsystems |
| Vimentin | P | Immunocorp |

*Protease digestion if formalin-fixed.

different organs and sites. In ABC of the thyroid, staining for TG in conjunction with calcitonin is useful to differentiate medullary carcinoma from tumors of follicular cell origin (see Fig. 10.13). Medullary carcinoma may show extensive follicle formation, complicating cytological interpretation. When the staining for TG is focal, the possibility of entrapped normal follicular cells or passive absorption of TG by neoplastic cells should be considered.[166]

# VIMENTIN

See "Intermediate Filaments." Page 65.

# REFERENCES

1. Achtstätter T, Moll R, Moore B, Franke WW: Cytokeratin polypeptide patterns of different epithelia of the human male urogenital tract: Immunofluorescence and gel electrophoretic studies. *J Histochem Cytochem* 33:415–426, 1985.
2. Albores-Saavedra J, Nadji M, Civantos F, Morales AR: Thyroglobulin in carcinoma of the thyroid: An immunohistochemical study. *Hum Pathol* 14:62–66, 1983.
3. Altmannsberger M, Osborn M, Treuner J, et al: Diagnosis of human childhood rhabdomyosarcoma by antibodies to desmin, the structural protein of muscle specific intermediate filaments. *Virchows Arch {B}* 39:203–215, 1982.
4. Altmannsberger M, Weber K, Droste R, Osborn M: Desmin is a specific marker for rhabdomyosarcoma of human and rat origin. *J Pathol* 118:85–95, 1985.
5. Ang LC, Taylor AR, Bergin D, Kaufmann JCE: An immunohistochemical study of papillary tumors in the central nervous system. *Cancer* 65:2712–2719, 1990.
6. Angeletti RH: Chromogranins and neuroendocrine secretion. *Lab Invest* 55:387–390, 1986.
7. Angeli S, Koelma IA, Fleuren GJ, Van Steenis GJ: Malignant melanoma in fine needle aspirates and effusions; an immunocytochemical study using monoclonal antibodies. *Acta Cytol* 32:707–712, 1988.
8. Axe S, Kuhajda FP: Esthesioneuroblastoma; intermediate filaments, neuroendocrine and tissue-specific antigens. *Am J Clin Pathol* 88:139–145, 1987.
9. Azumi N, Battifora H: The distribution of vimentin and keratin in epithelial and nonepithelial neoplasms; a comprehensive immunohistochemical study on formalin- and alcohol-fixed tumors. *J Clin Pathol* 88:286–296, 1987.
10. Azumi N, Ben-Ezra J, Battifora H: Immunophenotypic diagnosis of leiomyosarcoma and rhabdomyosarcoma with monoclonal antibodies to muscle-specific actin and desmin in formalin-fixed tissue. *Mod Pathol* 1:469–474, 1988.
11. Battifora H, Kopinski MI: Distinction of mesothelioma from adenocarcinoma; an immunohistochemical approach. *Cancer* 55:1679–1685, 1985.
12. Bergeron C, Shatz P, Margolese R, Major P, Ferenczy A: Immunohistochemical studies with monoclonal antibodies B72.3 and MA5 on histologic and cytologic specimens from benign and malignant breast lesions. *Anal Quant Cytol Histol* 11:33–41, 1989.
13. Bergh J, Esscher T, Steinholtz L, Nilsson K, Påhlman S: Immunocytochemical demonstration of neuron-specific enolase (NSE) in human lung cancers. *Am J Clin Pathol* 84:1–7, 1985.
14. Blobel GA, Gould VE, Moll R, et al: Coexpression of neuroendocrine markers and epithelial cytoskeletal proteins in bronchopulmonary neuroendocrine neoplasms. *Lab Invest* 52:39–51, 1985.
15. Bonetti F, Colombari R, Manfrin E, et al: Breast carcinoma with positive results for melanoma marker (HMB-45); HMB-45 immunoreactivity in normal and neoplastic breast. *Am J Clin Pathol* 92:491–495, 1989.
16. Broers JLV, Carney DN, De Ley L, Vooijs GP, Ramaekers FCS: Differential expression of intermediate filament proteins distinguishes classic from variant small-cell lung cancer cell lines. *Proc Natl Acad Sci USA* 82:4409–4413, 1985.
17. Broers JLV, Ramaekers FCS, Rot MK, et al: Cytokeratins in different types of human lung cancer as monitored by chain-specific monoclonal antibodies. *Cancer Res* 48:3221–3229, 1988.

18. Brown DC, Theaker JM, Banks PM, Gatter KC, Mason DY: Cytokeratin expression in smooth muscle and smooth muscle tumours. *Histopathology* 11:477–486, 1987.
19. Bussolati G, Gugliotta P, Sapina A, Eusebi V, Lloyd RV: Chromagranin-reactive endocrine cells in argyrophilic carcinomas ("carcinoids") and normal tissue of the breast. *Am J Pathol* 120:186–192, 1985.
20. Bussolati G, Papotti M, Sapino A, et al: Endocrine markers in argyrophilic carcinomas of the breast. *Am J Surg Pathol* 11:248–256, 1987.
21. Chejfec G, Falkmer S, Grimelius L, et al: Synaptophysin; a new marker for pancreatic neuroendocrine tumors. *Am J Surg Pathol* 11:241–247, 1987.
22. Chen Y, Zhang Y: Use of monoclonal antibodies to glial fibrillary acidic protein in the cytologic diagnosis of brain tumors. *Acta Cytol* 33:922–928, 1989.
23. Choe B-K, Pontes EJ, Rose NR, Henderson MD: Expression of human prostatic acid phosphatase in a pancreatic islet cell carcinoma. *Invest Urol* 15:312–318, 1978.
24. Cibas ES, Corson JM, Pinkus GS: The distinction of adenocarcinoma from malignant mesothelioma in cell blocks of effusions: The role of routine mucin histochemistry and immunohistochemical assessment of carcinoembryonic antigen, keratin protein, epithelial membrane antigen, and milk fat globule-derived antigen. *Hum Pathol* 18:67–74, 1987.
25. Cohen C, Bentz MS, Budgeon LR: Prostatic acid phosphatase in carcinoid and islet cell tumors. *Arch Pathol Lab Med* 107:277, 1983.
26. Collins VP: Monoclonal antibodies to glial fibrillary acidic protein in the cytologic diagnosis of brain tumors. *Acta Cytol* 28:401–406, 1984.
27. Cooper D, Schermer A, Sun T-T: Biology of disease; classification of human epithelia and their neoplasms using monoclonal antibodies to keratins: Strategies, applications, and limitations. *Lab Invest* 52:243–256, 1985.
28. Corwin DJ, Gown AM: Review of selected lineage-directed antibodies useful in routinely processed tissues. *Arch Pathol Lab Med* 113:645–652, 1989.
29. Dabbs DJ, Geisinger KR, Norris HT: Intermediate filaments in endometrial and endocervical carcinomas; the diagnostic utility of vimentin patterns. *Am J Surg Pathol* 10:568–576, 1986.
30. Debus E, Moll R, Franke WW, Weber K, Osborn M: Immunohistochemical distinction of human carcinomas by cytokeratin typing with monoclonal antibodies. *Am J Pathol* 114:121–130, 1984.
31. DeLellis RA, Kwan P. Technical consideration in the immunohistochemical demonstration of intermediate filaments. *Am J Surg Pathol* 12(suppl 1):17–23, 1988.
32. Delsol G, Gatter KC, Stein H, et al: Human lymphoid cells may express epithelial membrane antigens; implications for diagnosis of human neoplasms. *Lancet* 2:1124–1128, 1984.
33. Denk H, Krepler R, Artlieb V, et al: Proteins of intermediate filaments; an immunohistochemical and biochemical approach to the classification of soft tissue tumors. *Am J Pathol* 110:193–208, 1983.
34. Domagala W, Halczy-Kowalik L, Weber K, Osborn M: Coexpression of glial fibrillary acid protein, keratin and vimentin; a unique feature useful in the diagnosis of pleomorphic adenoma of the salivary gland in fine needle aspiration biopsy smears. *Acta Cytol* 32:403–408, 1988.
35. Domagala W, Lubinski J, Lasota J, et al: Decisive role of intermediate filament typing of tumor cells in the differential diagnosis of difficult fine needle aspirates. *Acta Cytol* 31:253–266, 1987.

36. Domagala W, Lubinski J, Lasota J, et al: Neuroendocrine (Merkel-cell) carcinoma of the skin; cytology, intermediate filament typing and ultrastructure of tumor cells in fine needle aspirates. *Acta Cytol* 31:267–275, 1987.
37. Domagala W, Lubinski J, Weber K, Osborn M: Intermediate filament typing of tumor cells in fine needle aspirates by means of monoclonal antibodies. *Acta Cytol* 30:214–224, 1986.
38. Domagala W, Weber K, Osborn M: Diagnostic significance of coexpression of intermediate filaments in fine needle aspirates of human tumors. *Acta Cytol* 32:49–59, 1988.
39. Drier JK, Swanson PE, Cherwitz DL, Wick MR: S100 protein immunoreactivity in poorly differentiated carcinoma; immunohistochemical comparison with malignant melanoma. *Arch Pathol Lab Med* 111:447–452, 1987.
40. Drut R: Neuron-specific enolase-positive rosettes in nephroblastoma: A possible diagnostic pitfall in aspiration cytology. *Diagn Cytopathol* 3:74–76, 1987.
41. Duray PH, Palazzo J, Gown AM, Ohuchi N: Melanoma cell heterogeneity; a study of two monoclonal antibodies compared with S-100 protein in paraffin sections. *Cancer* 61:2460–2468, 1988.
42. Dwarakanath S, Lee AK, DeLellis RA, et al: S-100 protein positively in breast carcinomas: A potential pitfall in diagnostic immunohistochemistry. *Hum Pathol* 18:1144–1148, 1987.
43. Elias AD, Cohen BF, Bernal SD: Keratin subtypes of small cell lung cancer. *Cancer Res* 48:2724–2729, 1988.
44. Ellis DW, Leffers S, Davies JS, Ng ABP: Multiple immunoperoxidase markers in benign hyperplasia and adenocarcinoma of the prostate. *Am J Clin Pathol* 81:279–284, 1984.
45. Epenetos AA, Travers P, Gatter KC, et al: An immunohistological study of testicular germ cell tumors using two different monoclonal antibodies against placental alkaline phosphatase. *Br J Cancer* 49:11–15, 1984.
46. Epstein JI, Kuhajda FP, Lieberman PH: Prostate-specific acid phosphatase immunoreactivity in adenocarcinoma of the urinary bladder. *Hum Pathol* 17:939–942, 1986.
47. Erlandson RA: Cytoskeletal proteins including myofilaments in human tumors. *Ultrastruct Pathol* 13:155–186, 1989.
48. Eusebi V, Carcangiu ML, Dina R, Rosai J: Keratin-positive epithelioid angiosarcoma of thyroid; a report of four cases. *Am J Surg Pathol* 14:737–747, 1990.
49. Feiner HD, Gonzalez R: Carcinoma of the prostate with atypical immunohistological features; clinical and histologic correlates. *Am J Surg Pathol* 10:765–770, 1986.
50. Finley JL, Silverman JF, Dabbs DJ: Fine-needle aspiration cytology of pulmonary carcinosarcoma with immunocytochemical and ultrastructural observations. *Diagn Cytopathol* 4:239–243, 1988.
51. Fisher HP, Altmannsberger M, Weber K, Osborn M: Keratin polypeptides in malignant epithelial liver tumors; differential diagnosis and histogenetic aspects. *Am J Pathol* 127:530–537, 1987.
52. Fishleder A, Tubbs RR, Levin HS: An immunoperoxidase technique to aid in the differential diagnosis of prostate carcinoma. *Cleve Clin Q* 48:331–335, 1981.
53. Franke WW, Schmid E, Winter S, Osborn M, Weber K: Widespread occurrence of intermediate-sized filaments of the vimentin-type in cultured cells from diverse vertebrates. *Exp Cell Res* 123:25–46, 1979.
54. Gal R, Aronof A, Gertzmann H, Kessler E: The potential value of the demonstration

of thyroglobulin by immunoperoxidase techniques in fine needle aspiration cytology. *Acta Cytol* 31:713–716, 1987.
55. Gal R, Gukovsky-Oren S, Lehman JM, Schwartz P, Kessler E: Cytodiagnosis of a spindle-cell tumor of the breast using antisera to epithelial membrane antigen. *Acta Cytol* 31:317–321, 1987.
56. Gould VE: Synaptophysin; a new and promising pan-neuroendocrine marker. *Arch Pathol Lab Med* 111:791–794, 1987.
57. Gould VE, Lee I, Wiedenmann B, et al: Synaptophysin: A novel marker for neurons, certain neuroendocrine cells, and their neoplasms. *Hum Pathol* 17:979–983, 1986.
58. Gould VE, Wiedenmann B, Lee I, et al: Synaptophysin expression in neuroendocrine neoplasms as determined by immunocytochemistry. *Am J Pathol* 126:243–257, 1987.
59. Gown AM, Vogel AM: Monoclonal antibodies to human intermediate filament proteins. III. Analysis of tumors. *Am J Clin Pathol* 84:413–424, 1985.
60. Gown AM, Vogel AM, Hoak D, Gough F, McNutt MA: Monoclonal antibodies specific for melanocytic tumors distinguish subpopulations of melanocytes. *Am J Pathol* 123:195–203, 1986.
61. Gray MH, Rosenberg AE, Dickersin GR, Bhan AK: Cytokeratin expression in epithelioid vascular neoplasms. *Hum Pathol* 21:212–217, 1990.
62. Gray MH, Rosenberg AE, Dickersin GR, Bhan AK: Glial fibrillary acidic protein and keratin expression by benign and malignant nerve sheath tumors. *Hum Pathol* 20:1089–1096, 1989.
63. Haimoto H, Hosoda S, Kato K: Differential distribution of immunoreactive S100-α and S100-β proteins in normal nonnervous human tissues. *Lab Invest* 57:489–498, 1987.
64. Hagn C, Schmid KW, Fischer-Colbrie R, Winkler H: Chromogranin A, B and C in human adrenal medulla and endocrine tissues. *Lab Invest* 55:405–411, 1986.
65. Haimoto H, Takahashi Y, Koshikawa T, Nagura H, Kato K: Immunohistochemical localization of γ-enolase in normal human tissues other than nervous and neuroendocrine tissues. *Lab Invest* 52:257–263, 1985.
66. Hammar SP, Hallman KO: Unusual primary lung neoplasms: Simple cell and undifferentiated lung carcinomas expressing only vimentin. *Ultrastruct Pathol* 14:407–422, 1990.
67. Heitz PU, von Herbay G, Klöppel G, et al: The expression of subunits of human chorionic gonadotropin (hCG) by nontrophoblastic, nonendocrine, and endocrine tumors. *Am J Clin Pathol* 88:467–472, 1987.
68. Herman CJ, Moesker O, Kant A, et al: Is renal cell (Grawitz) tumor a carcinosarcoma? Evidence from analysis of intermediate filament types. *Virchows Arch {B}* 44:73–83, 1983.
69. Herrera GA, Turbat-Herrera EA, Lott RL: S-100 protein expression by primary and metastatic adenocarcinomas. *Am J Clin Pathol* 89:168–176, 1988.
70. Hoffman PN, Lasek RJ: The slow component of axonal transport; identification of major structural peptides of the axon and their generality among mammalian neurons. *J Cell Biol* 66:351–366, 1975.
71. Höfler H, Kerl H, Lackinger E, Helleis G, Denk H: The intermediate filament cytoskeleton of cutaneous neuroendocrine carcinoma (Merkel cell tumour); immunohistochemical and biochemical analysis. *Virchows Arch {A}* 406:339–350, 1985.
72. Höfler H, Walter GF, Denk H: Immunohistochemistry of folliculo-stellate cells in

normal human adenohypophyses and in pituitary adenomas. *Acta Neuropathol (Berl)* 65:35–40, 1984.
73. Holm R, Sobrinho-Simões M, Nesland JM, Sambade C, Johannessen JV: Medullary thyroid carcinoma with thyroglobin immunoreactivity; a special entity? *Lab Invest* 57:258–268, 1987.
74. Holthöfer H, Miettinen A, Paasivuo R, et al: Cellular origin and differentiation of renal carcinomas; a fluorescence microscopic study with kidney-specific antibodies, antiintermediate filament antibodies, and lectins. *Lab Invest* 49:317–326, 1983.
75. Hoog A, Gould VE, Grimelius L, et al: Tissue fixation methods alter the immunohistochemical demonstrability of synaptophysin. *Ultrastruct Pathol* 12:673–678, 1988.
76. Irie T, Watanabe H, Kawaoi A, Takeuchi J: Alpha-fetoprotein (AFP), human chorionic gonadotropin (HCG), and carcinoembryonic antigen (CEA) demonstrated in the immature glands of mediastinal teratocarcinoma; a case report. *Cancer* 50:1160–1165, 1982.
77. Iwa N, Yutani C, Ishibashi-Ueda H, Katayama Y: Immunocytochemical demonstration of glial fibrillary acidic protein in imprint smears of human brain tumors. *Diagn Cytopathol* 4:74–77, 1988.
78. Jasani B, Edwards RE, Thomas ND, Gibbs AR: The use of vimentin antibodies in the diagnosis of malignant mesothelioma. *Virchows Arch {A}* 406:441–448, 1985.
79. Johnson TL, Lloyd RV, Burney RE, Thompson NW: Hurthle cell thyroid tumors; an immunohistochemical study. *Cancer* 59:107–112, 1987.
80. Johnston WW: Applications of monoclonal antibodies in clinical cytology as exemplified by studies with monoclonal antibody B72.3; the George N. Papanicolaou award lecture. *Acta Cytol* 31:537–556, 1987.
81. Johnston WW, Szpak CA, Lottich SC, Thor A, Schlom J: Use of a monoclonal antibody (B72.3) as a novel immunohistochemical adjunct for the diagnosis of carcinomas in fine needle aspiration biopsy specimens. *Hum Pathol* 17:501–513, 1986.
82. Kahn HJ, Marks A, Thom H, Baumal R: Role of antibody to S100 protein in diagnostic pathology. *Am J Clin Pathol* 79:341–347, 1983.
83. Katz RL, Raval P, Brooks TE, Ordoñez NG: Role of immunocytochemistry in diagnosis of prostatic neoplasm by fine needle aspiration biopsy. *Diagn Cytopathol* 1:28–32, 1985.
84. Kawahara E, Oda Y, Ooi A, et al: Expression of glial fibrillary acidic protein (GFAP) in peripheral nerve sheath tumors; a comparative study of immunoreactivity of GFAP, vimentin, S-100 protein, and neurofilament in 38 schwannomas and 18 neurofibromas. *Am J Surg Pathol* 12:115–120, 1988.
85. Kawai T, Torikata C, Suzuki M: Immunohistochemical study of pulmonary adenocarcinoma. *Am J Clin Pathol* 89:455–462, 1988.
86. Khoury N, Raju U, Crissman JD, et al: A comparative immunohistochemical study of peritoneal and ovarian serous tumors, and mesotheliomas. *Hum Pathol* 21:811–819, 1990.
87. Kimura N, Sasano N, Yamada R, Satoh J: Immunohistochemical study of chromogranin in 100 cases of pheochromocytoma, carotid body tumour, medullary thyroid carcinoma and carcinoid tumour. *Virchows Arch {A}* 413:33–38, 1988.
88. Kline TS, Lundy J, Lozowski M: Monoclonal antibody B72.3; an adjunct for evaluation of suspicious aspiration biopsy cytolgy from the breast. *Cancer* 63:2253–2256, 1989.
89. Kobayashi TK, Gotoh T, Kamachi M, Watanabe S, Sawaragi I: Immunocytochemical presentation of alpha-fetoprotein-producing gastric cancer in ascitic fluid: A case study. *Diagn Cytopathol* 4:116–120, 1988.

90. Kondo K, Mukai K, Sato Y, et al: An immunohistochemical study of thymic epithelial tumors. III. The distribution of interdigitating reticulum cells and S-100 β-positive small lymphocytes. Am J Surg Pathol 14:1139–1147, 1990.
91. Kurman RJ, Ganjei P, Nadji M: Contribution of immunocytochemistry to the diagnosis and study of ovarian neoplasms. Int J Gynecol Pathol 3:3–26, 1984.
92. Lawson CW, Fisher C, Gatter KC: An immunohistochemical study of differentiation in malignant fibrous histiocytoma. Histopathology 11:375–383, 1987.
93. Leader M, Collins PM, Henry K: Anti-α1-antichymotrypsin staining of 194 sarcomas, 38 carcinomas, and 17 malignant melanomas, its lack of specificity as a tumour marker. Am J Surg Pathol 11:133–139, 1987.
94. Leff EL, Brooks JSJ, Trojanowski JQ: Expression of neurofilament and neuron-specific enolase in small cell tumors of skin using immunohistochemistry. Cancer 56:625–631, 1985.
95. Lehto V-P, Miettinen M, Dahl D, Virtanen I: Bronchial carcinoid cells contain neural-type intermediate filaments. Cancer 54:624–628, 1984.
96. Leong AS-Y, Kan AE, Milios J: Small round cell tumors in childhood: Immunohistochemical studies in rhabdomyosarcoma, neuroblastoma, Ewing's sarcoma, and lymphoblastic lymphoma. Surg Pathol 2:5–17, 1989.
97. Leong AS-Y, Milios J: An assessment of a melanoma-specific antibody (HMB-45) and other immunohistochemical markers of malignant melanoma in paraffin-embedded tissues. Surg Pathol 2:137–145, 1989.
98. Leong AS-Y, Parkinson R, Milios J: "Thick" cell membranes revealed by immunocytochemical staining: A clue to the diagnosis of mesothelioma. Diagn Cytopathol 6:9–13, 1990.
99. Liem KH, Yen S-H, Salomon GD, Shelanski ML: Intermediate filaments in nervous tissues. J Cell Biol 79:637–645, 1978.
100. Listrom MB, Dalton LW: Comparison of keratin monoclonal antibodies MAK-6, AE1:AE3, and CAM-5.2. Am J Clin Pathol 88:297–301, 1987.
101. Lloyd RV: Immunohistochemical localization of chromogranin in normal and neoplastic endocrine tissues. Pathol Annu 22(Part 2):69–90, 1987.
102. Lloyd RV, Cana M, Rosa P, Annette H, Huttner WB: Distribution of chromogranin A and secretogranin I (chromogranin B) in neuroendocrine cells and tumors. Am J Pathol 130:296–304, 1988.
103. Lloyd RV, Mervak T, Schmidt K, Warner TFCS, Wilson BS: Immunohistochemical detection of chromogranin and neuron-specific enolase in pancreatic endocrine neoplasms. Am J Surg Pathol 8:607–614, 1984.
104. Lloyd RV, Sisson JC, Shapiro B, Verhofstad AAJ: Immunohistochemical localization of epinephrine, norepinephrine, catecholamine-synthesizing enzymes, and chromogranin in neuroendocrine cells and tumors. Am J Pathol 125:45–54, 1986.
105. Lloyd RV, Wilson BS, Kovacs K, Ryan N: Immunohistochemical localization of chromogranin in human hypophyses and pituitary adenomas. Arch Pathol Lab Med 109:515–517, 1985.
106. Loeffel SC, Gillespie SY, Mirmiran SA, et al: Cellular immunolocalization of S100 protein within fixed tissue sections by monoclonal antibodies. Arch Pathol Lab Med 109:117–122, 1985.
107. Lundy J, Kline TS, Lozowski M, Chao S: Immunoperoxidase studies by monoclonal antibody B72.3 applied to breast aspirates: Diagnostic considerations. Diagn Cytopathol 4:95–98, 1988.

108. Lundy J, Lozowski M, Mishriki Y: Monoclonal antibody B72.3 as a diagnostic adjunct in fine needle aspirates of breast masses. *Ann Surg* 203:399–402, 1986.
109. Manivel JC, Jessurun J, Wick MR, Dehner LP: Placental alkaline phosphatase immunoreactivity in testicular germ-cell neoplasms. *Am J Surg Pathol* 11:21–29, 1987.
110. Marshall RJ, Herbert A, Braye SG, Jones DB: Use of antibodies to carcinoembryonic antigen and human milk fat globule to distinguish carcinoma, mesothelioma, and reactive mesothelium. *J Clin Pathol* 37:1215–1221, 1984.
111. Martin SE, Moshiri S, Thor A, et al: Identification of adenocarcinoma in cytospin preparations of effusions using monoclonal antibody B72.3. *Am J Clin Pathol* 86:10–18, 1986.
112. May EE, Perentes E: Anti-Leu 7 immunoreactivity with human tumours: Its value in the diagnosis of prostatic adenocarcinoma. *Histopathology* 11:295–304, 1987.
113. McNutt MA, Bolen JW, Gown AM, Hammar SP, Vogel AM: Coexpression of intermediate filaments in human epithelial neoplasms. *Ultrastruct Pathol* 9:31–43, 1985.
114. Mendelsohn G, Bigner SH, Eggleston JC, Baylin SB, Wells SA Jr: Anaplastic variants of medullary thyroid carcinoma; a light-microscopic and immunohistochemical study. *Am J Surg Pathol* 4:333–341, 1980.
115. Miettinen M: Synaptophysin and neurofilament proteins as markers for neuroendocrine tumors. *Arch Pathol Lab Med* 111:813–818, 1987.
116. Miettinen M: Immunoreactivity for cytokeratin and epithelial membrane antigen in leiomyosarcoma. *Arch Pathol Lab Med* 112:637–640, 1988.
117. Miettinen M: Keratin subsets in spindle cell sarcomas; keratins are widespread but synovial sarcoma contains a distinctive keratin polypeptide pattern and desmoplakins. *Am J Pathol* 138:505–513, 1991.
118. Miettinen M, Franssila K: Immunochemical spectrum of malignant melanoma; the common presence of keratins. *Lab Invest* 61:623–628, 1989.
119. Miettinen M, Franssila K, Lehto V-P, Paasivuo R, Virtanen I: Expression of intermediate filament proteins in thyroid gland and thyroid tumors. *Lab Invest* 50:262–270, 1984.
120. Miettinen M, Lehto V-P, Badley RA, Virtanen I: Alveolar rhabdomyosarcoma; demonstration of the muscle type of intermediate-filament protein, desmin, as a diagnostic aid. *Am J Pathol* 108:246–251, 1982.
121. Miettinen M, Lehto V-P, Dahl D, Virtanen I: Varying expression of cytokeratin and neurofilaments in neuroendocrine tumors of the human gastrointestinal tract. *Lab Invest* 52:429–436, 1985.
122. Miettinen M, Rapola J: Immunohistochemical spectrum of rhabdomyosarcoma and rhabdomyosarcoma-like tumors; expression of cytokeratin and the 68-kD neurofilament protein. *Am J Surg Pathol* 13:120–132, 1989.
123. Moll R, Franke WW: Cytoskeletal differences between human neuroendocrine tumors: A cytoskeletal protein of molecular weight 46,000 distinguishes cutaneous from pulmonary neuroendocrine neoplasms. *Differentiation* 30:165–175, 1985.
124. Moll R, Franke WW, Schiller DL, Geiger B, Krepler R: The catalog of human cytokeratins: Patterns of expression in normal epithelia, tumors and cultured cells. *Cell* 31:11–24, 1982.
125. Moll R, Levy R, Czernobilsky B, et al: Cytokeratins of normal epithelia and some neoplasms of the female genital tract. *Lab Invest* 49:599–610, 1983.
126. Moll R, Osborn M, Hartschuh W, et al: Variability of expression and arrangement of

cytokeratin and neurofilaments in cutaneous neuroendocrine carcinomas (Merkel cell tumors): Immunocytochemical and biochemical analysis of twelve cases. *Ultrastruct Pathol* 10:473–495, 1986.

127. Morohoshi T, Kanda M, Horie A, et al: Immunocytochemical markers of uncommon pancreatic tumors; acinar cell carcinoma, pancreatoblastoma, and solid cystic (papillary-cystic) tumor. *Cancer* 59:739–747, 1987.

128. Muensch HA, Maslow WC, Azama F, et al: Placental-like alkaline phosphatase; re-evaluation of the tumor marker with exclusion of smokers. *Cancer* 58:1689–1694, 1986.

129. Mukai K, Schollmeyer J, Rosai J: Immunohistochemical localization of actin; applications in surgical pathology. *Am J Surg Pathol* 5:91–97, 1981.

130. Mukai M, Torikata C, Iri H, et al: Expression of neurofilament triplet proteins in human neural tumors, an immunohistochemical study of paraganglioma, ganglioneuroma, ganglioneuroblastoma, and neuroblastoma. *Am J Pathol* 122:28–35, 1986.

131. Murthy L, Kapila K, Verma K: Immunoperoxidase detection of carcinoembryonic antigen in fine needle aspirates of breast carcinoma; correlation with studies of tissue sections. *Acta Cytol* 32:60–62, 1988.

132. Nadji M, Tabei SZ, Castro A, et al: Prostatic-specific antigen: An immunohistologic marker for prostatic neoplasms. *Cancer* 48:1229–1232, 1981.

133. Nadji M, Tabei SZ, Castro A, Chu TM, Morales AR: Prostatic origin of tumors; an immunohistochemical study. *Am J Clin Pathol* 73:735–739, 1980.

134. Nakajima T, Watanabe S, Sato Y, et al: An immunoperoxidase study of S-100 protein distribution in normal and neoplastic tissues. *Am J Surg Pathol* 6:715–727, 1982.

135. Nelson WG, Battifora H, Santana H, Sun T-T: Specific keratins as molecular markers for neoplasms with a stratified epithelial origin. *Cancer Res* 44:1600–1603, 1984.

136. Nguyen G-K, Kline TS: *Essentials of Aspiration Biopsy Cytology*. New York, Igaku-Shoin, 1991. Pp. 152–164.

137. Norton AJ, Thomas JA, Isaacson PG: Cytokeratin-specific monoclonal antibodies are reactive with tumours of smooth muscle derivation. An immunocytochemical and biochemical study using antibodies to intermediate filament cytoskeletal proteins. *Histopathology* 11:487–499, 1987.

138. Nuti M, Mottolese M, Viora M, et al: Use of monoclonal antibodies to human breast-tumor-associated antigens in fine-needle aspiration cytology. *Int J Cancer* 37:493–498, 1986.

139. O'Connor DT, Deftos LJ: Secretion of chromogranin A by peptide-producing endocrine neoplasms. *N Engl J Med* 314:1145–1151, 1986.

140. Ordóñez NG: The immunohistochemical diagnosis of mesothelioma; differentiation of mesothelioma and lung adenocarcinoma. *Am J Surg Pathol* 13:276–291, 1989.

141. Ordóñez NG, Sneige N, Hickey RC, Brooks TE: Use of monoclonal antibody HMB-45 in the cytologic diagnosis of melanoma. *Acta Cytol* 32:684–688, 1988.

142. Osborn M, Altmannsberger M, Shaw G, Schauer A, Weber K: Various sympathetic derived human tumors differ in neurofilament expression; use in diagnosis of neuroblastoma, ganglioneuroblastoma and pheochromocytoma. *Virchows Arch [B]* 40:141–156, 1982.

143. Osborn M, Dirk T, Käser H, Weber K, Altmannsberger M: Immunohistochemical localization of neurofilaments and neuron-specific enolase in 29 cases of neuroblastoma. *Am J Pathol* 122:433–442, 1986.

144. Otis CN, Carter D, Cole S, Battifora H: Immunohistochemical evaluation of pleural mesothelioma and pulmonary adenocarcinoma; a bi-institutional study of 47 cases. Am J Surg Pathol 11:445–456, 1987.

145. Papsidero LD, Croghan GA, Asirwatham J, et al: Immunohistochemical demonstration of prostate-specific antigen in metastases with the use of monoclonal antibody F5. Am J Pathol 121:451–454, 1985.

146. Pavelic ZP, Petrelli NJ, Herrera L, et al: D-14 monoclonal antibody to carcinoembryonic antigen: Immunohistochemical analysis of formalin-fixed, paraffin-embedded human colorectal carcinoma, tumors of non-colorectal origin and normal tissues. J Cancer Res Clin Oncol 116:51–56, 1990.

147. Pelosi G, Bonetti F, Colombari R, Bonzanini M, Iannucci A: Use of monoclonal antibody HMB-45 for detecting malignant melanoma cells in fine needle aspiration biopsy samples. Acta Cytol 34:460–462, 1990.

148. Perentes E, Rubinstein LJ: Recent applications of immunoperoxidase histochemistry in human neuro-oncology; an update. Arch Pathol Lab Med 111:796–812, 1987.

149. Pinkus GS, Kurtin PJ: Epithelial membrane antigen-A diagnostic discrimination in surgical pathology: Immunohistochemical profile in epithelial, mesenchymal, and hematopoeitic neoplasms using paraffin sections and monoclonal antibodies. Hum Pathol 16:929–940, 1985.

150. Pinto MM: An immunoperoxidase study of S-100 protein in neoplastic cells in serous effusions; use as a marker for melanomas. Acta Cytol 30:240–244, 1986.

151. Pohoresky J, Yazdi HM: Value of B72.3 staining in "suspicious" fine needle aspiration biopsies of the breast. Acta Cytol 35:254–255, 1991.

152. Pollen JJ, Dreilinger A: Immunohistochemical identification of prostatic acid phosphatase and prostate specific antigen in female periurethral glands. Urology 23:303–304, 1984.

153. Quinlan RA, Schiller DL, Hatzfeld M, et al: Patterns of expression and organization of cytokeratin intermediate filaments. Ann NY Acad Sci 455:282–306, 1985.

154. Ramaekers F, Haag D, Jap P, Vooijs PG: Immunochemical demonstration of keratin and vimentin in cytologic aspirates. Acta Cytol 28:385–392, 1984.

155. Rosai J: *Ackerman's Surgical Pathology,* ed. 7. St. Louis, CV Mosby, 1989.

156. Said JW: Immunohistochemical localization of keratin proteins in tumor diagnosis. Hum Pathol 14:1017–1019, 1983.

157. Said JW, Vimadalal S, Nash G, et al: Immunoreactive neuron-specific enolase, bombesin, and chromogranin as markers for neuroendocrine lung tumors. Hum Pathol 16:236–240, 1985.

158. Schlaepfer W, Freeman LA: Neurofilament proteins of rat peripheral nerve and spinal cord. J Cell Biol 78:653–662, 1978.

159. Schmidt RA, Cone R, Haas JE, Gown AM: Diagnosis of rhabdomyosarcomas with HHF35, a monoclonal antibody directed against muscle actins. Am J Pathol 131:19–28, 1988.

160. Schmitt FC, Bacchi CE: S-100 protein: Is it useful as a tumor marker in diagnostic immunocytochemistry? Histopathology 15:281–288, 1989.

161. Schröder S, Dockhorn-Dworniczak B, Kastendieck H, Böcker W, Franke WW: Intermediate-filament expression of thyroid gland carcinomas. Virchows Arch {A} 409:751–766, 1986.

162. Sheahan K, O'Brien MJ, Burke B, et al: Differential reactivities of carcinoembryonic

antigen (CEA) and CEA-related monoclonal and polyclonal antibodies in common epithelial malignancies. *Am J Clin Pathol* 94:157–164, 1990.

163. Shevchuk MM, Romas NA, Ng PY, Tannenbaum M, Olsson CA: Acid phosphatase localization in prostatic carcinoma; a comparison of monoclonal antibody to heteroantisera. *Cancer* 52:1642–1646, 1983.

164. Shoup SA, Johnston WW, Siegler HF, et al: A panel of antibodies useful in the cytologic diagnosis of metastatic melanoma. *Acta Cytol* 34:385–391, 1990.

165. Sloane JP, Ormerod MG: Distribution of epithelial membrane antigen in normal and neoplastic tissues and its value in diagnostic tumor pathology. *Cancer* 47:1786–1795, 1981.

166. Stanta G, Carcangiu ML, Rosai J: The biochemical and immunohistochemical profile of thyroid neoplasia, in Rosen PP, Fechner RE (eds): *Pathology Annual*. (Part 1). Norwalk, Appleton & Lange, 1988. Pp 129–157.

167. Stefansson K, Wollmann R, Jerkovic M: S-100 protein in soft-tissue tumors derived from schwann cells and melanocytes. *Am J Pathol* 106:261–268, 1982.

168. Stein BS, Petersen RO, Vangore S, Kendall AR: Immunoperoxidase localization of prostate-specific antigen. *Am J Surg Pathol* 6:553–557, 1982.

169. Stirling JW, Henderson DW, Spagnolo DV, Whitaker D: Unusual granular reactivity for carcinoembryonic antigen in malignant mesothelioma. *Hum Pathol* 21:678–679, 1990.

170. Stroup RM, Pinkus GS: S-100 immunoreactivity in primary and metastatic carcinoma of the breast: A potential source of error in immunodiagnosis. *Hum Pathol* 19:949–953, 1988.

171. Swanson PE, Scheithauer BW, Manivel JC, Wick MR: Epithelial membrane antigen reactivity in mesenchymal neoplasms: An immunohistochemical study of 306 soft tissue sarcomas. *Surg Pathol* 2:313–322, 1989.

172. Szpak CA, Johnston WW, Roggli V, et al: The diagnostic distinction between malignant mesothelioma of the pleura and adenocarcinoma of the lung as defined by a monoclonal antibody (B72.3). *Am J Pathol* 122:252–260, 1986.

173. Tahara E, Ito H, Taniyama K, Yokozaki H, Hata J: Alpha 1-antitrypsin, alpha 1-antichymotrypsin, and alpha 2-macroglobulin in human gastric carcinomas: A retrospective immunohistochemical study. *Hum Pathol* 15:957–964, 1984.

174. Takashi M, Haimoto H, Murase T, Mitsuya H, Kato K: An immunochemical and immunohistochemical study of S100 protein in renal cell carcinoma. *Cancer* 61:889–895, 1988.

175. Tavassoli FA, Jones MW, Majeste RM, Bratthauer GL, O'Leary TJ: Immunohistochemical staining with monoclonal Ab B72.3 in benign and malignant breast disease. *Am J Surg Pathol* 14:128–133, 1990.

176. Taylor CR: *Immunomicroscopy: A Diagnostic Tool for the Surgical Pathologist*. Philadelphia, WB Saunders, 1986.

177. Thomas P, Battifora H, Manderino GL, Patrick J: A monoclonal antibody against neuron-specific enolase, immunohistochemical comparison with polyclonal antiserum. *Am J Clin Pathol* 88:146–152, 1987.

178. To A, Dearnaley DP, Ormerod MG, Canti G, Coleman DV: Epithelial membrane antigen; its use in the cytodiagnosis of malignancy in serous effusions. *Am J Clin Pathol* 78:214–219, 1982.

179. Trojanowski JQ, Lee VM-Y, Schlaepfer WW: An immunohistochemical study of hu-

man central and peripheral nervous system tumors, using monoclonal antibodies against neurofilaments and glial filaments. *Hum Pathol* 15:248–257, 1984.

180. Tron V, Wright JL, Churg A: Carcinoembryonic antigen and milk-fat globule protein staining of malignant mesothelioma and adenocarcinoma of the lung. *Arch Pathol Lab Med* 111:291–293, 1987.

181. Truong LD, Rangdaeng S, Cagle P, et al: The diagnostic utility of desmin; a study of 584 cases and review of the literature. *Am J Clin Pathol* 93:305–314, 1990.

182. Tsukada T, McNutt MA, Ross R, Gown AM: HHF35, a muscle actin-specific monoclonal antibody. II. Reactivity in normal, reactive, and neoplastic human tissues. *Am J Pathol* 127:389–402, 1987.

183. Tsukada T, Tippens D, Gordon D, Ross R, Gown AM: HHF35, a muscle-actin-specific monoclonal antibody. I. Immunocytochemical and biochemical characterization. *Am J Pathol* 126:51–60, 1987.

184. Uchida T, Shimoda T, Miyata H, et al: Immunoperoxidase study of alkaline phosphatase in testicular tumor. *Cancer* 48:1455–1462, 1981.

185. Upton MP, Hirohashi S, Tome Y, et al: Expression of vimentin in surgically resected adenocarcinomas and large cell carcinomas of lung. *Am J Surg Pathol* 10:560–567, 1986.

186. Vandekerckhove J, Weber K: At least six different actins are expressed in a higher mammal: An analysis based on the amino acid sequence of the amino-terminal tryptic peptide. *J Mol Biol* 126:783–802, 1978.

187. Van Der Kwast TH, Versnel MA, Delahaye M, et al: Expression of epithelial membrane antigen on malignant mesothelioma cells; an immunocytochemical and immunoelectron microscopic study. *Acta Cytol* 32:169–174, 1988.

188. Van Eyken P, Sciot R, Paterson A, et al: Cytokeratin expression in hepatocellular carcinoma: An immunohistochemical study. *Hum Pathol* 19:562–568, 1988.

189. Van Muijen GNP, Ruiter DJ, Van Leeuwen C, et al: Cytokeratin and neurofilament in lung carcinomas. *Am J Pathol* 116:363–369, 1984.

190. van Nagell JR Jr, Goldenberg DM: Carcinoembryonic antigen staining of endometrial and endocervical carcinomas. *Lancet* 1:213, 1980.

191. Viale G, Dell'Orto P, Coggi G, Gambacorta M: Coexpression of cytokeratin and vimentin in normal and diseased thyroid glands; lack of diagnostic utility of vimentin immunostaining. *Am J Surg Pathol* 13:1034–1040, 1989.

192. Vinores SA, Bonnin JM, Rubinstein LJ, Marangos PJ: Immunohistochemical demonstration of neuron-specific enolase in neoplasms of the CNS and other tissues. *Arch Pathol Lab Med* 108:536–540, 1984.

193. Waldherr R, Schwechheimer K: Co-expression of cytokeratin and vimentin intermediate-sized filaments in renal cell carcinoma; comparative study of the intermediate-sized filament distribution in renal cell carcinomas and normal human kidney. *Virchows Arch {A}* 408:15–27, 1985.

194. Walts AE, Said JW: Specific tumor markers in diagnostic cytology; immunoperoxidase studies of carcinoembryonic antigen, lysozyme and other tissue antigens in effusions, washes and aspirates. *Acta Cytol* 27:408–416, 1983.

195. Walts AE, Said JW, Shintaku IP: Epithelial membrane antigen in the cytodiagnosis of effusions and aspirates: Immunocytochemical and ultrastructural localization in benign and malignant cells. *Diagn Cytopathol* 3:41–49, 1987.

196. Warhol MJ, Longtine JA: The ultrastructural localization of prostatic specific antigen

and prostatic acid phosphatase in hyperplastic and neoplastic human prostates. *J Urol* 134:607–613, 1985.

197. Weiler R, Fischer-Colbrie R, Schmid KW, et al: Immunological studies on the occurance and properties of chromogranin A and B and secretogranin II in endocrine tumors. *Am J Surg Pathol* 12:877–884, 1988.

198. Weiss SW, Bratthauer GL, Morris PA: Postirradiation malignant fibrous histocytoma expressing cytokeratin; indications for the immunodiagnosis of sarcomas. *Am J Surg Pathol* 12:554–558, 1988.

199. Weiss SW, Langloss JM, Enzinger FM: Value of S-100 protein in the diagnosis of soft tissue tumors with particular reference to benign and malignant schwann cell tumors. *Lab Invest* 49:299–308, 1983.

200. Wick MR, Swanson PE, Scheithauer BW, Manivel JC: Malignant peripheral nerve sheath tumor, an immunohistochemical study of 62 cases. *Am J Clin Pathol* 87:425–433, 1987.

201. Wilander E, Påhlman S, Sällström J, Lindgren A: Neuron-specific enolase expression and neuroendocrine differentiation in carcinomas of the breast. *Arch Pathol Lab Med* 111:830–832, 1987.

202. Wilson BS, Lloyd RV: Detection of chromogranin in neuroendocrine cells with a monoclonal antibody. *Am J Pathol* 115:458–468, 1984.

203. Wong MA, Yazdi HM: Hepatocellular carcinoma versus carcinoma metastatic to the liver; value of stains for carcinoembryonic antigen and naphthylamidase in fine needle aspiration biopsy material. *Acta Cytol* 34:192–196, 1990.

204. Wu Y-J, Parker LM, Binder NE, et al: The mesothelial keratins: A new family of cytoskeletal proteins identified in cultured mesothelial cells and nonkeratinizing epithelia. *Cell* 31:693–703, 1982.

205. Yam LT, Janckila AJ, Lam WKW, Li C-Y: Immunohistochemistry of prostatic acid phosphatase. *Prostate* 2:97–107, 1981.

206. Zarbo RJ, Gown AM, Nagle RB, Visscher DW, Crissman JD: Anomalous cytokeratin expression in malignant melanoma: One- and two-dimensional Western blot analysis and immunohistochemical survey of 100 melanomas. *Mod Pathol* 3:494–501, 1990.

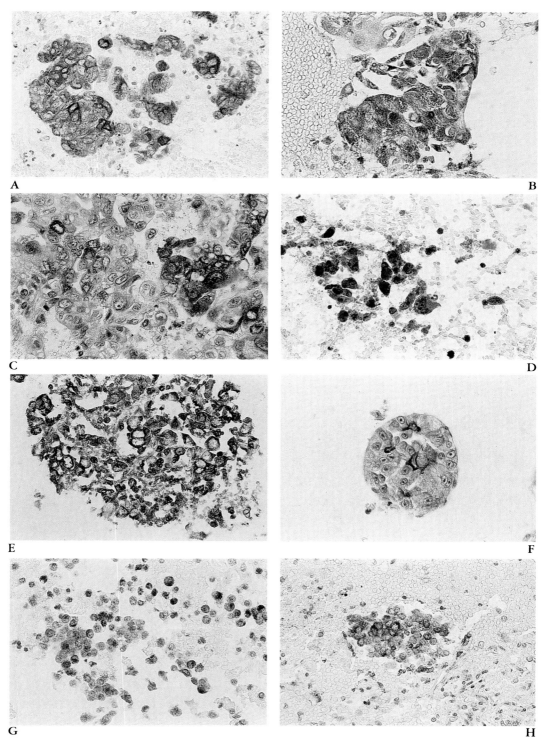

**Plate 5.1.** Immunoreactivity of different neoplasms in ABC specimens. **A.** Leiomyosarcoma (actin). **B.** Hepatocellular carcinoma (AFP). **C.** Germ-cell tumor (AFP). **D.** Medullary carcinoma of the thyroid (calcitonin). **E.** Metastatic adenocarcinoma to liver. Note diffuse cytoplasmic staining (CEA). **F.** Hepatocellular carcinoma. Note predominantly canalicular staining (CEA). **G.** Carcinoid tumor (chromogranin). **H.** Small-cell neuroendocrine carcinoma (chromogranin). (A–H) × 256.

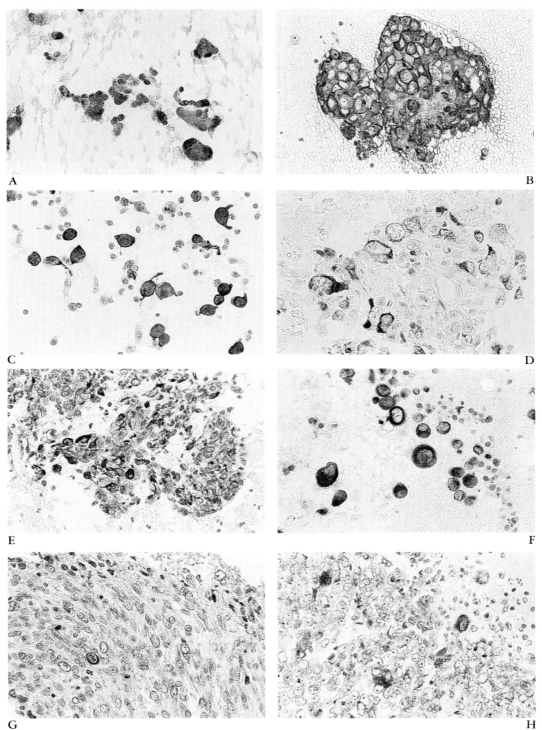

**Plate 5.2.** Immunoreactivity of different neoplasms in ABC specimens. **A.** Medullary carcinoma of the thyroid (chromogranin). **B.** Adenocarcinoma (EMA). **C.** Metastatic carcinoma to lymph node (cytokeratin). **D.** Anaplastic carcinoma (cytokeratin). **E.** Leiomyosarcoma (vimentin). **F.** Malignant melanoma (vimentin). **G.** Leiomyosarcoma (desmin). **H.** Malignant melanoma (HMB-45). (A–H) × 256.

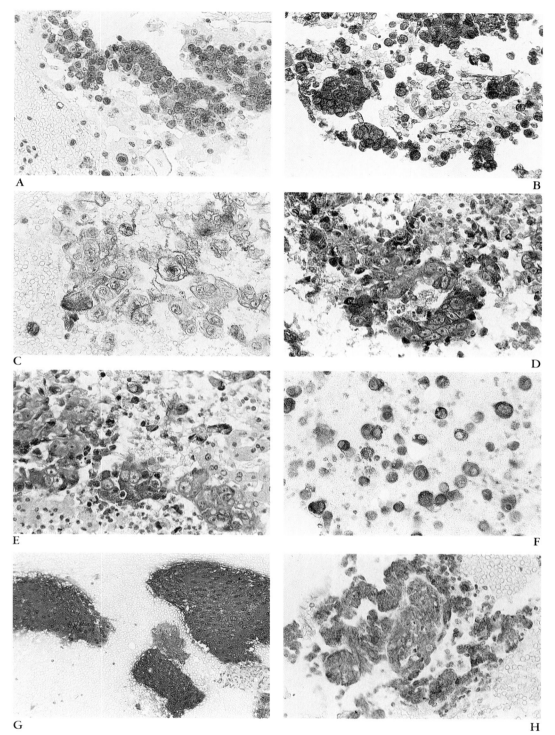

Plate 5.3. Immunoreactivity of different neoplasms in ABC specimens. **A.** Carcinoid tumor (NSE). **B.** Small-cell neuroendocrine carcinoma (NSE). **C.** Germ-cell tumor (PLAP). **D.** Prostatic carcinoma (PrAP). **E.** Prostatic carcinoma (PSA). **F.** Malignant melanoma (S-100). **G.** Schwannoma (S-100). **H.** Carcinoid tumor (synaptophysin). (A–F,H) × 256; (G) × 160.

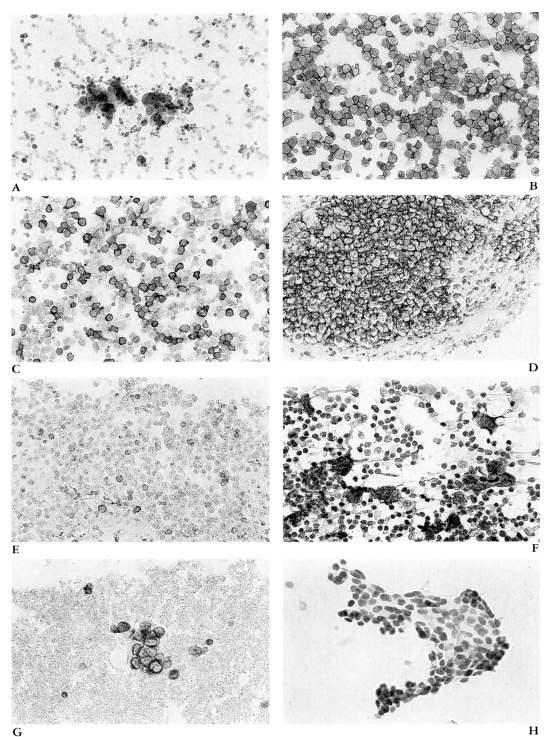

**Plate 5.4.** Immunoreactivity of different neoplasms in ABC specimens. **A.** Follicular neoplasm (thyroglobulin). **B.** Non-Hodgkin's lymphoma (LCA). **C.** Non-Hodgkin's B-cell lymphoma (lambda light chain). **D.** Non-Hodgkin's B-cell lymphoma (L26). **E.** Reactive T cells in a B-cell non-Hodgkin's lymphoma (UCHL1). **F.** Ki-1–positive large-cell lymphoma (Ki-1 antigen). **G.** Prostatic carcinoma (Leu-7). **H.** Breast carcinoma. Note nuclear staining (estrogen receptor). (A,D–H) × 256; (B,C) × 160.

# 6
# Principles of Electron Microscopy

Identification of diagnostic features in aspiration biopsy cytology (ABC) specimens maximally tests the skills of electron microscopists. Because the spectrum of differentiation and gene expression in tumors has become evident partly by electron microscopy, awareness is needed of both normal and altered cellular features to use electron microscopy effectively for diagnostic purposes. There are extensive articles and books on the diagnostic applications of electron microscopy;[18,27,35,40,46,60,78] however, all that is really necessary to diagnose successfully many problem cases is knowledge of a limited number of cellular characteristics, organizational features, or organelles. A primary purpose of this chapter is to describe and illustrate how electron microscopy can be applied in ABC. The challenge in surgical pathology is the use of the electron microscopy to elicit the tissue or organ source for metastatic tumors;[28,45,51] determining the unknown primary using ABC specimens is even more challenging, and the feasibility requires ongoing study.

Fortunately, most ABC of tumors reviewed by cytopathologists have such characteristic or distinctive cytological features, or are reasonably well differentiated, that they pose little or no diagnostic problem. With increasing anaplasia, however, specific and therefore diagnostic tumor characteristics become more difficult to recognize. Also, common cytoplasmic features can be shared by histogenetically diverse tumors and these may be difficult to differentiate; examples include tumor cells with clear, granular, or vacuolated cytoplasm. In both circumstances (i.e., with look-alike tumors and the undifferentiated to poorly differentiated categories of sarcomas and carcinomas), electron microscopy could provide the fine structural and organizational details of the tumor cells that enable a specific diagnosis or at least narrow the list of possible diagnoses. Even confirming that the lesion is undifferentiated can be valuable information for clinicians.

The exquisite resolution provided by electron microscopy at an almost limitless magnification has decided advantages. The cellular and structural details that are not otherwise evident in routinely prepared cytological smears may be crucial in solving a diagnostic dilemma. The ability to detect secretory products that are often

TABLE 6.1. Indications for Electron Microscopy of Needle Aspiration Biopsies

| |
|---|
| Determining the type of undifferentiated or poorly differentiated neoplasm |
| Specific diagnosis of poorly differentiated carcinomas or sarcomas |
| Determination of the primary site in metastatic tumors |
| Differentiating primary from metastatic tumors |
| Identifying neuroendocrine tumors |
| Confirmation of light microscopic diagnosis |

in too limited amounts to be detected by immunocytochemistry, for example, is instrumental in diagnosing benign and malignant neuroendocrine neoplasms. At the ultrastructural level, other secretory products and additional organelles can be essential in differentiating certain carcinomas and therefore can assist in determining the primary site for some metastatic neoplasms or even distinguishing primary from metastatic tumors. The spectrum of indications for ancillary assessments that can be readily applied to ABC specimens were outlined in Tables 1.1 and 1.2; more specific uses for electron microscopy are provided in Table 6.1.

The clinical setting dictates which tests will be used, because determination of the exact cytological diagnosis may well decide the therapeutic approach. There also may be cases where the cytological diagnosis is readily determined from the aspirate, but confirmation of the light microscopic diagnosis would assist clinicians. In the vast majority of ABC specimens, particularly from such sites or organs as breast, salivary gland, testis, thyroid, prostate, and lymph nodes, there is no general role for electron microscopy. As will be detailed in subsequent sections, ultrastructural studies are best reserved for aspirates from deep lesions obtained by radiological or ultrasound imaging; the indications outlined in Table 6.1 usually apply in these circumstances. The final decision to perform ultrastructural examination of a particular aspirate rests with cytopathologists. For effective use of this imporant diagnostic procedure, however, cooperation between radiologists, cytopathologists, and electron microscopists is necessary for the proper handling and interpretation of needle aspirates.

# DIAGNOSTIC ARCHITECTURAL FEATURES

As important as the enhanced cellular details obtainable by increased magnification are to electron microscopists for diagnostic purposes, the organization of the tumor cells is of equal value. Cellular relationships are a vital element for tissue diagnosis in surgical pathology. The many fragments of tumor cells in ABC specimens also can be used to obtain this type of information. Occasionally, the amount of tissue that can be obtained with a 22-gauge needle for this purpose may surpass expectations. Reference to ABC specimens as minisurgical biopsies relies on this fact. The following two cases will illustrate the quantity and quality of tissue fragments that can be obtained by needle aspiration biopsy (NAB), and the nature of architectural details that can be important diagnostically.

A 56-year-old woman who had previously undergone a mastectomy for an infiltrating carcinoma of the breast presented with right chest-wall pain and dyspnea. A chest radiograph revealed pleural effusion and a diffuse thickening of the right pleura. NAB of the pleura (using a 22-gauge needle) resulted in the tissue fragments, illustrated in Figure 6.1A, which were available in the epon block. Not only can the papillary-type of epithelial growth pattern be readily appreciated, but the biopsy nature of the aspirate is evident from the intact fibrovascular stroma. The size of the accumulated fragments is self evident (Fig. 6.1A), as is the superb degree of tissue preservation (Fig. 6.1B) despite 20 to 40 minutes in a balanced electrolyte solution (see Chapter 2) prior to fixation in glutaraldehyde. The papillary growth pattern could represent metastatic breast carcinoma or diffuse epithelial mesothelioma. Apical surface microvilli were numerous and reasonably long. The many cytoplasmic organelles and the few secretory granules in the apical regions of some tumor cells (Fig. 6.1B), however, showed that the lesion was a metastatic adenocarcinoma. Immunocytochemistry performed on the cell block section revealed a positive staining for carcinoembryonic antigen, confirming this interpretation.

Many smears prepared from ABC specimens of deep sites contain fragments of tissue whose general architecture provides valuable clues about the diagnosis. When the cytology of such material suggests a site of origin other than the tissue from which it was obtained, then diagnostic problems or even errors in interpretation can arise. The next case illustrates how electron microscopy can provide the diagnostic clues to prevent this situation.

A 76-year-old man had a chest radiograph performed because of persisting cough after a respiratory tract infection. The hilum of the upper lobe of the left lung contained a 2-cm diameter lesion (Fig. 6.2A,B). NAB was performed and produced an adequate smear that contained cellular material (see Fig. 6.2C) and what appeared to be psammoma bodies. The provisional diagnosis of a metastatic papillary carcinoma of the thyroid was reasonable. Electron microscopy of similar fragments, however, revealed that the tumor cells were not regularly aligned on the surface of fibrovascular tissue as is expected in a thyroid carcinoma. There were small glandular structures internally, whereas other tumor cells with attenuated cytoplasmic processes enclosed basal lamina-lined and glycosaminoglycan-containing intercellular spaces (Fig. 6.2D). The latter produced the expansion of the papillary processes seen by light microscopy (Fig. 6.2C), and their contents accounted for the psammoma-like structures. The two cell-types (i.e., luminal cells and the cells enclosing the localized extracellular materials) revealed this tumor to be an adenoid cystic carcinoma. There was no evidence for a primary salivary gland tumor and no history of surgery for such a lesion, suggesting that this tumor represented a primary adenoid cystic carcinoma of bronchial gland origin. Because of the patient's age, absence of symptoms, and lack of evidence for bronchial obstruction, surgery was deferred. He died 18 months later of causes unrelated to the lung tumor.

This case also highlights the role that diagnostic characteristics revealed by electron microscopy can have in clinical management of individual cases. Another feature that may be apparent in ABC, cell-to-cell relationships, also provides valuable diagnostic clues in the ultrastructural evaluation of tumors. Fortunately, many ABC specimens are of sufficient magnitude to reveal this feature.

The characteristic cellular relationships in squamous-cell carcinomas, which even

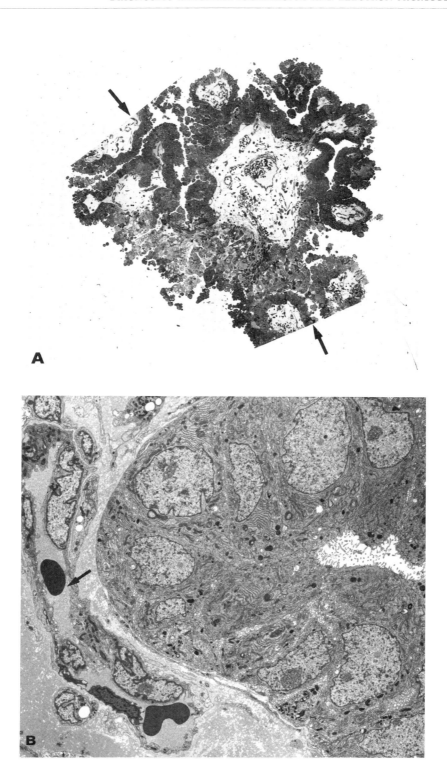

in low magnification, survey-type micrographs help establish diagnoses, are evident in three ABC specimens of lung tumors (Fig. 6.3A–D). This pattern relies on the multiple gap-like intercellular spaces produced by apposed, blunt cytoplasmic extensions placed at intervals along adjacent cell borders. This process is somewhat exaggerated in one example (Fig. 6.3A). The length and complexity of these cytoplasmic processes seem dependent on the degree of differentiation of the squamous-cell carcinoma. In a poorly differentiated tumor, the processes are simple, broad, and at times lack desmosomes (Fig. 6.3D), whereas in the better differentiated tumor the desmosomes are plentiful, well developed, and often placed at the sides of the cytoplasmic extensions (Fig.6.3A–C). It is the combination of alternating intercellular gaps and abutting processes that results in the classic "intercellular bridges" seen by light microscopy. Even when this feature is not discernible by light microscopy, it may be observed readily by electron microscopy. Figure 8.16 also illustrates the diagnostic utility of cellular relationships in neoplasms such as schwannomas.

# DIAGNOSTIC CELLULAR FEATURES

The asset underlying ultrastructural evaluation of biopsy specimens, whether in cytology or surgical pathology, is the ability to identify certain organelles and secretory products (Table 6.2) that cannot be appreciated readily in routine diagnostic preparations. Thus, recognition of these cellular features, and the spectrum of their differentiation and structure in the neoplastic state, is crucial. Fortunately, identification of at least the more common tumors depends only on knowing several cellular features (Table 6.2). Learning how to apply these features in diagnostic problems will enable more precise diagnoses in an appreciable number of cases and will assist in resolving the diagnostic problems listed in Table 6.1.

## JUNCTIONS

Qinonez and Simon[76] provide a practical guide to the nomenclature and diagnostic application of intercellular junctions in malignant tumors. Specializations of the cell surface such as desmosomes, junctional complexes, paired subplasmalemmal densities, and subplasmalemmal linear densities (Table 6.3) are the most useful.[39-41,68,76] Of these junctions, desmosomes and the tight junction component of junctional complexes are the more frequently found and the most diagnostically

---

Fig. 6.1. NAB of pleura, metastatic breast carcinoma. **A.** Needle rinse tissue fragments processed without centrifugation, includes both epithelium and stromal tissues. These tissues fill a considerable portion of the epon block face (edges of the plastic section denoted by arrows). Toluidine blue stain, × 100. **B.** Ultrastructurally, the tissue fragments have an intact architecture consisting of rough endoplasmic reticulum-rich glandular epithelium and an underlying capillary (arrow). × 2,200. (From Dardick I, et al: A quantitative comparison of light and electron microscopic diagnoses in specimens obtained by fine-needle aspiration biopsy. *Ultrastruct Pathol* 15:105–129, 1991)

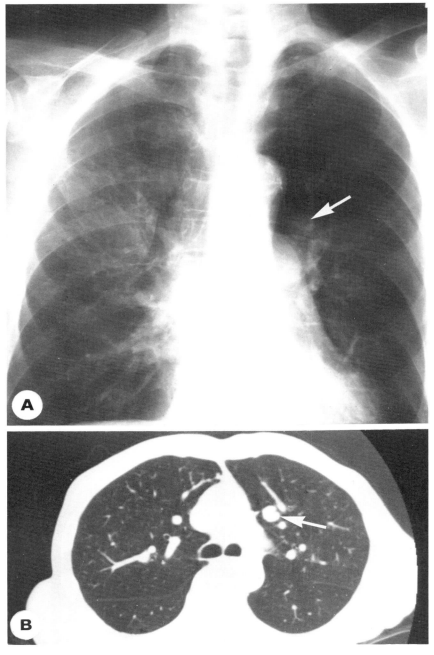

Fig. 6.2. NAB of lung, adenoid cystic carcinoma. A. Chest radiograph with discrete hilar nodule (arrow). B. Computed tomographic scan localizes the tumor (arrow) for subsequent NAB.

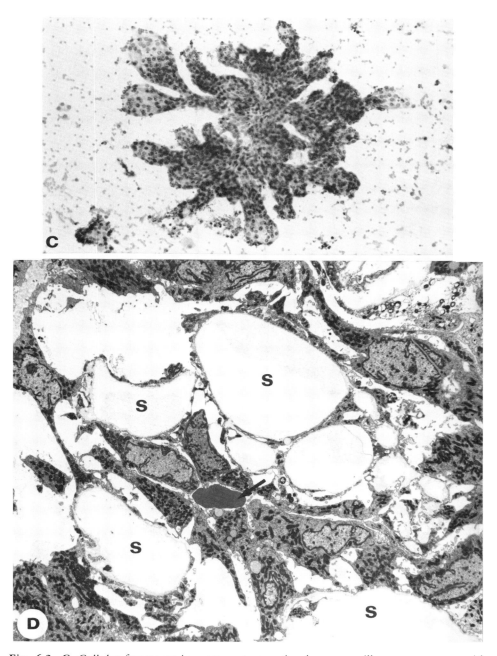

Fig. 6.2. C. Cellular fragments in a smear preparation have a papillary arrangement with the terminal segments having a bulbous appearance. × 150. D. An extravasated red blood cell (arrow) lies in a microlumen. External to the luminal cells are tumor cells with narrow cytoplasmic processes forming numerous basal lamina-lined "cystic" spaces (S). This combination of features is diagnostic of adenoid cystic carcinoma. × 2,200. (From Dardick I, et al: A quantitative comparison of light and electron microscopic diagnoses in specimens obtained by fine-needle aspiration biopsy. *Ultrastruct Pathol* 15:105–129, 1991)

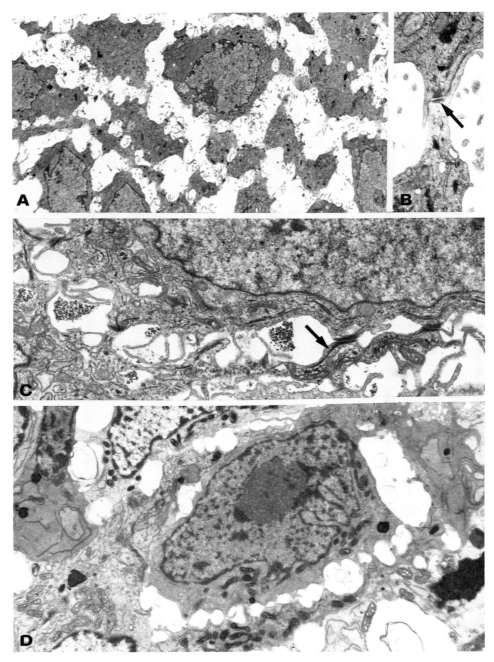

Fig. 6.3. **A.** NAB of a 2 cm left lung tumor diagnosed as well-differentiated squamous-cell carcinoma. Tumor cells are considerably separated and have many fine processes projecting from the surface. × 2,100. **B.** The tumor in A. Occasional broader intercellular "bridges" occur and are connected by desmosomes (arrow). × 11,000. **C.** NAB of a metastatic squamous-cell carcinoma to the scapula. Apposing blunt cytoplasmic processes, joined by desmosomes with narrow curved bundles of tonofilaments (arrow), intervene between widened extracellular zones into which project a few fine cellular processes. × 16,800. **D.** NAB of a poorly differentiated squamous-cell lung carcinoma. The cell-to-cell growth pattern is similar to that seen in A and C, but the intercellular "bridges" do not develop desmosomes. × 10,400.

TABLE 6.2. Cellular Organelles of Value in Diagnostic Electron Microscopy

Intercellular junctions
Microvilli
Secretory products
   Glycocalyx and glycocalyceal bodies
   Mucin
   Glycogen
   Neurosecretory granules
   Lipid
   Melanosomes
Basal lamina
Cytoplasmic filaments
   Cytokeratins
   Vimentin
   Desmin
   Actin/myosin complexes
Intracytoplasmic lumens

useful in epithelial neoplasms, whereas the latter two forms are generally found in mesenchyme-derived tumors.[39-41,68,76]

Fully differentiated desmosomes, with their paired densely stained internal plaques and associated intermediate filaments (with or without the intermediate line in the intercellular gap) (Fig. 6.4), are usually associated with better-differentiated epithelial tumors such as squamous-cell carcinoma, adenocarcinoma, urothelial neoplasms, meningioma, mesothelioma, adenomatoid tumor and thymoma, and with the soft-tissue tumors, synovial and epithelioid sarcomas.[14,40,41] A distinct advantage of electron microscopy is that, occasionally, even in poorly differentiated (Fig. 6.5) or anaplastic tumors, desmosomes that are surprisingly well developed may be identifiable. In this circumstance, the diagnostic possibilities are narrowed or a diagnosis that was suspected from the interpretation of the cytological features and clinical setting is confirmed. One practical marker of squamous-cell carcinoma in poorly differentiated lesions is attenuated desmosomes with short, comma-like

TABLE 6.3. Definitions for Intercellular Junctions and Tumors

Desmosome: tonofilament bundles inserting into densely staining plaques, associated with apposed portions of cell membrane, which are separated by approximately a 25-nm extracellular space that is bisected by a narrow linear density.
Tight junction: segments of adjacent cell surface of varying length that have fused; the resulting junction has a width less than the combined unit membranes.
Paired subplasmalemmal density (desmosome-like): plaque-like to somewhat fuzzy–appearing thickenings on the inner surface of adjacent portions of cell membranes; such poorly formed or "primitive" junctions are not associated with intermediate filaments or with an intermediate line.
Subplasmalemmal linear density: short segments of the inner surface of the cell membrane bearing a strip of dense staining material, often accompanied by a bar-like zone of moderately staining material adjacent to the outer aspect of the cell membrane.

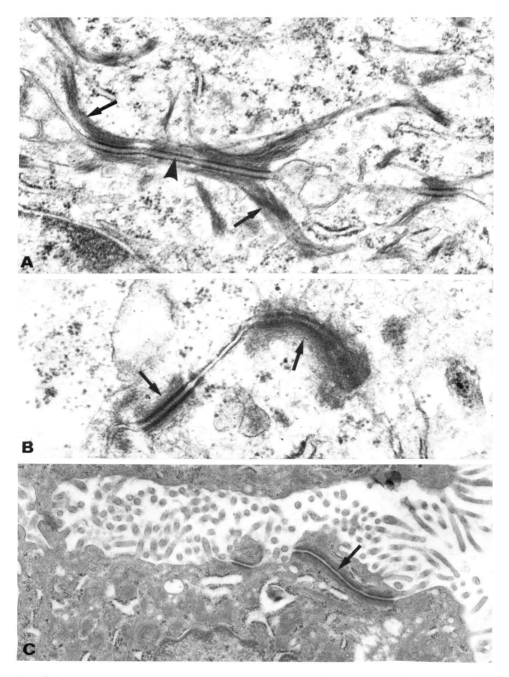

Fig. 6.4. **A.** Metastatic squamous-cell carcinoma in parotid lymph node. Well-formed desmosomes with a granular intercellular zone (arrowhead) lacking an intermediate line, and narrow, uniformly thick dense plaques from which project curved, comma-like, cytokeratin intermediate filaments (arrows). × 16,500. **B.** NAB of a thymoma. Desmosomes with a similar basic structure, but the intermediate filaments (arrows) are less structured and tend to distribute along the length of the junction. × 16,500. **C.** NAB of a diffuse epithelial mesothelioma of pleura. These desmosomes are even simpler with a narrower, less obvious band of intermediate filaments (arrow). × 8,700.

PRINCIPLES OF ELECTRON MICROSCOPY 99

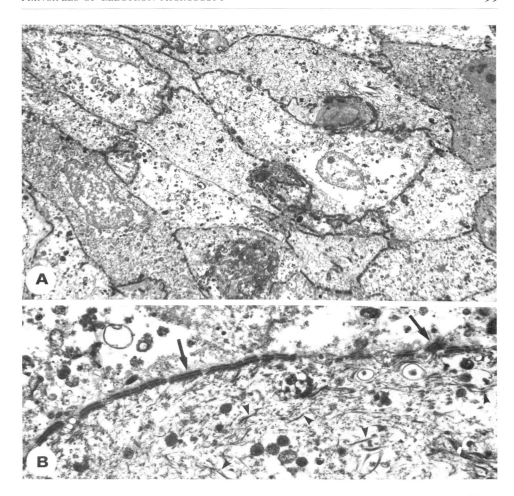

Fig. 6.5. ABC interpreted as large-cell carcinoma and as squamous-cell carcinoma of lung ultrastructurally. **A.** In this poorly differentiated neoplasm, the cells are approximated and have prominent cellular membranes. Cytoplasmic clear regions represent extracted glycogen. × 1,650. **B.** Intercellular junctions (arrows), responsible for the "thick" cell membrane, are particularly numerous and reasonably structured. Tonofilaments are present (arrowheads). × 6,000.

"tails" of tonofilaments projecting at an angle from a plaque (Figs. 6.3C, 6.4A). Occasionally this type of desmosome can be found in adenocarcinomas but with other features of glandular differentiation.

In epithelial tumors, usually intercellular junctions are less frequently found and simpler in form, but some may still have features suggesting a desmosome (Fig. 6.6A–C). These intercellular junctions often lack the intermediate line and have less prominent staining of the plaques associated with the cytoplasmic face of the cell membranes. Although frequently there is a reduced complement or very few intermediate (cytokeratin and/or vimentin) filaments, there may be sufficient filaments to recognize the junction as a desmosome (Fig. 6.6C). With further anaplasia

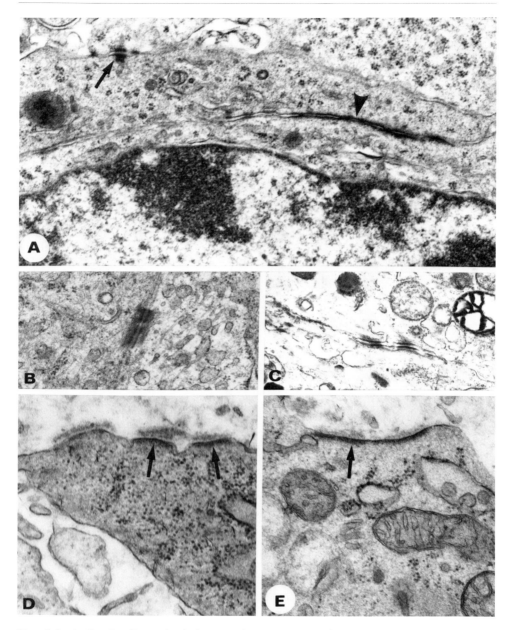

Fig. 6.6. **A.** Small-cell anaplastic lung carcinoma. A small desmosome (arrow) has projecting intermediate filaments, but the longer, simpler junction (arrowhead) is of the paired subplasmalemmal density type. × 16,000. **B.** Metastatic colonic adenocarcinoma to ovary. Dense plaques are less apparent, but the adjacent intermediate filaments are relatively thick. × 26,000. **C.** NAB of an adenocarcinoma presenting as an intra-abdominal mass. Simple desmosomes have poorly developed plaques and limited intermediate filaments. × 10,500. **D., E.** Malignant fibrous histiocytoma. Subplasmalemmal linear densities (arrows), each associated with an extracellular band of granular material that is less well formed in E. Both × 16,000.

(Fig. 6.6A), junctions in epithelial tumors may closely resemble the paired subplasmalemmal densities.

A variety of benign and malignant soft-tissue tumors can exhibit paired subplasmalemmal densities (Table 6.3), where two cell membranes are in close apposition (Fig. 6.6A), and subplasmalemmal linear densities (Table 6.3), where the structure is hemidesmosome-like and involves only one cell membrane (Fig. 6.6D,E).[40,68] In both cases, they consist of focal, narrow, and dense-staining regions on the cytoplasmic face of the cell membrane. In paired subplasmalemmal densities, these plaque-like zones have many short filaments producing a "fuzzy" appearance (Fig. 6.6A), a feature that is distinct from the homogeneously staining plaques of the true desmosome (Fig. 6.4). The absence of tonofilaments from paired subplasmalemmal densities (desmosome-like) is a further feature distinguishing this type of intercellular junction from the true desmosome.[40,41] The subplasmalemmal linear densities associated with normal and neoplastic mesenchymal cells have varying development of a thin band of granular material usually separated by a narrow lucent zone from the cell membrane and its associated dense-staining inner region (Fig. 6.6D,E). Because paired and unpaired subplasmalemmal linear densities are an important electron microscopic clue to the sarcomatous nature of poorly differentiated, anaplastic, and spindle-cell tumors, familiarity with this form of junction is essential.

The classic terminal bar or junctional complex, joining the apical regions of glandular epithelial cells, has a tight junction (Table 6.3), below which is a sequence of an intermediate junction and a desmosome.[40,41] An apically situated junction closely approximating that of normal tissues can be found in adenomas and adenocarcinomas (Figs. 6.7, 6.8); they are often altered in malignant tumors. In both well- and poorly differentiated adenocarcinomas, a few simple to many complexly organized tight junctions may be present (i.e., the intermediate junction with its adjacent desmosome may be missing) (Fig. 6.7B), associated with a small to reasonably sized lumen. In other examples, the tight junction may have an adjacent desmosome (Fig. 6.7A). In anaplastic lesions, a few tight junctional complexes even without lumens should begin a further search for other markers of glandular differentiation such as mucous secretory granules, lumens, and microvilli. Both tight junctions and junctional complexes can be identified in diffuse epithelial mesotheliomas and biphasic synovial sarcomas.

## MICROVILLI

The form, number, and character of microvilli are all useful diagnostic features not only for separating such neoplasms as mesothelioma and adenocarcinoma, but also to help determine primary sites for metastatic adenocarcinomas. Although definitive statements are often made concerning the type of microvilli in different tumors, it is essential to be aware of the range of development of this oganelle to recognize essential diagnostic criteria in ABC specimens of metastatic tumors. One method to gain this experience is to study primary tumors of various organs ultrastructurally.

In terms of form and frequency, it is unusual for either renal[63] or intestinal[52] adenocarcinomas to develop the "brush-border" type of microvilli that are so characteristic of their respective normal epithelia. In a review of 156 primary and 69 metastatic renal cell carcinomas, Mackay and associates[63] observed that "microvilli

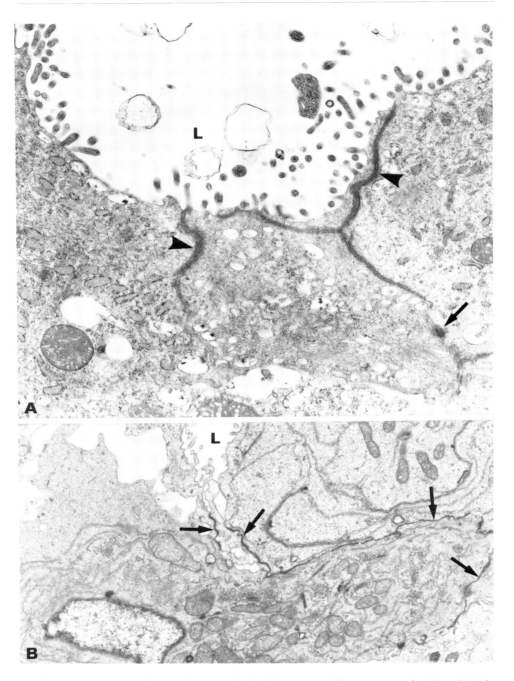

Fig. 6.7. **A.** Prostatic adenocarcinoma. A glandular lumen (L) is associated with relatively long densely staining tight junctions (arrowheads), one of which is associated with a desmosome (arrow). × 32,500. **B.** NAB of a pulmonary adenocarcinoma. Intercellular lumens (L) were infrequent and poorly developed, but tight junctions (arrows) were numerous and more often unassociated with lumens. × 6,000.

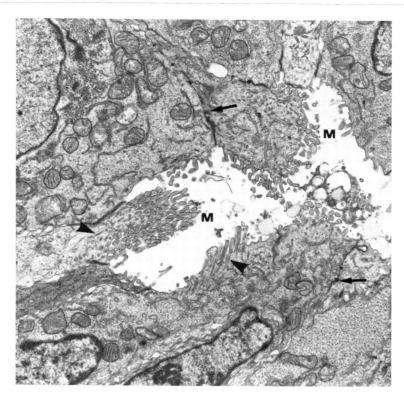

**Fig. 6.8.** NAB of metastatic colonic adenocarcinoma to bone (tibia). Apical regions of almost every tumor cell have tight junctions of varying length, but only some have adjacent desmosomes (arrows). Many tumor cell apices bear well-formed microvilli (M), some of which have microfilamentous core rootlets (arrowheads). × 2,900.

are quite variable, ranging from a profusion of straight or slightly curved protrusions into the lumen of a gland to a few short projections from a free cell surface or into a narrow gap between neighboring cells." This description can apply even to moderately and well-differentiated renal cell carcinomas (Fig. 6.9A) and, in our experience, a "brush-border" pattern is the least frequent finding, but something approaching this pattern does occur (Figs. 6.9B, 10.17C). This finding applies equally to gastrointestinal carcinomas (Fig. 6.8; 6.12, 6.15A, 6.18A). Small putative glandular lumens crammed with microvilli also must be recognized in ABC specimens (Figs. 6.10, 10.17C).

Microvilli of both lung (Fig. 6.10; Fig. 6.11A) and gastrointestinal tract (Figs. 6.8, 6.12, 6.15A,B, 6.18A) adenocarcinomas may or may not exhibit a central core of microfilaments. The feature of microvilli (Figs. 6.8, 6.12, 6.15B) that produces a distinctive, but not necessarily exclusive, marker for the colorectal region, duodenum, stomach, esophagus, pancreas, gall bladder, and common bile duct is the identification of discrete bundles of microfilaments extending down from microvilli into the underlying cell cytoplasm.[29,52,65,66,80] Using immunofluorescent techniques, normal microvilli (and those in tumors) contain a central zone of actin filaments.

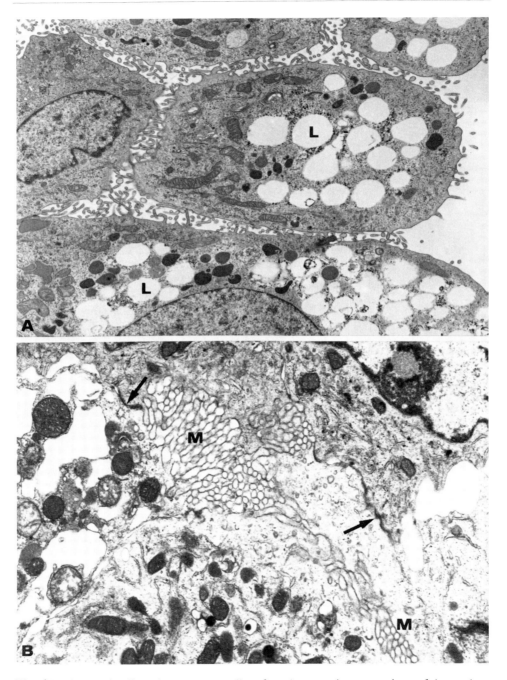

Fig. 6.9. **A.** Renal-cell carcinoma; most cell surfaces have moderate numbers of short microvilli and the cytoplasm containing lipid droplets (L). × 6,200. **B.** NAB of a large renal mass with a cytological differentiatial diagnosis of carcinoma or lymphoma. Packed microvilli (M) with a fairly orderly arrangement and associated tight junctions (arrows) established the diagnosis of renal adenocarcinoma. × 10,500.

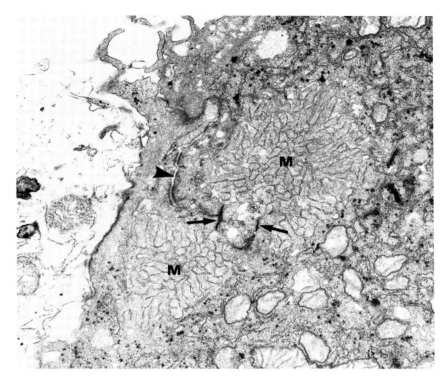

**Fig. 6.10.** NAB of a 6-cm mass in the left lower lobe of lung. Intercellular lumen (identified at least on one aspect by tight junctions [arrows] and an intercellular space [arrowhead]) entirely filled with microvilli (M) assisted in establishing the diagnosis of adenocarcinoma. × 22,000.

In the gastrointestinal tract, the actin-rich core of micovilli is particularly prominent ultrastructurally, and the microfilaments extend down into the apical cytoplasm to a varying extent, but usually for a considerable distance; the latter feature is termed *microvillous core rootlets*.[66] They must be recognized in both cross-section, (where they produce a regular punctate appearance to the apical cytoplasm) (Figs. 6.8, 6.12B,D), and in longitudinal cuts (Figs. 6.8, 6.12A,C, 6.15B). The rootlets are particularly valuable; Hickey and Seiler[52] observed that they can be present even when microvilli are almost absent (Fig. 6.12D) and that they may be more obvious in malignant than in normal epithelium. A few short extensions of filaments from microvilli may be seen in a variety of adenocarcinomas other than from the gastrointestinal tract, most often in primary lung tumors, but particularly long rootlets are the hallmark of colorectal neoplasms.[52,80]

After finding that microvillous core rootlets were present in 16 of 67 (24%) pulmonary adenocarcinomas, Engstrand and colleagues[29] noted that "in neither a qualitative nor a quantitative sense can the presence of microvillous core rootlets or glycocalyceal bodies be an absolute criteria to permit the pathologist to distinguish a primary from metastatic tumor in lung." Other reports, however, have noted a lower incidence[52,80] or the infrequency of microvillous core rootlets in pulmonary

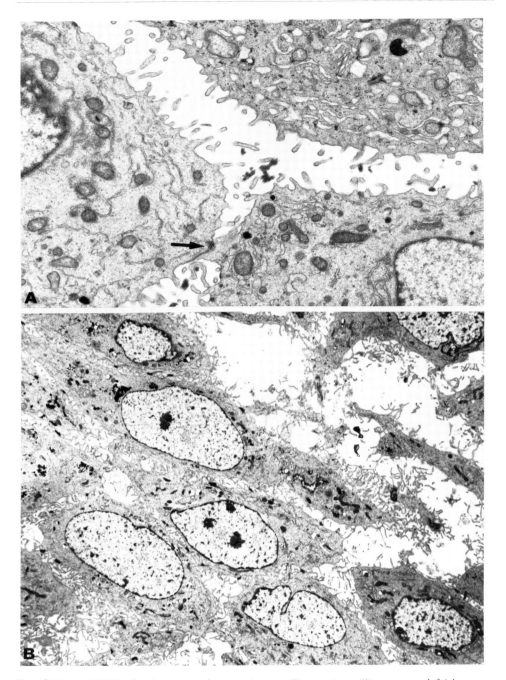

Fig. 6.11. A. NAB of pulmonary adenocarcinoma. Short microvilli are spaced fairly regularly on most cell surfaces. A small desmosome (arrow) joins two tumor cells. × 16,800. B. NAB of squamous-cell carcinoma of lung illustrating that not all surface projections, which can be numerous in this type of tumor, are true microvilli. Typical desmosome-associated "bridges" and tonofilaments are present in other fields. Microvillus-like processes are also seen in Figure 6.3A–C. × 2,200.

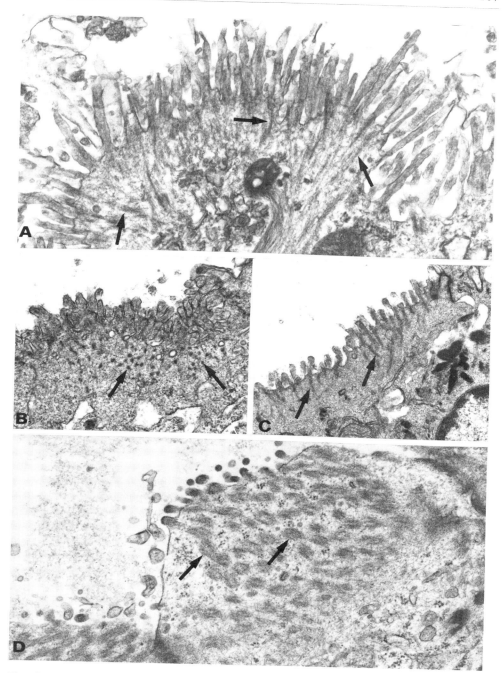

**Fig. 6.12.** NABs of metastatic gastric **A.** and colonic **B.–D.** adenocarcinomas. Most microvilli have a central core of microfilaments, but the main feature is the extent to which these filaments extend down into the apical cytoplasm as microvillous core rootlets (arrows). Such rootlets are seen both in a longitudinal (A,C) and a cross-sectional (B,D) plane. In D, they are particularly prominent, even where microvilli are poorly differentiated. (A) × 22,200; (B) × 16,800; (C) × 13,200; (D) × 27,200.

adenocarcinomas.[47,50] For example, Hickey and Seiler[52] observed that although 87% of lung adenocarcinomas and 94% of colorectal adenocarcinomas had filaments present within the microvilli, only 13% of the lung tumors exhibited rootlets compared with 100% for the colorectal tumors. In surveying our own experience of 65 ABC specimens of primary pulmonary adenocarcinomas (adenocarcinoma, not otherwise specified, n = 44; bronchioloalveolar, n = 3; large cell, n = 18) studied ultrastructurally over a three-year period,[23] none of the cases exhibited microvillous core rootlets. Perhaps the size of the specimen obtained by NAB precluded an adequate survey, but this finding only highlights the focal differentiation of this marker in lung carcinomas compared with colorectal carcinomas. It at least reduces concern for misdiagnosing lung carcinomas as metastatic gastrointestinal adenocarcinomas. Seiler and associates[80] came to the same conclusion.

The region of the apical cytoplasm containing the core rootlets is also rich in actin filaments (the terminal web), which provides an equally useful cellular marker for colorectal tumors. This material usually excludes other organelles, such as mitochondria and endoplasmic reticulum, and produces an apical "clear" zone (Fig. 6.12B–D).[52] In comparative ultrastructural studies, Hickey and Seiler[52] and Seiler and associates[80] found that a combination of microvillous features, densely staining apical secretory granules, glycocalyceal bodies, and the apical cytoplasmic clearing assisted in differentiating lung and colorectal cancers. The absence of such features and the appearance and arrangement of the microvilli in a metastatic carcinoma of rib (Fig. 6.13) were the principal reasons for suggesting that this lesion was likely of pulmonary origin. In a woman with a previous history of breast cancer and a metastatic carcinoma to the tibia identified by ABC (Figs. 6.8, 10.3), subsequent electron microscopy of the ABC specimen revealed microvilli that identified the metastatic tumor as an unsuspected primary colonic carcinoma.

Microvilli can have sufficiently characteristic forms and arrangements that help distinguish certain neoplasms. Classically, apical cell surfaces of the tumor cells in diffuse epithelial mesotheliomas display a high concentration of relatively long, narrow, and curvaceous microvilli (Figs. 6.14A, 10.10D). A distinctive feature is their uniform appearance and length. When microvilli are present to this degree, they are a key ultrastructural element in the diagnosis of mesothelioma. Caution needs to be exercised, however, because adenocarcinomas also can have plentiful and somewhat long microvilli.[95,97]

Generalizations about microvillous form and length as a diagnostic parameter are difficult. Some studies comparing mesotheliomas and adenocarcinomas (lung, breast, and upper gastrointestinal tract) observed that the former had significantly longer and more complex-appearing microvilli than the latter.[96,98] A separate report comparing mesothelioma with adenocarcinomas of the ovary and endometrium noted no difference in microvillous length, but commented that ovarian adenocarcinomas (Fig. 6.16B) were more likely to have branched microvilli projecting from apical surface extensions;[97] Hammar and associates[45] also commented on the frequency of this feature in serous cystadenocarcinoma of the ovary. Unfortunately, branched microvilli do occur in mesotheliomas (Fig. 6.14A).[98]

By measuring microvillous length objectively, clearly considerable overlap can exist between diffuse epithelial mesotheliomas and adenocarcinomas.[13,95,96,98] Indeed, in many subtypes of mesotheliomas (e.g., well differentiated, desmoplastic, sarcomatoid, and poorly differentiated [also termed *transitional*]), microvilli can be

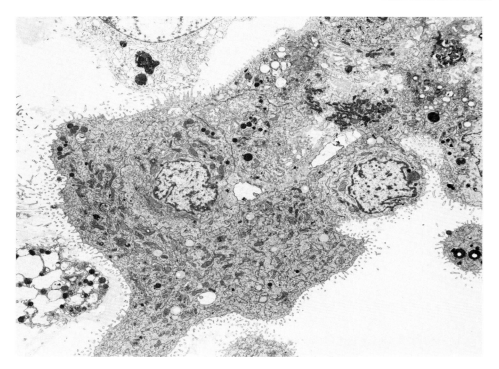

Fig. 6.13. NAB of bone (rib) tumor. Adequate tissue fragment of cohesive tumor cells with many relatively short, fairly regularly distributed microvilli lacking filament cores and rootlets suggested a pulmonary origin for this metastatic adenocarcinoma. A large-cell carcinoma of lung was subsequently diagnosed. × 2,200.

short, sparse, or absent.[6,7,19,21,95,96] Still, the form and length of microvilli are practical criteria to assist in differentiating diffuse epithelial mesotheliomas of pleura from peripheral adenocarcinomas of the lung.[98]

Except for the gastrointestinal tract, there are generally no particularly distinctive features of microvilli in the various adenocarcinomas. On the basis of our experience, however, lung tumors (even when metastatic) can often be recognized by their somewhat orderly arrangement of short, blunt microvilli. The microvilli from neighboring cells may roughly alternate and produce an interlocking, "zipper-like" appearance (Fig. 6.14A), particularly when glandular lumens are small or when microvilli are present on the lateral aspects of adjacent tumor cells. As previously detailed, microvillous core rootlets are infrequently found and poorly formed in lung adenocarcinomas.[50] Microvillous-like cytoplasmic extensions or filipodia should not be mistaken for microvilli. These extensions are frequently found in squamous-cell carcinomas, where they may protrude into the widened extracellular spaces between the intercellular "bridges" (Figs. 6.3A, 6.14B). Anemone (villiform or filiform) cell tumors, by definition, also express profuse numbers of narrow cellular processes often circumferentially arranged on the cell surface in sectional profiles.[24,56,90]

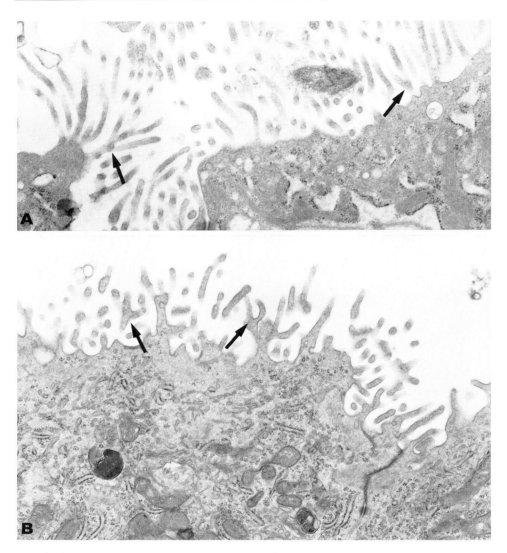

Fig. 6.14. A. NAB of pleural mesothelioma. × 40,000. B. Papillary serous cystadenocarcinoma of ovary. × 30,000. Multiple and occasionally branching (arrows) microvilli project from short, irregularly shaped cytoplasmic extensions in both neoplasms.

## Secretory Products

### Glycocalyx and Glycocalyceal Bodies

The specialized secretory material, glycocalyx, is a subtle but useful marker of glandular differentiation. The subtlety arises from the higher magnifications required to document its presence and because it may only be observed focally. The "fuzzy" coating of radiating, fine fibrils attached to the apical cell surface, projecting microvilli, and secretory products such as glycocalyceal bodies helps to identify adenocarcinomas, particularly from such sites as the gastrointestinal tract, lung, and mucinous lesions of the ovary and cervix.

Glycocalyceal bodies (Fig. 6.15) consist of clusters of a few to many vesicular structures wedged between the microvilli or lying within the glandular lumen, often mixed with the principal mucinous secretory product of the tumor cells.[65,66] They are generally round, membrane-bound structures and have a diameter approximately 20% of that for a microvillus,[66,72] and are of practical use during the investigation of lesions with unknown primary sites.[44,51,65,66] Although almost universally present in colorectal carcinomas, they also can be found in gastric, esophageal (Fig. 6.15A), pancreatic (Fig. 6.15B), bile duct, and pulmonary (Fig. 6.15C) adenocarcinomas, and mucinous cystadenocarcinomas of the ovary.[66] Approximately 70% of lung carcinomas are reported to display glycocalyceal bodies.[52] There is, however, some contradictory evidence; Engstrand and associates[29] found only 9% of lung adenocarcinomas with glycocalyceal bodies, and they were infrequently noted by Herrera and associates[47,50] and in our own series of ABC specimens of lung adenocarcinomas (n = 65). When glycocalyceal bodies are detected in such carcinomas, they do not compare quantitatively with those evident in gastrointestinal carcinomas. Therefore, glycocalyceal bodies and the microvillous core rootlets previously described are useful markers for differentiating between pulmonary and gastrointestinal carcinomas in appropriate clinical settings.

Mesothelioma is the other lesion to consider when a particularly prominent layer of glycocalyx entangles the microvilli in the occasional well-differentiated diffuse epithelial form.[21,44] Although other ultrastructural features of mesothelioma, such as microvilli (Fig. 6.14A), tonofilaments, and desmosomes (Fig. 6.4C), may be characteristic, the presence of glycocalyx should not sway the observer to conclude that the tumor is an adenocarcinoma. Glycocalyceal bodies and apical cytoplasmic secretory granules, which can be present in adenocarcinomas, will be absent in mesotheliomas.[45]

## *Mucin*

Mucin is an invaluable marker for the identification of glandular differentiation. At times, mucin is readily recognizable, for example in goblet cells, but in other circumstances it presents in a variety of guises. This variability has been well illustrated by the range of differentiation of mucin-type granules in pulmonary adenocarcinomas.[10,16,45,49,88] Descriptive segregation of secretory granules in pulmonary adenocarcinomas by Taccagni and associates[88] should prove useful. On the basis of material from pleural effusions or aspiration biopsies of lung lesions, these authors observed six types of secretory products in the tumor cells: "mature and immature mucus granules, myelinated and lamellar granules, myelinated and partially lamellar granules, osmiophilic nonlamellar granules, neurosecretory-like granules, and not otherwise specified (indeterminate) secretory granules." Figures 6.16 and 6.17 provide examples of some of these forms and the range of mucinous secretory granules.

Epithelial cell secretory granules may be small, infrequent, and darkly staining so that they resemble neurosecretory granules (Fig. 6.18A). Mucin droplets, however, will be localized next to luminal surfaces and not basally situated or in cytoplasmic processes like most neuroendocrine neoplasms (Figs. 6.22, 6.23). Such localization may prove useful in the subtyping of metastatic "large-cell" or undiffer-

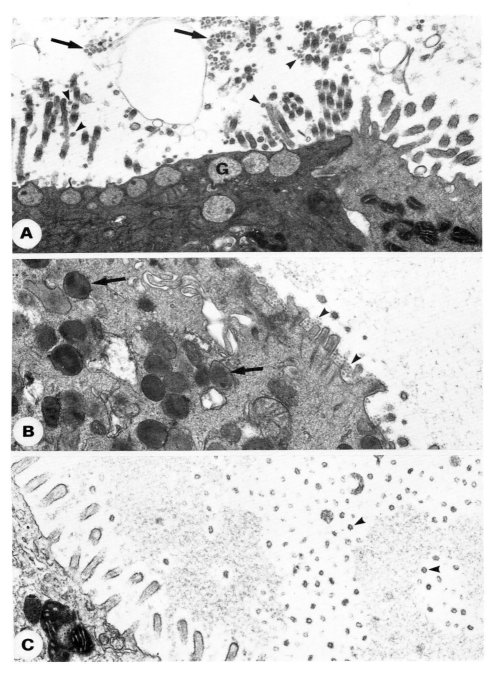

Fig. 6.15. **A.** NAB of metastatic adenocarcinoma to bone (rib) in a patient with a previous esophageal carcinoma. Groups of glycocalyceal bodies lie within the lumen (arrow) and are associated with microvilli (arrowheads). Note mucous granules (G) below apical surface. × 16,800. **B.** NAB of primary pancreatic adenocarcinoma. Glycocalyceal bodies (arrowheads) mainly associated with the microvilli that are covered by wisps of glycocalyx. Secretory granules (arrows) are denser staining in this case. × 13,200. **C.** NAB of supraclavicular lymph node with metastatic pulmonary adenocarcinoma. Glycocalyceal bodies (arrowheads) are primarily within the lumen mixed with the more granular mucin. × 16,800.

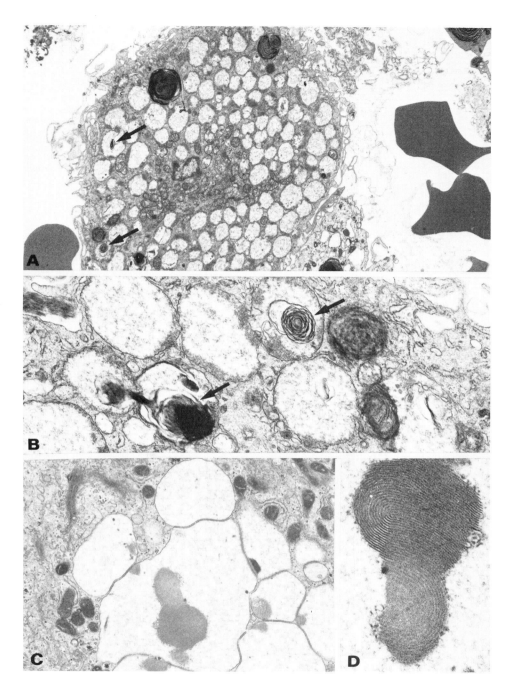

Fig. 6.16. NAB of pulmonary adenocarcinomas, morphological variation of secretory products. **A.** Bronchioloalveolar carcinoma with mucin-type secretory granules containing mainly finely granular material, but some osmiophilic contents in other granules (arrows). × 4,000. **B.** Same case as in A, with higher magnification showing myelinated secretory products (arrows) in some mucin-type granules. × 16,800. **C.** An adenocarcinoma NOS with accumulations of denser staining product in some mucin granules. × 10,400. **D.** Higher magnification reveals the lamellar and "finger-print" structure to a portion of the secretory material. × 26,800.

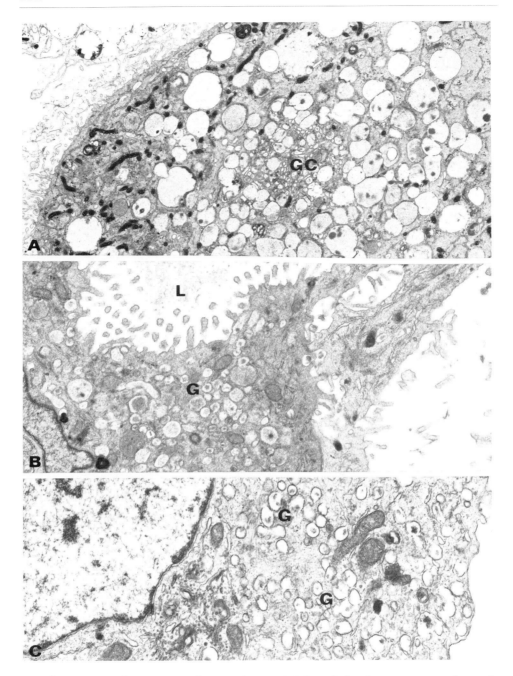

**Fig. 6.17.** NAB of pulmonary adenocarcinomas NOS, variation in secretory products. **A.** Central Golgi complex (GC) with surrounding pale mucous secretory granules, many of which have one or more eccentric darker staining zones. × 5,200. **B.** An intracytoplasmic lumen (L) is associated with variably sized secretory granules (G) that differ in internal structure and staining quality. × 10,400. **C.** Mucin-type granules (G) are smaller, more focal, fewer in number, and poorly formed in this example. × 13,200.

entiated carcinomas. Mucin also may accumulate in somewhat large amounts in a focal area within the cytoplasm, displacing the nucleus and creating what is apparent in cytological preparations as a signet-ring cell. In the example in Figure 6.18B, such a feature allowed identification of an unsuspected primary gastric carcinoma by examination of a metastatic adenocarcinoma presenting as a nodule in the skin of the groin. Note should be taken of the range of morphology and stainability of the mucus-type secretory droplets (Figs. 6.16 to 6.18); some may be relatively large and still quite electron-dense (Fig. 6.18C).

Zymogen-type secretory granules of pancreatic (Fig. 6.19) or salivary gland neoplasms also need to be considered. Practical aspects of secretory granule identification and useful tables of differential features are available in a review of the application of electron microscopic characteristics in tumor diagnosis by Russo and associates.[79]

## Glycogen

In the cytoplasm, glycogen has the ultrastructural appearance of irregularly shaped particles of two types (Fig. 6.20). Although a few scattered glycogen granules are present in the cytoplasm of many types of normal and tumor cells, they become obvious and of diagnostic value when synthesized in amounts that occupy a considerable proportion of the cytoplasm (Fig. 6.20A). The beta-type (Fig. 6.20B) is seen as individual particles approximately 15 to 30 nm in diameter, whereas while the alpha-type (Fig. 6.20C) is a rosette-like arrangement of multiple beta particles (typically seen in normal hepatocytes). Synthesis and storage of prominent amounts of intracytoplasmic glycogen is a particularly useful diagnostic feature of certain tumors such as Ewing's sarcoma, rhabdomyosarcoma (Fig. 6.20B), chondrosarcoma, renal-cell carcinoma (Fig 6.20A), and seminoma. In each of these tumors, other ultrastructural characteristics are present that combine with the glycogen to help in establishing a diagnosis. It should be stressed that glycogen may be minimal or absent in some examples. Many other types of clear-cell epithelial neoplasms, such as hepatocellular carcinoma; clear-cell tumors of lung, chordoma, oncocytoma, and epithelial-myoepithelial carcinomas of salivary gland; and mesothelioma can also have a considerable complement of glycogen. The ultrastructural study of a needle aspirate of one reported sacrococcygeal chordoma did not reveal excessive deposits of glycogen in this example but did have the typical features of physaliphorous cells.[74]

## Neurosecretory Granules

By definition, neuroendocrine cells are identifiable by their cytoplasmic content of small, membrane-bound secretory granules of somewhat varying size and morphology. The contents of the granules include chromogranins, nucleotides, biological amines, and the many peptide hormones and neurotransmitters specific for the function of each type of neuroendocrine cell.[73] Both the morphological and functional characteristics of the various neuroendocrine cells can be reflected in the respective benign and malignant tumors.

Ghadially[40] provided an excellent overview of structural aspects of the neuroendocrine granule and the application of this organelle to tumor diagnosis. Table 6.4 provides a description of the different forms of neurosecretory granules.

The various neuroendocrine tumors, such as small-cell anaplastic (oat-cell) carcinoma; carcinoid and atypical carcinoid tumors; pheochromocytoma, paraganglioma, and carotid body tumor; islet-cell tumor; pituitary adenoma; medullary carcinoma; and Merkel-cell carcinoma reproduce most of these types of granules, but the majority of such tumors synthesize membrane-bound granules with a narrow electron lucent band surrounding the central densely staining core (type B in Table 6.4 and the secretory granules in Fig. 6.21A). Some neuroendocrine tumor cells only contain granules that are homogeneously electron-dense, whereas others contain a mixture of type A and B granules. Only some islet-cell tumors (insulinomas) and pheochromocytomas secrete granules with a morphology that is specific enough to be diagnostic; type E in the former (Fig. 6.21B) and type F in the latter (see Fig. 6.21C). Very occasionally a mid-gut or bronchial carcinoid will display neurosecretory granules of the type seen in the normal enterochromaffin (argentaffin) cell of the gastrointestinal tract (type D in Table 6.4).

Identification of neurosecretory granules has an important role in tumor diagnosis in ABC specimens (see "Neuroendocrine Neoplasms" section in Chapter 8). As noted previously, the range of organs and sites associated with neuroendocrine tumors shows the frequency with which this type of tumor must be considered

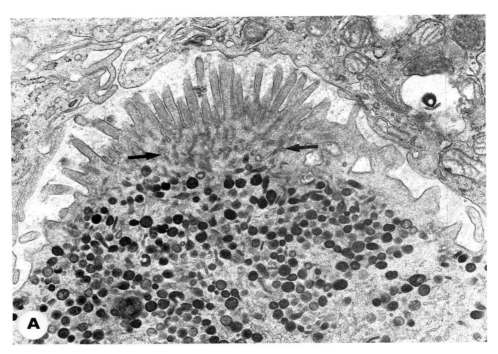

Fig. 6.18. A. Metastatic gastric adenocarcinoma with many small, round, moderately to densely staining secretory granules below the microvillous core rootlets (arrows). × 17,000.

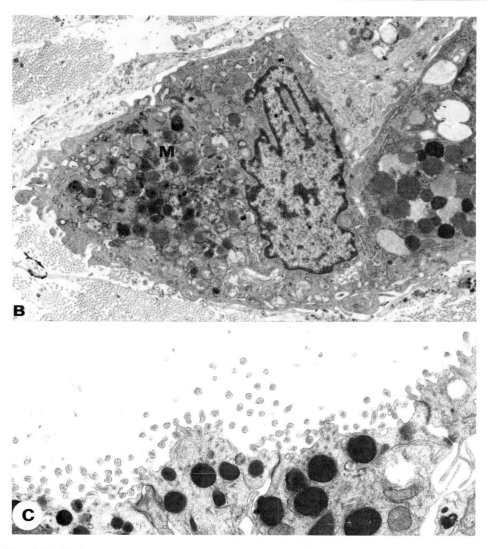

Fig. 6.18. B. Same tumor as in A, with a signet-ring cell formed by the extensive accumulation of mucous droplets (M) and displacement of the nucleus. × 9,000. C. NAB of a pulmonary adenocarcinoma with apical homogeneously and darkly staining mucinous secretory granules. × 13,200.

in the differential diagnosis of ABC specimens. In some cases, neuroendocrine neoplasms, both benign and malignant, may be unsuspected from the cytological assessment but may be obvious ultrastructurally.

The differential diagnosis of small-cell carcinoma from lymphoma rests primarily on observing the characteristic secretory granules, because other features such as desmosomes and tonofilaments may be undetectable or poorly developed in such carcinomas. Small-cell anaplastic carcinomas may have few if any granules in the

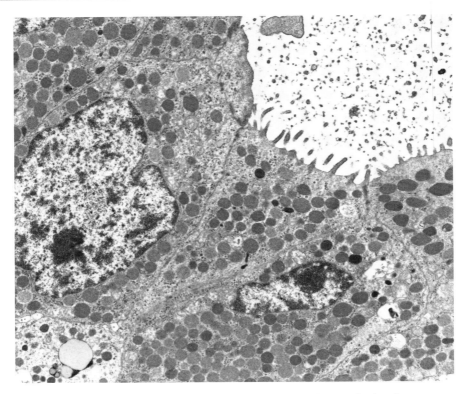

Fig. 6.19. Acinar cell carcinoma of the pancreas with extensive synthesis of zymogen-type granules occupying considerable portions of the cytoplasm. Staining intensity also varies in these granules. × 8,900.

main portion of cell cytoplasm, because the neurosecretory granules (often still limited quantitatively) may be concentrated at the ends of cytoplasmic processes, wedged between the tumor cells (Fig. 6.22). At times, the aspirate may consist only of small clumps or single-tumor cells, some of which may be fragmented due to the extreme sensitivity of these cells to crushing, anoxia, and possibly even the negative pressure resulting during the aspiration biopsy. It is, however, often still possible to identify small packets of cytoplasm containing a few neurosecretory granules among fragmented or degenerating cells (Fig. 6.23). Such localization increases the diagnostic confidence in those cases where initially only a few small uniformly electron-dense granules are present in the cytoplasm and could be interpreted as or in fact be lysosomes. Lysosomes, however, are usually present as a small group in the main cytoplasm of the tumor cell and are more heterogeneous in shape and variable in the staining of their contents than neurosecretory granules.[40]

The importance of recognizing neuroendocrine lesions in ABC specimens is evident from our experience with the ultrastructural examination of lung and liver biopsies (Table 6.5). In the pulmonary aspirates, 21.6% were neuroendocrine lesions of which the majority were small-cell anaplastic carcinomas. In liver aspirates, nearly one-third were metastatic neuroendocrine carcinomas, again mainly of pulmonary origin (Table 6.5). Because some of these cases included neuroendocrine

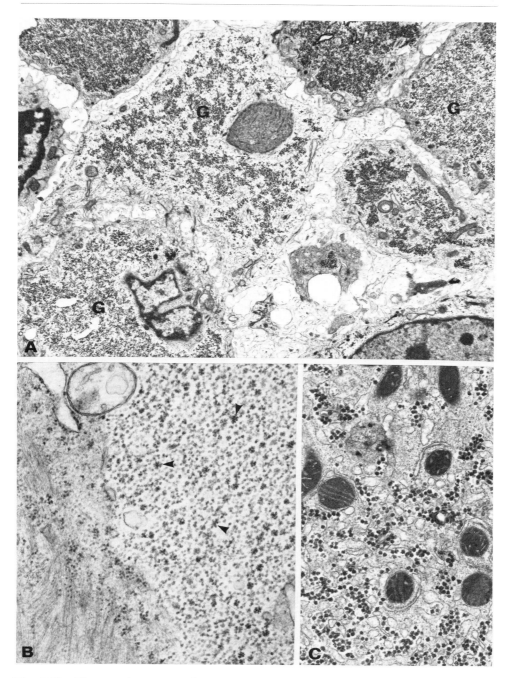

**Fig. 6.20.** Glycogen in tumor cells. **A.** NAB of renal-cell carcinoma (metastatic to lung) with differing but considerable storage of particulate glycogen (G) in individual tumor cells. × 5,200. **B.** Rhabdomyosarcoma (thick and thin filaments in the lower left) with many paler staining beta-type glycogen particles contrasting with the larger, darker staining alpha form (arrowheads). × 13,200. **C.** NAB of hepatocellular carcinoma with groups of individual alpha-type glycogen particles among the mitochondria and endoplasmic reticulum. × 18,000.

TABLE 6.4. Descriptive Morphology of Neuroendocrine Granules[*]

A. Round, homogeneously electron-dense, 100–250 nm diameter
B. Central dense core, peripheral lucent "halo" of 100–250 nm diameter
C. Uniformly electron-dense but 200–400 nm in diameter
D. Pleomorphic-shaped, uniformly densely staining (enterochromaffin cells are of this type)
E. Electron-dense crystalline core with a larger electron-lucent background (beta granules of islet cells)
F. Electron-dense core eccentrically placed in an electron-lucent background of 200–450 nm diameter (noradrenaline)

*Adapted from Ghadially FN: *Diagnostic Electron Microscopy of Tumours*, 2nd ed. London, Butterworths, 1985. Used with permission.

lesions in the differential diagnosis or this diagnosis was unsuspected, identification of specific granules by electron microscopy rapidly clarified the diagnostic dilemma or confirmed the light microscopic interpretation.

## *Lipid*

The occasional occurrence of a cytoplasmic lipid droplet is not uncommon in epithelial and sarcomatous tumor cells and even neoplastic lymphocytes. Lipid, therefore, is nonspecific in these circumstances. Still, there are a few situations where lipid synthesis and storage may be of sufficient magnitude to be diagnostic. If the cytoplasm contains many generally uniform-sized nonmembrane-bound lipid droplets, then the differential diagnosis must include nonneoplastic macrophages (histiocytes) associated with an inflammatory or reactive process; neoplastic histiocytes, as in malignant fibrous histiocytoma;[25,37,59] and even liposarcoma (Fig. 6.24A).[9,25,94] In the latter, however, there is generally more variation in lipid droplet size, at least in some tumor cells differentiated as lipoblasts (Fig. 6.24B). A good rule for the diagnosis of liposarcoma is to search for the few tumor cells in which one or two droplets are particularly larger than the rest; this finding suggests fusion of some droplets and a mimicry of the development of the mature adipocyte. Lipoblasts in which the lipid droplets are many and tend to fill the cytoplasm may exhibit a scalloped appearance of the nucleus.[25]

Prominent lipid droplets are also a feature of some renal-cell carcinomas (Fig. 6.9A) and are a key element with glycogen and microvilli in establishing this diagnosis by electron microscopy in metastatic lesions with an unknown primary site. Adrenal cortical (Fig. 10.20) and other steroid synthesizing neoplasms are another class of lesions in which lipid may be a prominent feature. Recognition of the range of electron density of lipid droplets is an essential aspect of the use of lipid droplets as a diagnostic feature. The osmophilia of lipid is greater the higher the content of unsaturated fatty acids and their degree of unsaturation.[40]

## *Melanosomes*

Because malignant melanoma is so frequently included in the differential diagnosis of poorly differentiated and anaplastic tumors and commonly presents in an amela-

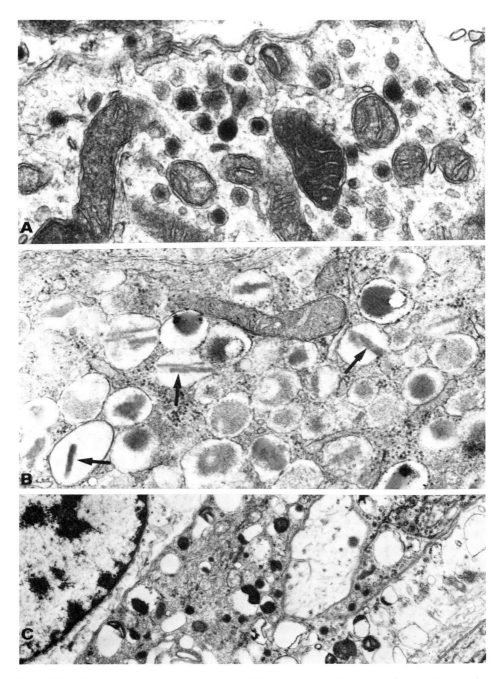

**Fig. 6.21.** Neurosecretory granules. **A.** NAB of a carcinoid tumor of lung. Reasonably uniform-sized cytoplasmic granules with a pale outer zone separating the enclosing membrane from the central core. × 40,000. **B.** Insulinoma with plate-like crystalloids (arrows) within some secretory granules and rounder accumulations of granular-appearing secretory product in other granules. × 21,500. **C.** Pheochromocytoma with cytoplasmic granules formed by densely staining, eccentrically placed portions and relatively large electron-lucent regions. × 16,000.

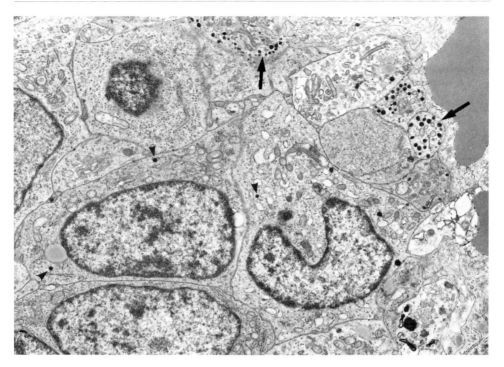

**Fig. 6.22.** Metastatic small-cell anaplastic carcinoma to bone (proximal femur). Tumor cells, with sparse cytoplasmic organelles, have only a few secretory granules in the main portion of the cytoplasm (arrowheads); the majority are located in the cytoplasmic processes (arrows). × 13,300.

notic form, awareness of the ultrastructural characteristics of typical (Fig. 6.25) and atypical or aberrant (Fig. 6.26) melanosomes is essential. This knowledge can be gained from several excellent reviews of normal and aberrant melanosomes.[31,40,67,75] It is also recommended that metastatic melanomas from patients with a previously resected primary tumor be examined ultrastructurally to obtain personal experience with the range of melanosome morphology, particularly the aberrant forms. The latter are an integral feature of most amelanotic lesions.[31,67,77] In fact, it has been our experience that typical melanosomes are absent in many metastatic melanomas. Infrequently, in some cases, which by light microscopy and clinical presentation are likely metastatic malignant melanomas, melanosomes in any form may be absent by electron microscopy; this is a limitation of this technique that must be recognized. Investigators should rely on the immunocytochemical findings in such cases.[82]

In some melanotic and amelanotic melanomas, there are typical or nearly typical melanosomes resembling the types II and III stages of normal melanogenesis occurring in epidermal melanocytes.[31,40,75] The elliptical organelles with striated crystalline form or combinations of filamentous rods and beaded or coiled inclusion bodies closely resembling the type II melanosomes are illustrated in Figure 6.25, with similar cytoplasmic structures that have partial or complete melanin deposition (i.e., stages III and IV) (Fig. 6.25A). A confident diagnosis of metastatic malignant

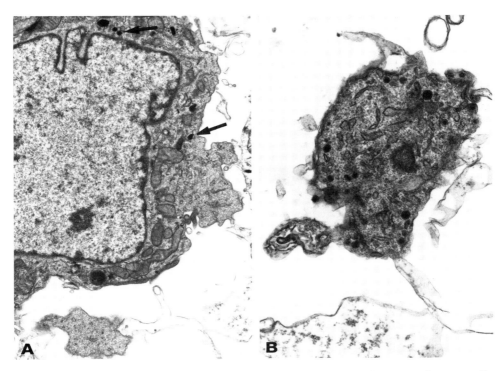

Fig. 6.23. NAB of small-cell anaplastic carcinoma of lung. **A.** The few isolated tumor cells identified had an occasional small, darkly staining cytoplasmic granule (arrows). × 10,400. **B.** Among the considerable cellular debris are cytoplasmic processes containing increased numbers of neuroendocrine-type granules, a feature increasing the confidence for the favored diagnosis. × 16,800.

TABLE 6.5. Neuroendocrine Tumors in Needle Aspiration Biopsies of the Lung and Liver

|  | No. | Percent |
|---|---|---|
| Lung |  |  |
|   Nonneuroendocrine | 87 | 78.4 |
|     Squamous cell |  |  |
|     Adenocarcinoma |  |  |
|     Large cell |  |  |
|   Neuroendocrine | 24 | 21.6 |
|     Small cell (n = 18) |  |  |
|     Carcinoid (n = 5) |  |  |
|     Large cell (n = 1) |  |  |
|   Total | 111 | 100.0 |
| Liver |  |  |
|   Nonneuroendocrine | 29 | 67.4 |
|     Primary and metastatic |  |  |
|   Neuroendocrine | 14 | 32.6 |
|     Pulmonary (n = 12) |  |  |
|     Pancreatic (n = 2) |  |  |
|   Total | 43 | 100.0 |

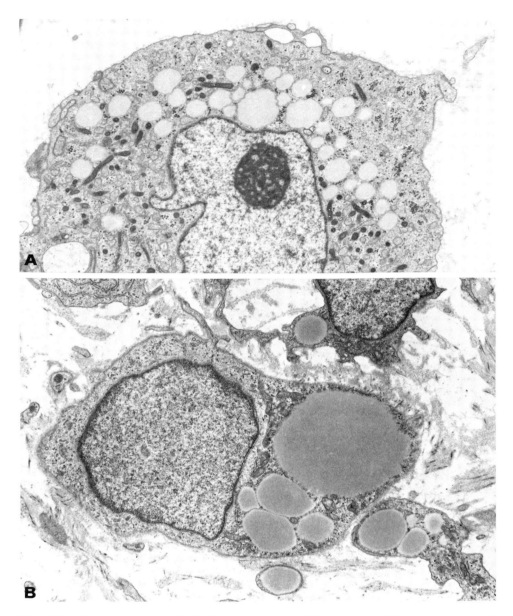

Fig. 6.24. Lipid droplets. A. NAB of a primary myxoid liposarcoma of rib with a tumor cell containing individual nonmembrane-enclosed globules of lightly contrasted lipid material. × 6,400. B. A lipoblast from another liposarcoma shows the apparent, but characteristic, fusion of smaller, glassy-appearing lipid droplets into a larger one. × 7,500.

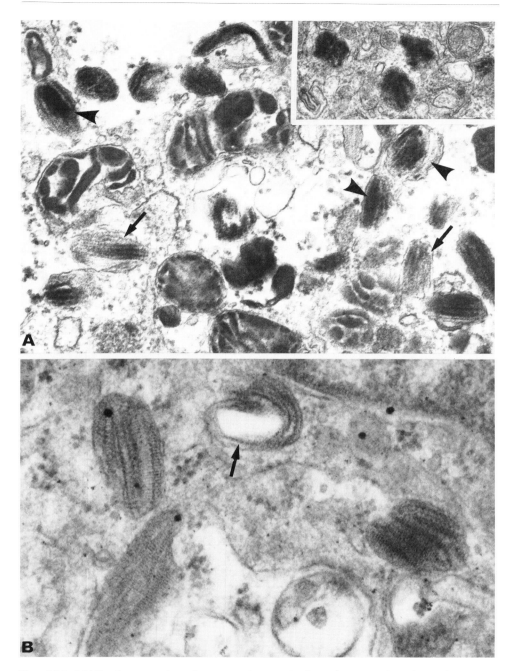

Fig. 6.25. NAB of metastatic malignant melanoma in the soft tissue of the buttock. **A.** The cytoplasm contains both typical stages II (arrows) and III melanosomes (arrowheads). These melanosomes are slightly smaller than the adjacent mitochondria (M). × 60,000. Inset: NAB of liver metastases from a primary ocular melanoma show melanosomes with a full complement of melanin (stage IV). × 30,000. **B.** Stage II melanosomes from the case in A have longitudinal filaments cross-linked to produce a finely striated appearance. The arrow points to an aberrant melanosome. × 90,000.

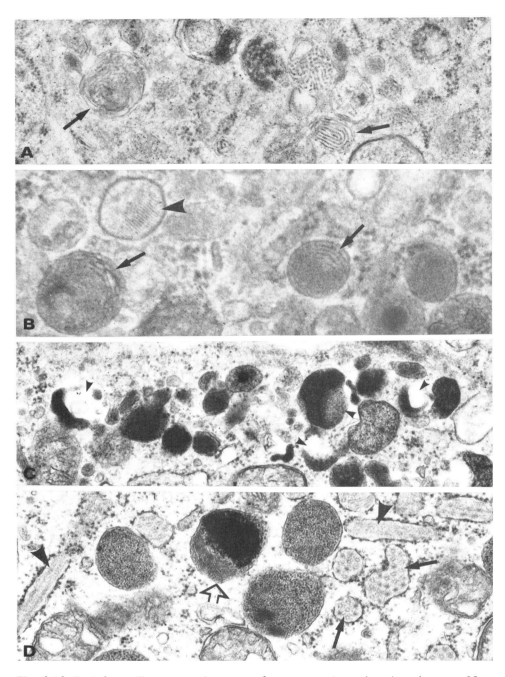

**Fig. 6.26.** Each frame illustrates melanosomes from metastatic amelanotic melanomas. Note the variety of morphology that aberrant or atypical melanosomes can assume and the spectrum of form even in any one example. **A., B.** Some melanosomes contain stacks of straight or curved filaments (arrows), others more granular materials. One has a typical laminated appearance (arrowhead). (A) × 46,000; (B) × 96,000. **C.** Aberrant melanosomes contain mixtures of lucent and darkly staining regions (small arrowheads), whereas others have more homogeneous pigment content. × 46,000. **D.** Note the presence of rigid microtubules, both in cross-section (arrows) and longitudinally (arrowheads), within cisternae of rough endoplasmic reticulum. Some melanosomes have a uniform fine granularity, whereas one is partially melanized but still has a lamellar structure internally (open arrows). × 46,000.

melanoma ideally requires even a few of these forms. Clearly, however, many metastatic melanomas contain mainly aberrant forms of melanosomes that are just as valid for diagnostic purposes (Fig. 6.26). It is important to note that melanosomes can be found in a range of nonmelanocytic tumors,[87] but the majority of these melanosomes are not a consideration in the differential diagnosis of metastatic neoplasms.

On the basis of ultrastructural studies of cell cultures established from human malignant melanomas[36] and detailed surveys of metastases from unequivocal cases of amelanotic melanomas, the range of morphology of atypical or aberrant melanosomes and the considerable percentage of cases in which only these forms are present has become evident. Figure 6.26 illustrates some examples of aberrant melanosomes; unfortunately, their forms are legion.[31,67,77] Probably the most common types are membrane-bound vesicles with one or more internal layers of semicircular membranes or membranous rings with or without multiple tiny granules of varying staining quality. Others consist almost entirely of microgranules and small vesicular or filamentous structures, whereas in other cases the melanosomes are represented by granules that are primarily composed of electron-dense material with eccentrically placed and irregularly shaped lucent zones. An article by Mazur and Katzenstein[67] provides a more complete description of aberrant melanosomes.

A small proportion of malignant melanomas have an additional useful diagnostic feature: the presence of microtubules within profiles of rough endoplasmic reticulum (Fig. 6.26D).[61]

In metastatic tumors suspected of being malignant melanoma in cytopathological smears or histological sections, ultrastructural studies can involve a prolonged and tedious search, often at high magnifications, in an attempt to find the ellusive typical type II or III melanosome. By recognizing the more frequent aberrant melanosomes and knowing that they qualify diagnostically, this process can be circumvented and the correct diagnosis rendered with confidence.

## Basal Lamina

The basal or external lamina is a polysaccharide-, collagen-, and other protein-rich layer applied to basal, focal, or circumferential surfaces of normal and tumor cells. In spindle cell and poorly differentiated tumor cells, the amount and distribution of basal lamina can be of particular assistance no only in differentiating carcinomas from sarcomas but also in distinguishing between various sarcomas. Reduplication of basal lamina or excessive deposition of basal lamina materials between tumor cells is an additional significant diagnostic feature.

The ultrastructural detection of focal or continuous basal lamina in relation to small groups of tumor cells is a more likely finding in adenocarcinomas than in squamous-cell carcinomas. In the appropriate clinical setting, this finding can help distinguish between a metastatic tumor or a new primary lesion in aspiration biopsy specimens. In epithelial tumors, excessively reduplicated basal lamina is a common feature of neoplastic myoepithelial cells in both benign and malignant tumors and, in lesions of the head and neck or in metastatic tumors, should signal the possibility that the tumor is of salivary gland origin. Similarly, clump-like and thickened sinuous depositions of basal lamina materials can be a helpful diagnostic feature in epithelial tumors and even certain sarcomas.

We reported an anterior mediastinal embryonal carcinoma diagnosed by ABC in which the amount and distribution of basal lamina in the surgically resected neoplasm also showed a component of endodermal sinus tumor.[12] This latter finding is a characteristic of both gonadal and extragonadal endodermal sinus (yolk sac) tumors,[85,86,93] and it can be detected in ABC specimens as recently demonstrated by Akhtar and associates[3] and as evident in Figure 8.22. This feature was absent in the ABC of another germ-cell tumor in the anterior mediastinum, diagnosed as seminoma by light microscopy, that did have ultrastructural evidence for additional components of embryonal carcinoma, choriocarcinoma, and teratoma.[89]

Of the sarcomas, basal lamina is associated with benign and malignant peripheral nerve-sheath tumors (both schwannomas and perineuromas), smooth-muscle tumors, hemangiopericytoma, angiosarcoma, rhabdomyosarcoma, and synovial sarcoma. This feature alone effectively separates out the fibromatoses and fibrosarcoma, malignant fibrous histiocytoma, and liposarcoma. Some electron microscopic studies suggest basal lamina is present in liposarcomas, but our personal experience and that of Vuzevski and van der Heul[94] indicate this material is at best infrequently evident.

Benign schwannomas generally have at least some attenuated cytoplasmic processes ensheathed by a narrow basal lamina exhibiting a lamina rara and lamina densa. This finding is illustrated in the ABC of a spindle-cell tumor of the chest wall (Fig. 8.16) and a case report of a facial nerve schwannoma considered a malignant mesenchymal tumor using conventional cytological preparations.[69] Although benign and low-grade schwannomas can have almost every process covered with a single layer or reduplicated layers of basal lamina (foci of extensive accumulations of basal lamina also can be present), this feature becomes focal or even absent with higher grade lesions.[26,33,53] Nerve sheath tumors composed of neoplastic perineurial cells also form basal lamina.[33,54]

The general statements presented regarding peripheral nerve-sheath tumors apply equally well to leimyomas and leiomyosarcomas, even in aspiration biopsies.[57,71] Benign and well-differentiated smooth-muscle tumors may display a well-defined typical basal lamina surrounding each tumor cell (Figs. 7.2B,C; 8.14F). With higher grade leiomyosarcomas, basal lamina is focal or absent.[35,57] Occasionally, basal lamina may be reduplicated or present as irregularly shaped clumps of basal lamina–like materials scattered between the tumor cells.

In hemangiopericytomas, basal lamina synthesis has a major role in the ultrastructural diagnosis of this soft-tissue tumor.[22] Again, it is more likely to be present as thickened masses of basal lamina materials partially or completely enclosing single or small groups of neoplastic pericytes and the many small capillary-like structures ramifying throughout the tumor. Two ABC studies of soft-tissue tumors examined ultrastructurally contained examples of hemangiopericytomas.[57,71]

In rhabdomyosarcoma, it is usually the large, elongated tumor cells with readily recognized myofilaments (often arranged in sarcomeres) that are enclosed by discrete basal lamina.[30] Basal lamina is not a useful diagnostic criterion in less-differentiated lesions because it is absent or infrequently focally expressed.[30] It is the epithelial component of synovial sarcomas that best develops basal lamina.[35]

## Cytoplasmic Filaments

All the various types of cytoplasmic filaments (i.e., actin, intermediate filaments, and myosin (Table 6.6)) have a role in tumor diagnosis, so their recognition in ABC

TABLE 6.6. Characteristics of Cytoplasmic Filaments

| Type | Size (nm) | Molecular Weight, (kd) | Tissue |
|---|---|---|---|
| Muscle-specific actin | 6–8 | 42 | Muscle cells, myoepithelium, and myofibroblasts |
| Intermediate filaments | 8–11 | | |
| Cytokeratin | | 40–68 | Epithelia |
| Vimentin | | 57–58 | Mesenchyme |
| Desmin | | 50–55 | Smooth, cardiac, and striated muscle cells |
| Glial fibrillary acidic protein | | 51–52 | Astrocytes |
| Neurofilaments | | 68, 150, 200 | Neurons |
| Myosin | 12–15 | . . . | Muscle cells |

specimens, usually involving their organization and structural characteristics,[32,40] is essential. General reviews that incorporate correlation of the immunocytochemistry and ultrastructure of cytoplasmic filaments are particularly useful.[8,17] In using intermediate filaments for diagnostic purposes, there is a widening range of normal tissues and neoplasms in which coexpression of two or more types occurs so that the original restriction of cytokeratins to epithelia, vimentin to mesenchyme, desmin to muscle, and glial fibrillary acidic protein to glia no longer applies completely.[17] In routinely stained electron micrographs, this aspect generally cannot be appreciated and the usual diagnostic characteristics of cytoplasmic filaments apply.

## Cytokeratin

High-molecular-weight cytokeratins, present in complex and stratified epithelia, are a key element in differentiating carcinoma from sarcoma and are found, ultrastructurally, in the cytoplasm as aggregates of intermediate-sized filaments referred to as tonofilaments. These usually have a slightly curved form, and may be scattered randomly in the cytoplasm or have a perinuclear distribution (Fig. 6.27). Low-molecular-weight cytokeratins, found in simple epithelia, are cytoskeletal filaments that are not readily apparent in routine electron micrographs as a specific structural component but are demonstrable using immunocytochemical techniques.

The detection of tonofilaments in poorly differentiated or spindle-cell tumors can be instrumental in establishing a correct diagnosis. Using this marker and desomosomes, nasopharyngeal carcinomas can be differentiated from lymphoma in biopsy specimens from the head and neck region, and the diagnosis of metastatic squamous-cell carcinoma can be established in a variety of clinical settings. In two series of ABC specimens of soft-tissue tumors examined by electron microscopy,[57,71] three pleomorphic and spindle-cell lesions diagnosed cytologically as sarcomas were in fact squamous-cell carcinomas. The role of ultrastructural study of ABC specimens in determining squamous differentiation is illustrated in Figure 10.1.

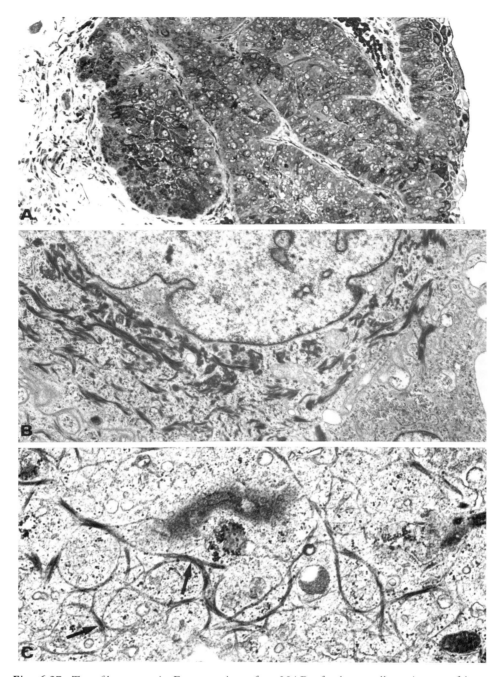

Fig. 6.27. Tonofilaments. **A.** Epon section of an NAB of a large-cell carcinoma of lung. Note the intact architecture of tumor cell clusters. × 200. **B.** Same case as in A. Ultrastructurally, the perinuclear zones of tumor cells in this squamous-cell carcinoma contain many darkly staining and compact linear to slightly curved bundles of tonofilaments (cytokeratin). × 10,400. **C.** Another NAB of squamous-cell carcinoma of lung. The tonofilaments are more delicate, curved to a greater degree, and have a characteristic arrangement with adjacent bundles (arrows). × 16,800.

Tumor cells exhibiting a few to moderate numbers of tonofilaments are not unusual in adenocarcinomas, which does not imply squamous metaplasia or a diagnosis of adenosquamous carcinoma because these diagnoses should be based primarily on light microscopic characteristics. In differentiating squamous carcinomas from adenocarcinomas ultrastructurally, note should be made of the generally fewer cytoplasmic organelles, particularly rough endoplasmic reticulum, and Golgi complexes in the former compared with the latter. For differential diagnostic purposes, it should be appreciated that cytokeratin and other intermediate filaments may occur as expanded focal accumulations within the cytoplasm; these have been termed *globular filamentous bodies*.[41] Such filament accumulations are most useful diagnostically in rhabdoid and Merkel-cell tumors, carcinoids, oncocytomas, and as the hyaline (plasmacytoid) cells of pleomorphic adenomas and myoepitheliomas.[4,5,8,11,20,32]

## Vimentin

Vimentin filaments may be visible and seemingly arranged with no discernible pattern, as in a metastatic meningioma of the lung confirmed by the electron microscopy of an ABC specimen (Fig. 6.28A). Even in this illustrative case, the vimentin filaments, cytoplasmic distribution is regularly spaced in a somewhat wavy pattern. When this feature, which is not uncommon, is coupled with a general whorled and focally concentric organization within the cytoplasm, an ultrastructural pattern emerges that is characteristic for this 10-nm filament.[32] Although coexpression of vimentin and cytokeratins can occur in a variety of epithelial neoplasms,[17] the focal accumulation of vimentin is more common in a variety of sarcomas.[91]

## Desmin

Smooth and skeletal muscle coexpress the intermediate filaments desmin and vimentin.[8,17] By electron microscopy, desmin is usually not discernible in normal muscle cells or their tumors except with the use of immunogold techniques[48] or the correlation of immunocytochemistry with routine electron micrographs in rhabdomyosarcomas and leiomyosarcomas.[8] Perhaps the difficulty in visualizing this filament in tumors relates to the localization of desmin in normal skeletal muscle to the Z disc.[92] Currently, desmin does not have a diagnostic role ultrastructurally.

## Actin/Myosin Complexes

This filament association is an integral component of the sarcomeres of skeletal and cardiac muscle, and both filament types are present in smooth muscle,[38] where special techniques are required to identify the myosin filaments. These polypeptides are essential tools for the immunocytochemical and electron microscopic identification of tumors of smooth or skeletal muscle histogenesis in adults and children.

Actin/myosin complexes as alternating, linear, rigid, thin (6–8 nm diameter) and thick (12–15 nm diameter) filaments, with Z-disc material, are the primary ultrastructural features used to identify rhabdomyosarcomas[27,30,32] among the various small round-cell tumors of children and pleomorphic and spindle-cell sarcomas of adults (see Chapter 8). These filaments may be evident within the cytoplasm of

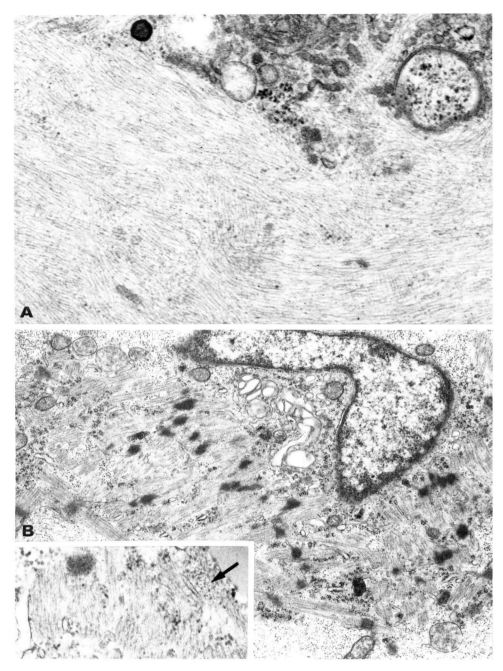

**Fig. 6.28.** Vimentin and myofilaments. **A.** NAB of a metastatic meningioma to lung in which the tumor cells have the fairly orderly and evenly spaced filaments characteristic of vimentin. × 46,500. **B.** A malignant mixed müllerian tumor of ovary with rhabdomyosarcomatous differentiation. Darkly staining foci of Z-band material have rigid thick and thin filaments emanating from opposite sides. At higher magnification (inset), the simulation of the specific inter-relationship of thin to thick filaments in normal skeletal muscle is best seen when these filaments are cross-sectioned (arrow). × 10,000; inset, × 40,000.

the tumor cells as well-organized bundles of thick and thin filaments and Z discs forming sarcomere-like structures, as scattered sheaths of thick and thin filaments of varying length with and without associated Z-disc material (Fig. 6.28B), or as clusters of parallel thick (myosin) filaments between which are rows of ribosomes (termed *ribosome/myosin complexes*[30]). The full ultrastructural spectrum of rhabdomyosarcoma, the diagnostic criteria, and the application of electron microscopy in the differential diagnosis of small, round-cell tumors of children and young adults are fully presented in a practical review article by Erlandson.[30] The segregation of rhabdomyosarcomas was based on these criteria in a series of ABC specimens of small-cell malignant tumors in children.[2]

In smooth-muscle tumors, whether benign or malignant, muscle-specific actin is the filament visualized ultrastructurally that is of diagnostic importance. Although desmin is a component of the normal smooth-muscle cell[38] and may be detectable in benign and low-grade malignant tumors, it is often not expressed in leiomyosarcomas:[8,32] this finding limits its usefulness even immunocytochemically. In well-differentiated leiomyosarcomas (Fig. 7.2), most tumor cells will have a rich complement of actin filaments arrayed in a parallel fashion, either filling much of the cytoplasm or as multiple aggregates of varying size.[62,64] Within the filaments, varying numbers of linear, short, dense bodies (containing alpha-actinin and other structural and contractile proteins) are present, and there are thin (30–40 nm in depth) electron-dense plaques associated with the inner surface of the cell membrane into which the various cytoplasmic filaments insert (Fig. 6.29). Less-differentiated leiomyosarcomas will have only scattered smaller aggregates of myofilaments, usually at the terminal ends of the tumor cell or at the periphery of their lateral margins, but some cells (and even a majority in some tumors) will be completely without filaments, at least as seen ultrastructurally.[64]

## Intracytoplasmic Lumens

True glandular or intercellular lumens are the hallmark of adenomas and adenocarcinomas, but they are also evident in mesotheliomas, transitional-cell carcinomas, and synovial sarcomas.[21,40,45,85] Two or more tumor cells form such lumens, which are associated with various combinations of apical junctional complexes, surface microvilli, and secretory products, both within the apical cytoplasm and the lumen. This combined evidence for glandular differentiation, with each type of organelle revealing varying degrees of structural form, is a particularly useful ultrastructural marker for identifying poorly differentiated adenocarcinomas, whether primary or metastatic.[44,45,50,62]

Equally helpful is the identification of intracellular lumens (intracytoplasmic lumens or neolumens) in primary or metastatic tumors. In electron micrographs, these lumens usually present as vacuole-like cytoplasmic regions of differing size lined by a unit membrane bearing varying numbers of microvilli (Figs. 6.30, 6.31); the absence of intercellular junctions of any type associated with the lumen is a key feature distinguishing these from true lumens. At other times, the intracytoplasmic lumen may be nearly or completely occupied by microvilli, producing a pattern seen in the true lumens illustrated in Figure 6.10. The lumen may contain obvious secretory products (Fig. 6.31) or glycocalyceal bodies, or both, and specific secretory granules may be evident in the cytoplasm next to the luminal surface. Microvilli

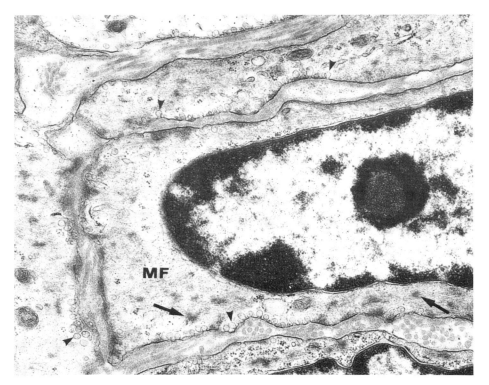

Fig. 6.29. Tumor cells from a leiomyoma of the bowel wall have a major portion of the cytoplasm occupied by microfilaments (MF), within which are variously sized and shaped dense bodies (arrows). Micropinocytotic vesicles (arrowheads) are associated with the cell membrane. × 20,000.

associated with intracytoplasmic lumens may possess the core of actin filaments extending down into the apical cytoplasm that strongly suggests a gastrointestinal carcinoma (Figs. 6.8, 6.12).

Intracytoplasmic lumens are most frequently noted in breast carcinoma (more often in lobular than ductal) and are particularly helpful in confirming a primary breast neoplasm in a poorly differentiated tumor involving an axillary lymph node.[44,70,84] Neolumens, however, are occasionally evident in primary lung carcinomas (Figs. 6.17B, 6.31A); gastrointestinal adenocarcinomas such as pancreas,[99] esophagus, (Fig. 6.30) and stomach;[40] mesotheliomas;[19,21,45] renal-cell adenomas;[55] and adenomatoid tumors.[15] With focal accumulations of mucus-type secretory granules, somewhat large intracytoplasmic lumens produce the signet-ring–type of tumor cell in certain adenocarcinomas, even the lung.[58] Despite the occurrence of intracytoplasmic lumens in a considerable list of normal, benign, and malignant tissues,[40] breast carcinomas continue to be the preeminent lesion in which this diagnostic feature presents.[70,84] In our series of ABC specimens of the lung, 3 of 78 cases (3.8%) diagnosed cytologically as large-cell and adenocarcinomas showed intracytoplasmic lumens.

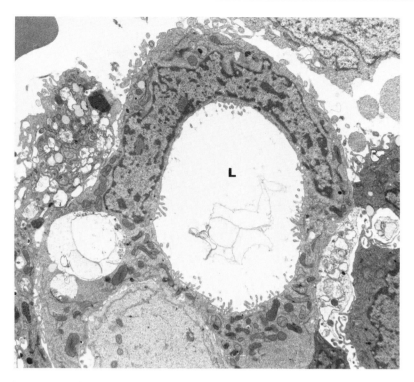

Fig. 6.30. Intracytoplasmic lumen in an NAB of an esophageal adenocarcinoma metastatic to rib. The lumen (L), with modest numbers of short microvilli, occupies a considerable portion of the cytoplasm and, with the displaced nucleus, is another mechanism for the formation of a signet-ring cell. × 8,000.

# BENIGN LESIONS

There are few, if any, specific indications for electron microscopy of benign lesions; however, electron microscopy can on occasion establish a benign diagnosis where benign, malignant, or reactive processes are all considerations based on the cytological criteria. The ABC specimen diagnosed simply as a spindle-cell tumor, which was readily recognizable as a schwannoma by electron microscopy, is detailed in Figure 8.16. Electron microscopy has diagnosed other schwannomas in ABC specimens.[69,100] Similarly, the degree of differentiation based on the number of neurosecretory granules can help distinguish between benign and malignant neuroendocrine tumors.[43]

Ferguson and associates[34] published a detailed ultrastructural analysis of 36 ABC specimens from a variety of benign breast lesions. The ability to detect myoepithelial cells in an ABC specimen of a case of adenosis tumor of the breast examined ultrastructurally suggests this feature is a potential diagnostic tool.[81]

Electron microscopy and immunocytochemistry of certain lesions suspected of being caused by bacterial, viral, or other organisms might benefit from sampling by

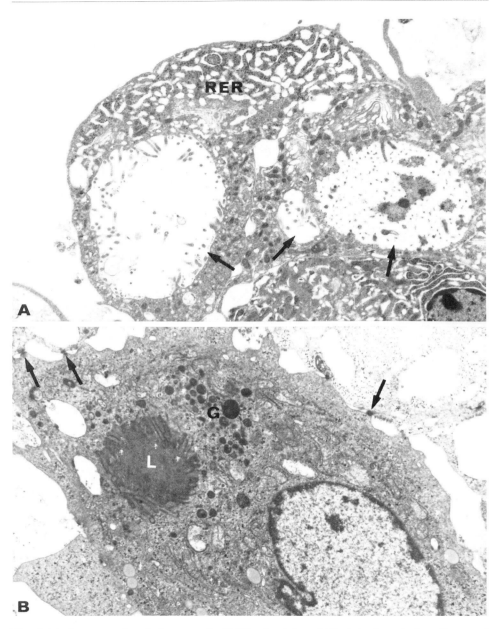

**Fig. 6.31.** Intracytoplasmic lumens. **A.** NAB of a right-middle lobe lung lesion diagnosed cytologically as "possible carcinoma." The presence of three intracytoplasmic lumens (arrows) and considerable rough endoplasmic reticulum (RER) established the diagnosis of adenocarcinoma. × 8,000. **B.** NAB of liver with metastatic adenocarcinoma of unknown primary site. An intracytoplasmic lumen (L) is filled with electron-dense secretory product similar to the contents of cytoplasmic granules (G). Small desmosomes (arrows) are also present. × 10,400.

NAB. These diagnostic techniques have an essential role in determining etiological agents in the viral encephalitides.[42] Recently, electron microscopy was used to confirm the cytological diagnosis of human papillomavirus infection.[83] An example of malacoplakia of the colon and retroperitoneum was diagnosed by ABC of the abdominal mass and the results of therapy followed by repeated ABC specimens.[1]

# REFERENCES

1. Akhtar M, Ali MA, Robinson C, Harfi H: Role of fine needle aspiration biopsy in the diagnosis and management of malacoplakia. *Acta Cytol* 29:457–460, 1985.
2. Akhtar M, Ali MA, Sabbah R, Bakry M, Nash JE: Fine-needle aspiration biopsy diagnosis of round cell malignant tumors of childhood: A combined light and electron microscopic approach. *Cancer* 55:1805–1817, 1985.
3. Akhtar M, Ali MA, Sackey K, Jackson D, Bakry M: Fine-needle aspiration biopsy diagnosis of endodermal sinus tumor: Histologic and ultrastructural correlations. *Diagn Cytopathol* 6:184–192, 1990.
4. Battifora H, Silva EG: The use of antikeratin antibodies in the immunohistochemical distinction between neuroendocrine (Merkel cell) carcinoma of the skin, lymphoma, and oat cell carcinoma. *Cancer* 58:1040–1046, 1986.
5. Berger G, Berger F, Bejui F, et al: Bronchial carcinoid with fibrillary inclusions related to cytokeratins: An immunohistochemical and ultrastructural study with subsequent investigation of 12 foregut APUDomas. *Histopathology* 8:245–257, 1984.
6. Bolen JW, Hammar SP, McNutt MA: Reactive and neoplastic serosal tissue: A light-microscopic, ultrastructural, and immunocytochemical study. *Am J Surg Pathol* 10:34–47, 1986.
7. Bolen JW, Hammar SP, McNutt MA: Serosal tissue: Reactive tissue as a model for understanding mesotheliomas. *Ultrastruct Pathol* 11:251–262, 1987.
8. Bolen JW, McNutt MA: Cytoskeletal intermediate filaments: Practical applications of intermediate filament analysis. *Ultrastruct Pathol* 11:175–189, 1987.
9. Bolen JW, Thorning D: Benign lipoblastoma and myxoid liposarcomas: A comparative light and electron microscopic study. *Am J Surg Pathol* 4:163–169, 1980.
10. Bolen JW, Thorning D: Histogenetic classification of pulmonary carcinoma: Peripheral adenocarcinoma studied by light microscopy, histochemistry, and electron microscopy. *Pathol Annu* 17:77–100, 1983.
11. Bonsib SM, Bromley C, Lager DJ: Renal oncocytoma: Diagnostic utility of cytokeratin-containing globular filamentous bodies. *Mod Pathol* 4:16–23, 1991.
12. Burns BF, Dardick I: Mixed germ cell tumour of the mediastinum (seminoma, embryonal carcinoma, choriocarcinoma and teratoma): Light and electron microscopic cytology and histological investigation. Letter to the case. *Pract Res Pathol* 185:511–513, 1989.
13. Burns TR, Greenberg SD, Mace ML, Johnson EH: Ultrastructural diagnosis of epithelial malignant mesothelioma. *Cancer* 56:2036–2040, 1985.
14. Burns TR, Johnson EH, Cartwright J Jr, Greenberg SD: Desmosomes of epithelial malignant mesothelioma. *Ultrastruct Pathol* 12:385–388, 1988.
15. Carlier MT, Dardick I, Lagace AF, Sreeram V: Adenomatoid tumor of uterus: Presentation in uterine curettings. *Int J Gynecol Pathol* 5:69–74, 1986.

16. Clayton F: Bronchioloalveolar carcinomas: Cell types, patterns of growth, and prognostic corelates. *Cancer* 57:1555–1564, 1986.

17. Coggi G, Dell'Orto P, Braidotti P, Coggi A, Viale G: Coexpression of intermediate filaments in normal and neoplastic human tissues: A reappraisal. *Ultrastruct Pathol* 13:501–514, 1989.

18. Dardick I: Diagnostic electron microscopy, in Gnepp DR (ed): *Pathology of the Head and Neck, Contemporary Issues in Surgical Pathology,* vol 10. New York, Churchill Livingstone, 1988. Pp 101–190.

19. Dardick I, Al-Jabi M, McCaughey WTE, et al: Ultrastructure of poorly differentiated diffuse epithelial mesotheliomas. *Ultrastruct Pathol* 7:151–160, 1984.

20. Dardick I, Cavell S, Boivin M, et al.: Salivary gland myoepithelioma variants: Histological, ultrastructural, and immunocytochemical features. *Virchows Arch {A}* 416:25–42, 1989.

21. Dardick I, Jabi M, McCaughey WTE, et al: Diffuse epithelial mesothelioma: a review of the ultrastructural spectrum. *Ultrastruct Pathol* 11:503–533, 1987.

22. Dardick I, Hammar SP, Scheithauer BW: Ultrastructural spectrum of hemangiopericytoma: A comparative study of fetal, adult and neoplastic pericytes. *Ultrastruct Pathol* 13:111–154, 1989.

23. Dardick I, Yazdi HM, Brosko C, Rippstein P, Hickey NM: A quantitative comparison of light and electron microscopic diagnoses in specimens obtained by fine needle aspiration biopsy. *Ultrastruct Pathol* 15:105–130, 1991.

24. Deck JHN, Ramjohn S, Dardick I: "Anemone" cell (villiform) tumor of the brain. *Ultrastruct Pathol* 14:87–94, 1990.

25. Dickersin GR: The contributions of electron microscopy in the diagnosis and histogenesis of controversial neoplasms. *Clin Lab Med* 4:123–164, 1984.

26. Dickersin GR: The electron microscopic spectrum of nerve sheath tumors. *Ultrastruct Pathol* 11:103–146, 1987.

27. Dickersin GR: *Diagnostic Electron Microscopy: A Text/Atlas.* New York, Igaku-Shoin, 1988.

28. Dvorak AM, Monahan RA: Metastatic adenocarcinoma of unknown primary site: Diagnostic electron microscopy to determine the site of tumor origin. *Arch Pathol Lab Med* 106:21–24, 1982.

29. Engstrand DA, England DM, Oberley TD: Limitations of the usefulness of microvillous ultrastructure in distinguishing between carcinoma primary in and metastatic to the lung. *Ultrastruct Pathol* 11:53–58, 1987.

30. Erlandson RA: The ultrastructural distinction between rhabdomyosarcoma and other undifferentiated "sarcomas." *Ultrastruct Pathol* 11:83–101, 1987.

31. Erlandson RA: Ultrastructural diagnosis of amelanotic malignant melanoma: Aberrant melanosomes, myelin figures or lysosomes? *Ultrastruct Pathol* 11:191–208, 1987.

32. Erlandson RA: Cytoskeletal proteins including myofilaments in human tumors. *Ultrastruct Pathol* 13:155–186, 1989.

33. Erlandson RA, Woodruff JM: Peripheral nerve sheath tumors: An electron microscopic study of 43 cases. *Cancer* 49:273–287, 1982.

34. Ferguson DJP, Wells CA, Crucioli V: Ultrastructural analysis of fine needle aspirates from benign breast lesions. *J Clin Pathol* 43:22–26, 1990.

35. Fisher C: The value of electronmicroscopy and immunohistochemistry in the diagnosis of soft tissue sarcomas: A study of 200 cases. *Histopathology* 16:441–454, 1990.

36. Foa C, Aubert C: Ultrastructural comparison between cultured and tumor cells of human malignant melanomas. *Cancer Res* 37:3957–3963, 1977.

37. Fukuda T, Tsuneyoshi M, Enjoji M: Malignant fibrous histiocytoma of soft parts: An ultrastructural quantitative study. *Ultrastruct Pathol* 12:117–129, 1988.

38. Gabella G: General aspects of the fine structure of smooth muscles, in Motta PM (ed): *Ultrastructure of Smooth Muscle: Electron Microscopy in Biology and Medicine*, vol 8. Boston, Kluwer Academic Publishers, 1990. Pp 1–22.

39. Ghadially FN: The role of electron microscopy in the determination of tumour histogenesis. *Diagn Histopathol* 4:245–262, 1981.

40. Ghadially FN: *Diagnostic Electron Microscopy of Tumours*, 2nd ed. London, Butterworths, 1985.

41. Ghadially FN: *Ultrastructural Pathology of the Cell and Matrix*, 3rd ed, vols 1 and 2. London, Butterworths, 1988.

42. Gosztonyi G, Cervós-Navarro J: Immunohistochemical and electron microscopic techniques in the diagnosis of viral encephalitides. *Pathol Res Pract* 183:223–252, 1988.

43. Gould VE, Lee I, Warren WK: Immunohistochemical evaluation of neuroendocrine cells and neoplasms of the lung. *Pathol Res Pract* 183:200–213, 1988.

44. Hammar S: Adenocarcinoma and large cell undifferentiated carcinoma of the lung.*Ultrastruct Pathol* 11:263–291, 1987.

45. Hammar S, Bockus D, Remington F: Metastatic tumors of unknown origin: An ultrastructural analysis of 265 cases. *Ultrastruct Pathol* 1:209–250, 1987.

46. Henderson DW, Papadimitriou JM, Coleman M: *Ultrastructural Appearances of Tumours: Diagnosis and Classification of Human Neoplasia by Electron Microscopy*, 2nd ed. Edinburgh, Churchill Livingstone, 1986.

47. Herrera GA: Readers' forum, letter to the editor. *Ultrastruct Pathol* 12:151, 1988.

48. Herrera GA: Ultrastructural postembedding immunogold labeling: Applications to diagnostic pathology. *Ultrastruct Pathol* 13:485–499, 1989.

49. Herrera GA, Alexander B, De Moraes HP: Ultrastructural subtypes of pulmonary adenocarcinoma: A correlation with patient survival. *Chest* 84:581–586, 1983.

50. Herrera GA, Alexander CB, Jones JM: Ultrastructural characteristics of pulmonary neoplasms. Part I—the role of electron microscopy in characterization of the most common epithelial neoplasms. *Surv Synth Pathol Res* 3:520–546, 1985.

51. Herrera GA, Reimann BEF: Electron miscroscopy in determining origin of metastatic adenocarcinomas. *South Med J* 77:1557–1566, 1984.

52. Hickey WF, Seiler MW: Ultrastructural markers of colonic adenocarcinoma. *Cancer* 47:140–145, 1981.

53. Hirose T, Sano T, Hizawa K: Heterogeneity of malignant schwannomas. *Ultrastruct Pathol* 12:107–116, 1988.

54. Hirose T, Sumitomo M, Kudo E, et al: Malignant peripheral nerve sheath tumor (MPNST) showing perineurial differentiation. *Am J Surg Pathol* 13:613–620, 1989.

55. Holm-Nielsen P, Olsen TS: Ultrastructure of renal adenoma. *Ultrastruct Pathol* 12:27–39, 1988.

56. Ishihara T, Takahashi M, Uchino F: A filiform large cell lymphoma in the spleen: A case report with immunohistochemical and electron microscopic study. *Ultrastruct Pathol* 14:193–199, 1990.

57. Kindblom LG: Light and electron microscopic examination of embedded fine-needle

aspiration biopsy specimens in the preoperative diagnosis of soft tissue and bone tumors. *Cancer* 51:2264–2277, 1983.

58. Kish JK, Ro JY, Ayala AG, McMurtrey MJ: Primary mucinous adenocarcinoma of the lung with signet-ring cells: A histochemical comparison with signet-ring cell carcinomas of other sites. *Hum Pathol* 20:1097–1102, 1989.

59. Lagacé R, Delage C, Seemayer TA: Myxoid variant of malignant fibrous histiocytoma: Ultrastructural observations. *Cancer* 43:526–534, 1979.

60. Mackay B, (ed): *Diagnostic Electron Microscopy: Clinics in Laboratory Medicine,* vol 7. Philadelphia, W.B Saunders, 1987.

61. Mackay B, Ayala AG: Intracisternal tubules in human melanoma cells. *Ultrastruct Pathol* 1:1–6, 1980.

62. Mackay B, Fanning T, Bruner JM, Steglich MC: Diagnostic electron microscopy using fine needle aspiration biopsies. *Ultrastruct Pathol* 11:659–672, 1987.

63. Mackay B, Ordóñez NG, Khoursand J, Bennington JL: The ultrastructural and immunocytochemistry of renal cell carcinoma. *Ultrastruct Pathol* 11:483–502, 1987

64. Mackay B, Ro J, Floyd C, Ordóñez NG: Ultrastructural observations on smooth muscle tumors. *Ultrastruct Pathol* 11:593–607, 1987.

65. Marcus PB: Glycocalyceal bodies and their role in tumor typing. *J Submicrosc Cytol* 13:483–500, 1981.

66. Marcus PB, Martin JH, Green RH, Krouse MA: Glycocalyceal bodies and microvillous core rootlets: Their value in tumor typing. *Arch Pathol Lab Med* 103:89–92, 1979.

67. Mazur MT, Katzenstein A-LA: Metastatic melanoma: The spectrum of ultrastructural morphology. *Ultrastruct Pathol* 1:337–356, 1980.

68. Mirra SS, Miles ML: Subplasmalemmal linear density: A mesodermal feature and a diagnostic aid. *Hum Pathol* 13:365–380, 1982.

69. Navas-Palacios JJ, de Agustin de Agustin PP, Alvarez de los Heros F, Perez-Barrios A, Alvarez-Vicent JJ: Ultrastructural diagnosis of facial nerve schwannoma using fine needle aspiration. *Acta Cytol* 27:441–445, 1983.

70. Nesland JM, Holm R, Lunde S, Johannessen JV: Diagnostic problems in breast pathology: The benefit of ultrastructural and immunocytochemical analysis. *Ultrastruct Pathol* 11:293–311, 1987.

71. Nordgren H, Akerman M: Electron microscopy of fine needle aspiration biopsy from soft tissue tumors. *Acta Cytol* 26:179–188, 1982.

72. Ozzello L, Savary M, Roethlisberger B: Columnar mucosa of the distal esophagus in patients with esophageal reflux. *Pathol Annu* 12:41–86, 1977.

73. Payne CM: Phylogenetic considerations of neurosecretory granule contents: Role of nucleotides and basic hormone/transmitter packaging mechanisms. *Arch Histol Cytol* 52(suppl):277–292, 1989.

74. Plaza JA, Ballestín C, Pérez-Barrios A, Martínez MA, de Agustín P: Cytologic, cytochemical, immunocytochemical and ultrastructural diagnosis of a sacroccygeal chordoma in a fine needle aspiration biopsy specimen. *Acta Cytol* 33:89–92, 1989.

75. Quevedo WC Jr, Fitzpatrick TB, Szabo G, Jimbow K: Biology of melanocytes, in Fitzpatrick TB, Eisen AZ, Wolff K, Freeberg IM, Austen KF, (eds): *Dermatology in General Medicine, Textbook and Atlas,* 3rd ed. New York, McGraw-Hill, 1987. Pp 224–251.

76. Quinonez G, Simon GT: Cellular junctions in a spectrum of human malignant tumors. *Ultrastruct Pathol* 12:389–405, 1988.

77. Rennison A, Duff C, McPhie JL: Electron microscopic identification of aberrant melanosomes using a combined dopa/Warthin-Starry technique. *J Pathol* 152:333–336, 1987.

78. Russo J, Sommers SC, (eds): *Tumor Diagnosis by Electron Microscopy*. Philadelphia, Field & Wood, 1986 (vol 1); 1988 (vol 2); 1990 (vol 3).

79. Russo J, Tait L, Russo IH: Current basis for the ultrastructural clinical diagnosis of tumors: A review. *J Electron Microsc Techn* 2:305–351, 1985.

80. Seiler MW, Reilova-Velez J, Hickey W, Bono L: Ultrastructural markers of large bowel cancer, in Wolman SR, Mastromarino AJ, (eds): *Progress in Cancer Research and Therapy*. New York, Raven Press, 1984. Pp 51–65.

81. Silverman JF, Dabbs DJ, Gilbert CF: Fine needle aspiration cytology of adenosis tumor of the breast: With immunocytochemical and ultrastructural observations. *Acta Cytol* 33:181–187, 1989.

82. Shoup AS, Johnston WW, Siegler HF, et al: A panel of antibodies useful in the cytologic diagnosis of metastatic melanoma. *Acta Cytol* 34:385–392, 1990.

83. Shroyer KR, Hosey J, Swanson LE, Woodward WD, Fennel RH: Cytological diagnosis of human papillomavirus infection: Spindle cell nuclei. *Diagn Cytopathol* 6:178–183, 1990.

84. Sobrinho-Simões M, Johannessen JV, Gould VE: The diagnostic significance of intracytoplasmic lumina in metastatic neoplasms. *Ultrastruct Pathol* 2:327–335, 1981.

85. Srigley JR, Mackay B, Toth P, Ayala A: The ultrastructure and histogenesis of male germ neoplasia with emphasis on seminoma with early carcinomatous features. *Ultrastruct Pathol* 12:67–86, 1988.

86. Srigley JR, Toth P, Edwards V: Diagnostic electron microscopy of male genital tract tumors. *Clin Lab Med* 7:91–115, 1987.

87. Szpak CA, Shelburne J, Linder J, Klintworth GK: The presence of stage II melanosomes (prelemanosomes) in neoplasms other then melanomas. *Mod Pathol* 1:35–43, 1988.

88. Taccagni G, Dell'Antonio G, Terreni MR, Cantaboni A: Heterogeneous subcellular morphology of lung adenocarcinoma cells: Identification of different cytotypes on cytologic material. *Ultrastruct Pathol* 14:65–80, 1990.

89. Taccagni GL, Parafioriti A, Dell'Antonio G, Crespi G: Mixed germ cell tumour of the mediastinum (seminoma, embryonal carcinoma, choriocarcinoma and teratoma): Light and electron microscopic cytology and histological investigation. *Pract Res Pathol* 185:506–510, 1989.

90. Taxy JB, Almanaseer IY: "Anemone" cell (villiform) tumors: Electron microscopy and immunohistochemistry of five cases. *Ultrastruct Pathol* 7:143–150, 1984.

91. Tsuneyoshi M, Daimaru Y, Hashimoto H, Enjoji M: The existence of rhabdoid cells in specific soft tissue sarcomas. *Virchows Arch {A}* 411:509–514, 1987.

92. Tokuyasu KT, Dutton AH, Singer SJ: Immunoelectron microscopic studies of desmin (skeletin) localization and intermediate filament organization in chicken skeletal muscle. *J Cell Biol* 96:1727–1735, 1983.

93. Ulbright TM, Roth LM, Brodhecker CA: Yolk sac differentiation in germ cell tumors: A morphologic study of 50 cases with emphasis on hepatic, enteric and parietal yolk sac features. *Am J Surg Pathol* 10:151–164, 1986.

94. Vuzevski VD, van der Heul RO: Comparative ultrastructure of soft-tissue myxoid tumors. *Ultrastruct Pathol* 12:87–105, 1988.

95. Warhol MJ, Corson JM: An ultrastructural comparison of mesotheliomas with adenocarcinomas of the lung and breast. *Hum Pathol* 16:50–55, 1985.
96. Warhol MJ, Hickey WF, Corson JM: Malignant mesothelioma: Ultrastructural distinction from adenocarcinoma. *Am J Surg Pathol* 6:307–314, 1982.
97. Warhol MJ, Hunter NJ, Corson JM: An ultrastructural comparison of mesotheliomas and adenocarcinomas of the ovary and endometrium. *Int J Gynecol Pathol* 1:125–134, 1982.
98. Wick MR, Loy T, Mills SE, Legier JF, Manivel JC: Malignant epithelioid pleural mesothelioma versus peripheral pulmonary adenocarcinoma: A histochemical, ultrastructural, and immunohistologic study of 103 cases. *Hum Pathol* 21:759–766, 1990.
99. Wills EJ, Carr S, Philips J: Electron microscopy in the diagnosis of percutaneous fine needle aspiration specimens. *Ultrastruct Pathol* 11:361–387, 1987.
100. Zbieranowski I, Bedard YC: Fine needle aspiration of schwannomas: Value of electron microscopy and immunocytochemistry in the preoperative diagnosis. *Acta Cytol* 33:381–384, 1989.

# 7

# Value and Limitations of Electron Microscopy

Many of the clinicopathological cases presented herein attest to the value of electron microscopy in the practice of cytopathology. These examples can, quite rightly, be construed simply as anecdotal, with the majority of aspiration biopsy cytology (ABC) specimens examined deriving no direct benefit from ultrastructural study. Some form of comparison between diagnoses rendered from the study of routine ABC preparations and diagnoses established following electron microscopy of the same ABC, either from the literature or an analysis of our own samples, is required to establish the real benefits of this procedure. In addition, assessment is also needed of the proportion of ABC samples that are inadequate for electron microscopy and the factors that limit the application or interpretation of ABC specimens by electron microscopy.

## QUANTIFYING THE UTILITY

Several studies have judged the level of agreement (generally high) between light and electron microscopy of cytological preparations.[1,4,5,11,12,14,16,17] In most such studies, a few to a series of cases are presented in which electron microscopy provided essential information or an unsuspected diagnosis. For example, in a correlative study of 43 poorly differentiated lung carcinomas, Mennemeyer and associates[11] showed that electron microscopy altered the diagnosis in 32.6%, whereas Sehested and colleagues[13] reported on 105 consecutive cases of transthoracic ABC specimens and noted that in the 67 samples in which light microscopy demonstrated tumor cells, electron microscopy contributed to "a more specific histogenetic morphological tumor diagnosis" in 11 cases, or 16.4%.

Only Wills and associates[16] and Dabbs and Silverman[4] systematically assessed the degree of contribution to ABC made by electron microscopic examinations in a range of tumors from a variety of sites. In terms of providing information about

**TABLE 7.1. Quantitative Comparison of Light and Electron Microscopic Diagnoses of ABC Specimens***

| Category | No. | Percent |
|---|---|---|
| Similar diagnosis by LM and EM | 59 | 29.5 |
| Additional information from EM | 30 | 15.0 |
| Nondiagnostic by LM and EM | 52 | 26.0 |
| Diagnostic by LM but not by EM | 59 | 29.5 |
| Totals | 200 | 100.0 |

*From Wills EJ, Carr S, Philips J: Electron microscopy in the diagnosis of percutaneous fine needle aspiration specimens. *Ultrastruct Pathol* 11:361–387, 1987. Used with permission.
LM = light microscopy. EM = electron microscopy.

technical details, general indications, diagnostic ultrastructural characteristics, and a comparative analysis of the role of electron microscopy in ABC, the study by Wills and co-authors[16] is the most complete and informative. Their data analysis is based on the initial 200 consecutive ABC specimens that were examined by electron microscopy (Table 7.1). Of these specimens, electron microscopy provided additional diagnostic information in 30 cases (15%). As the authors correctly state, the overall results are tempered by the many nondiagnostic cases that always occur in the early phases of establishing this type of technical service. If the aspiration biopsies that were inadequate for either light or electron microscopy and those in which no material was available for ultrastructural studies (their groups 3 and 4) are removed from their results (Table 7.1), then the contribution of electron microscopy would rise to 33.7% (30 of 89 cases).

Of 46 adequate specimens analyzed by Dabbs and Silverman,[4] electron microscopy revealed diagnostic features in 25 case (54.3%), whereas the ultrastructural and light microscopic diagnoses were the same in the remainder. Aspirates from lung and liver accounted for 25 of the 46 cases. In a recent series of 10 lung tumors, the seven ABC specimens with adequate tissue were all representative of the resected specimen.[6] Poorly differentiated tumors in adults and small-cell tumors of children would appear to benefit most from the application of electron microscopy to ABC specimens.[2]

A few quantitative studies of the role of electron microscopy in surgical pathology are available in which a reasonable number of samples (ranging from 181 to 259) were compared.[9,10,15] The extent of the contribution provided by electron microscopy in these three studies is 18%,[9] 38%,[10] and 57%,[15] respectively. The reports by Wills and associates[16] and Dabbs and Silverman[4] indicate that similar results can be anticipated using ABC specimens for electron microscopy. To verify this finding using comparative criteria similar to those employed for surgical pathology specimens by Lombardi and Orazi,[10] we carried out a prospective study of 222 ABC specimens over a 3-year period in which adequate material was available for electron microscopy and the cytopathologist suggested that ultrastructural features might provide useful diagnostic information.[5] We compared the initial, tentative, or interim diagnosis based on a cytopathologist's light microscopic assessment of each of the 222 ABC specimens; the final diagnosis was established when the electron

TABLE 7.2. Quantitative Assessment of the Role of Electron Microscopy in ABC*

| Group | Role | No. | Percent |
|---|---|---|---|
| 1 | Specific diagnosis unsuspected or not possible by LM | 16 | 7.7 |
| 2 | Confirmed LM diagnosis, selected a specific diagnosis from a set of differential diagnoses, or provided clinically relevant information | 44 | 19.4 |
| 3 | Additional diagnostic data but not clinically relevant | 45 | 20.3 |
| 4 | Light and electron microscopic diagnoses identical | 117 | 52.7 |
| Totals | | 222 | 100.0 |

*From Dardick I, Yazdi HM, Brosko C, Rippstein P, Hickey NM: A quantitative comparison of light and electron microscopic diagnoses in specimens obtained by fine needle aspiration biopsy. *Ultrastruct Pathol* 15:105–129, 1991.

LM = light microscopy.

microscopic features were integrated with the light microscopy features.[5] The results of segregating the 222 ABC specimens into 4 defined categories are given in Table 7.2. A major contribution was made by electron microscopy in 27% (groups 1 and 2). This result, made possible because electron microscopy detected specific organelles or tumor cell arrangement not possible by light microscopy, emphasizes the role that this diagnostic modality can have for cytopathologists. In group 3, we included ABC specimens, primarily from the lung and diagnosed as nonsmall-cell carcinoma by light microscopy, in which a more definitive diagnosis was established by electron microscopy (usually squamous-cell carcinoma or adenocarcinoma), but this distinction did not affect the clinical management of the patient.

Segregating the ABC specimens on the basis of the site aspirated, one from lung or pleura and the other from nonpulmonary sites (Table 7.3), established that the contribution of ultrastructural examinations to ABC specimens from the lung cases was 19.5% compared with 35.4% for the nonlung aspirates. This type of information assists in deciding the types of cases in which electron microscopy is more likely to help in the diagnostic process.

TABLE 7.3. Contribution of Electron Microscopy in Lung and Nonlung ABC specimens

| Site | Total (%) | Groups 1 and 2 No. (%) | Groups 3 and 4 No. (%) |
|---|---|---|---|
| Lung | 123 (55.4) | 24 (19.5) | 99 (80.5) |
| Nonlung* | 99 (44.6) | 35 (35.4) | 64 (64.6) |
| Combined | 222 (100.0) | | |

*Includes biopsies from kidney, mediastinum, pelvis, abdomen, liver, bone, lymph node, retroperitoneum, neck, and adrenal.

TABLE 7.4. Limitations of Ultrastructural Examination of ABC Specimens

| |
|---|
| Inadequate tumor tissue |
| Necrotic tissue |
| Normal or stromal tissue |
| Blood clot only |
| Sampling |
| Degree of differentiation |

# SPECIFIC LIMITATIONS

In the case of ultrastructural examination of ABC specimens, some limitations result directly from the collection technique and the skill of the operator, and are thus unique to this diagnostic procedure when compared with techniques generally employed in surgical pathology. The more common problems encountered in a comparative assessment of light and electron microscopic diagnoses in ABC for a three-year period[5] are provided in Table 7.4.

Due to the limited material available through a needle aspirate, there is the risk that sufficient material will be obtained for routine smears but that none will be available for ancillary studies. In critical diagnostic situations, this problem can be obviated by a second ABC, which may be done exclusively for electron microscopy or immunocytochemistry, or both. Sufficient material may appear to have been obtained for electron microscopy, but following embedding and sectioning, the semi-thin plastic sections may reveal only necrotic, normal, or stromal tissue. Blood clots can often be recognized in the glutaraldehyde fixative due to their red-brown coloration, compared with the whitish to grayish color of tumor tissue. ABC specimens from lung are the most frequent source of blood clots. Even this material, however, may be useful, and multiple portions should be embedded if this is the only material received. In some cases, adequate tissue fragments or small clusters of tumor cells may be apparent within the blood clot in the initial semi-thin plastic sections cut for review. In other situations, no tumor tissue may be seen in the original sections, but by cutting deeper in the blocks, multiple fragments that are diagnostically useful may eventually be obtained.

Review of a three-year series of a total of 279 ABC specimens examined by electron microscopy showed that 57 (20.4%) were inadequate specimens.[5] Occasionally, in cases in which a definitive diagnosis was crucial to the management of the patient, a second aspirate usually netted a useful sample. A breakdown of the rate of inadequate specimens in each of the three years (Table 7.5) establishes that this occurrence is fairly constant and is not likely due to inexperience of the physician performing the aspirate, but is an inherent problem due to the nature of the lesion and the relatively small sample obtainable using this technique. A previous analysis of ABC specimens studied by electron microscopy also experienced a 20% failure rate.[3] To gain additional and subtle diagnostic information from ABC specimens that present as diagnostic problems, such a failure rate is acceptable.

The false impression or even misdiagnosis that can result from the sampling of stromal tissues by ABC is well illustrated by the following example. A 78-year-old man presented clinically with a tumor mass within the abdomen. Cytology smears

TABLE 7.5. Frequency of Inadequate Aspiration Biopsy Specimens for Electron Microscopy

|  | Total | Inadequate | Percent |
|---|---|---|---|
| Year 1 | 72 | 12 | 16.7 |
| Year 2 | 118 | 29 | 24.6 |
| Year 3 | 89 | 16 | 18.0 |
| Combined | 279 | 57 | 20.4 |

of an aspiration biopsy of his mass, located in the retroperitoneum, revealed a cellular, spindle-shaped tumor that could not be further classified (Fig. 7.1A). We were misled by the ultrastructural aspects that suggested a diagnosis of malignant fibrous histiocytoma. Two cell types, both loosely organized and not revealing junctions or basal lamina, were identified by electron microscopy. One type was polygonal in shape with marked irregularities of the cell contour due to many looping processes that created intricate surface channels (Fig. 7.1B). These cells, interpreted as histiocytes, had abundant cytoplasm containing lysosomes and lipid droplets. The second population was formed by spindle cells with the features of fibroblasts (Fig. 7.1C). In retrospect, these cells were part of the reactive stromal tissue. Autopsy revealed that the tumor in the retroperitoneum was a malignant lymphoma.

The degree of tumor-cell differentiation can pose diagnostic difficulty. Although undifferentiated neoplasms are often quoted as diagnostic limitations of electron

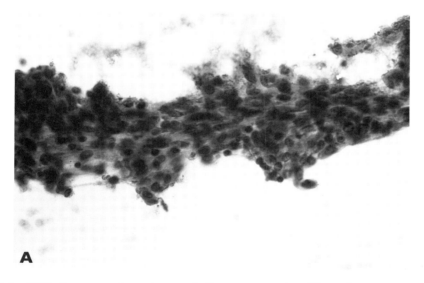

Fig. 7.1. ABC of a retroperitoneal mass. A. Smear preparation. Tissue fragments are highly cellular and composed of spindle cells. × 260.

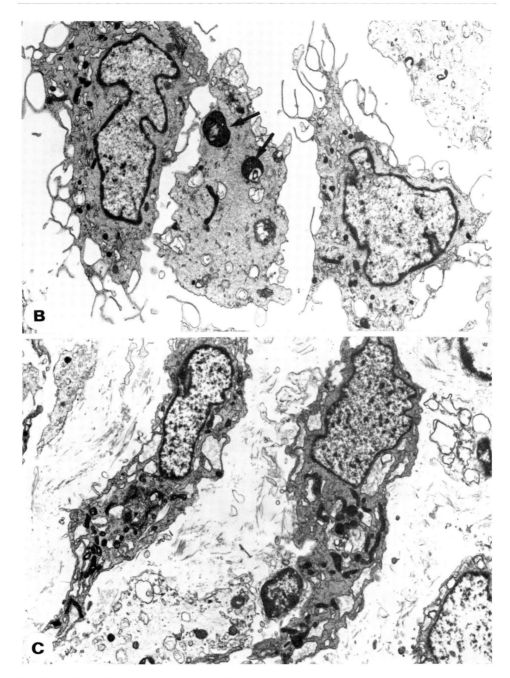

Fig. 7.1. B. Aspirate contains some cells with a complex surface formed by many looping cytoplasmic processes and lysosomes (arrows), features indicating histiocytic differentiation. × 8,000. C. Other cells, spindle-shaped and with a complement of dilated rough endoplasmic reticulum, are fibroblasts, × 6,000.

microscopy, in reality this is not the case. Direct confirmation of the anaplastic nature of a tumor is a positive and clinically helpful categorization. The only limitation in this situation may be the sample size, which may not allow the detection of foci or isolated cells that do express differentiation characteristics. An example would be the embryonal rhabdomyosarcoma in which there are only random tumor cells containing diagnostic features such as a few Z bands associated with a small number of thick and thin filaments or collections of filaments with their rows of aligned ribosomes.[7,8]

Occasionally, well-differentiated tumors may also present a diagnostic problem. The following case illustrates the difficulty distinguishing normal from neoplastic tissue by electron microscopy with confidence. A 10-cm mass was detected in the lower right abdomen of a 28-year-old male. The ABC obtained an adequate sample of cohesive, spindle cells with moderate amounts of fibrillar cytoplasm and elongated, blunt-ended nuclei (Fig. 7.2A). On the basis of these features and the positive immunostaining for vimentin and desmin, the lesion was diagnosed as a smooth-muscle tumor. Only a small group of tumor cells were evident in the semi-thin plastic sections and, ultrastructurally, these cells revealed a rippled, cellular surface, large numbers of cytoplasmic microfilaments, and external lamina (Fig. 7.2B). Because these cells so closely resembled normal smooth-muscle cells, which could have derived from the urinary bladder or a blood vessel, it was not possible to establish a neoplastic diagnosis. When the tumor was resected, electron microscopy confirmed the remarkable degree of cytological differentiation in this smooth-muscle tumor (Fig. 7.2C).

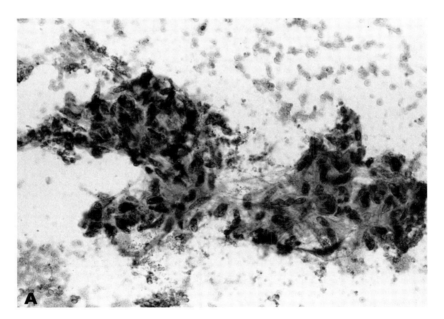

Fig. 7.2. Smooth-muscle tumor, ABC of a pelvic mass. **A**. Smear preparation. Tumor fragments are composed of spindle cells with rather plump, hyperchromatic nuclei with rounded ends. × 320.

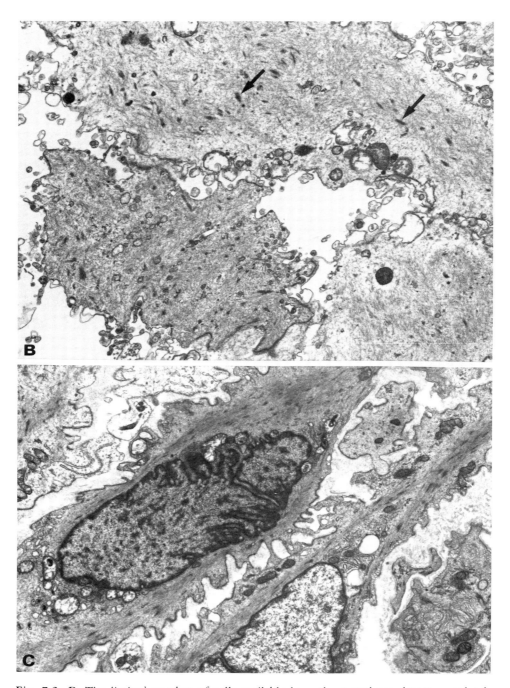

Fig. 7.2. B. The limited number of cells available have the cytoplasm almost completely occupied with microfilaments that have scattered dense bodies (arrows). × 11,000. C. The surgically resected tumor is formed of well-differentiated smooth-muscle cells. Note the wrinkled shape of both nuclear and cell surfaces due to the contraction of the cytoplasmic microfilaments. One tumor cell in (B) has the same feature. × 8,800. (From Dardick I, et al: A quantitative comparison of light and electron microscopic diagnoses in specimens obtained by fine-needle aspiration biopsy. *Ultrastruct Pathol* 15:105–129, 1991.)

Difficulties with the interpretation of ultrastructural findings in ABC specimens can be minimized if they are reviewed with the cytopathologist and are correlated with the light microscopy features. Misinterpretations may still occur.[2]

# REFERENCES

1. Berkman WA, Chowdhury L, Brown NL, Padleckas R: Value of electron microscopy in cytologic diagnosis of fine-needle biopsy. *Am J Radiol* 140:1253–1258, 1983.
2. Brooke PK, Wakely Jr PE, Frable WJ: Utility of electron microscopy as an adjunct in aspiration cytology diagnosis (abstract). *Lab Invest* 64:22A, 1991.
3. Collins VP, Ivarsson B: Tumor classification by electron microscopy of fine needle aspiration biopsy material. *Acta Pathol Microbiol Scand* [A] 89:103–105, 1981.
4. Dabbs DJ, Silverman JF: Selective use of electron microscopy in fine needle aspiration cytology. *Acta Cytol* 32:880–884, 1988.
5. Dardick I, Yazdi HM, Brosko C, Rippstein P, Hickey NM: A quantitative comparison of light and electron microscopic diagnoses in specimens obtained by fine needle aspiration biopsy. *Ultrastruct Pathol* 15:105–129, 1991.
6. Davidson DD, Conces DJ, Goheen M, Clark SA: Comparison of fine needle aspiration and surgical resection specimens for electron microscopy (abstract). *Acta Cytol* 33:735, 1989.
7. Erlandson RA: Cytoskeletal proteins including myofilaments in human tumors. *Ultrastruct Pathol* 13:155–186, 1989.
8. Erlandson RA: The ultrastructural distinction between rhabdomyosarcoma and other undifferentiated "sarcomas." *Ultrastruct Pathol* 11:83–101, 1987.
9. Fisher C, Ramsay AD, Griffiths M, McDougall J: An assessment of the value of electron microscopy in tumor diagnosis. *J Clin Pathol* 38:403–408, 1985.
10. Lombardi L, Orazi A: Electron microscopy in an oncologic institution. Diagnostic usefulness in surgical pathology. *Tumori* 74:531–535, 1988.
11. Mennemeyer R, Bartha M, Kidd CR: Diagnostic cytology and electron microscopy of fine needle aspirates of retroperitoneal lymph nodes in the diagnosis of metastatic pelvic neoplasms. *Acta Cytol* 23:370–373, 1979.
12. Nordgren H, Akerman M: Electron microscopy of fine needle aspiration biopsy from soft tissue tumors. *Acta Cytol* 26:179–188, 1982.
13. Sehested M, Francis D, Hainau B: Electron microscopy of transthoracic FABC specimens. *Acta Pathol Microbiol Immunol Scand* {A} 91:457–461, 1983.
14. Taccagni G, Cantaboni A, Dell'Antonio G, Vanzulli A, Del Mashio A: Electron microscopy of fine needle aspiration biopsies of mediastinal and paramediastinal lesions. *Acta Cytol* 32:868–879, 1988.
15. Williams MJ, Uzman BG: Uses and contributions of diagnostic electron microscopy in surgical pathology: A study of 20 Veterans' Administration Hospitals. *Hum Pathol* 15:738–745, 1984.
16. Willis EJ, Carr S, Philips J: Electron microscopy in the diagnosis of percutaneous fine needle aspiration specimens. *Ultrastruct Pathol* 11:361–387, 1987.
17. Yazdi HM, Dardick I: What is the value of electron microscopy in fine-needle aspiration biopsy? *Diagn Cytopathol* 4:177–182, 1988.

# 8

# Guidelines for Classification of Neoplasms

## UNDIFFERENTIATED OR POORLY DIFFERENTIATED (LOOK-ALIKE) NEOPLASMS

Determining the exact cell of origin of an undifferentiated or poorly differentiated neoplasm is the most difficult and challenging aspect of diagnosis of tumors by aspiration biopsy cytology (ABC). When the tumor is well differentiated or moderately differentiated, proper classification is greatly facilitated. By applying and interpreting well-established cytomorphological criteria, although subjective, most of these neoplasms can be classified. Experience in cytopathology certainly improves the accuracy of classification in better-differentiated neoplasms. The cytological appearance in various poorly differentiated or undifferentiated tumors, however, may be similar or identical and defy classification even by cytopathologists with considerable experience.

Multimodal investigations such as electron microscopy or detection of specific tumor markers, or both, enhance the accuracy of cytological interpretations and may result in a precise diagnosis. These techniques facilitate further classification of individual tumors in certain cases. Such diagnostic methods, however, should not force the classification if the evidence is scanty. For example, cytokeratin expression in an occasional cell or a few primitive cell junctions does not necessarily mean epithelial differentiation. There are certain tumors that are undifferentiated, indeed, a perfectly satisfactory diagnosis, and cannot be further classified despite all the modern diagnostic tools. Failure to recognize this fact may result in a wrong interpretation and undesirable consequences.

For obvious reasons, it is neither practical nor economical to perform an ultrastructural survey and immunocytochemical studies using multiple antibodies in all cases. One has to formulate a list of differential diagnoses based on light microscopy, clinical data (i.e., patient history, age, sex, presence of tumor elsewhere, and

TABLE 8.1. Undifferentiated Neoplasms—Patterns of Immunocytochemical Staining

| Support the Diagnosis of | Cytokeratin | Vimentin | LCA | HMB-45 |
|---|---|---|---|---|
| Carcinoma | + | − | − | − |
|  | +* | + | − | − |
| Malignant lymphoma | − | + | + | − |
|  | − | − | + | − |
| Malignant melanoma | − | + | − | + |
|  | − | − | − | + |
| Mesenchymal tumors | −* | + | − | − |

*With the exception of some mesenchymal tumors, including synovial sarcoma, epithelioid sarcoma, and smooth muscle neoplasms.
LCA = leukocyte common antigen.

history of previous tumor), and statistical probabilities. A panel of complementary antibodies should be used to address the possibilities. When there is no clue to the possible origin of the tumor cells, then it is better to start with a simple diagnostic panel consisting of antibodies to cytokeratins, vimentin, leukocyte common antigen (LCA), and melanoma-specific antibody (HMB-45) (Table 8.1). They can distinguish carcinomas, malignant lymphomas, malignant melanomas, and mesenchymal tumors or shorten the list of differential diagnoses by excluding some of the possibilities. Depending on the results, one may proceed to more "specific" antibodies in a particular sequence using some form of algorithm.

Sometimes it is only possible to categorize neoplastic cells, using descriptive terms, into one of four groups, such as large polygonal cells, pleomorphic cells, spindle cells, or small cells. The aspirate can then be investigated using a combined approach of initial diagnostic panels and algorithms of the type shown in Figures 8.1 to 8.4. Interpretation of the staining patterns should be tempered by awareness of the potential limitations and pitfalls of each antibody and expression of the same antigen by a variety of tumor types (see Chapters 4 and 5). It is the presence or absence of diagnostic subcellular features representing many of the organelles or synthetic cytoplasmic products outlined in Chapter 6 that account for the immunocytochemical findings and are equally useful for ultrastructural diagnosis. Although electron microscopy and immunocytochemistry are complementary diagnostic methods, electron microscopy has the advantage of being able to detect quantities of such organelles or patterns of cellular or tissue differentiation not detectable by light microscopy.

Assessment of diagnostic problems in ABC specimens by electron microscopy can also be systematic. Review of the toluidine blue–stained semi-thin plastic sections during selection of a block or blocks for thin sectioning usually permits a preliminary placement of the tumor into one of the large-, pleomorphic-, small- or spindle-cell groups. Parallel to the application of a panel of antibodies for diagnostic purposes, the presence or absence of certain key organelles or cellular characteristics provides a diagnosis of at least carcinoma, lymphoma, sarcoma, or melanoma in most cases (Table 8.2). Similarly, review and analysis of the ultrastructural cytological and organizational features of an undifferentiated or poorly differentiated

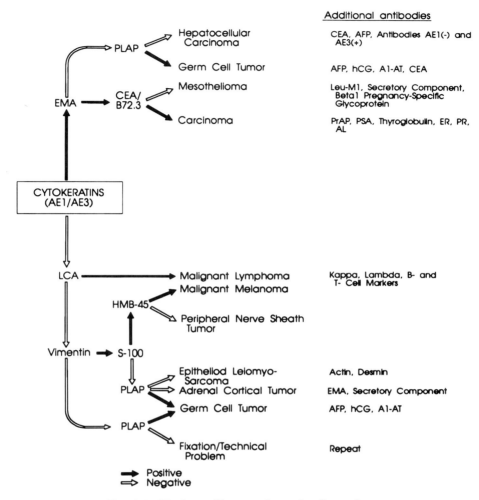

Fig. 8.1. Workup of large, polygonal-cell neoplasms.

tumor should allow a more specific diagnosis of large or pleomorphic- (Table 8.3), spindle- (Table 8.4), or small- (Table 8.6) cell tumors. The application and general usefulness of electron microscopy in poorly differentiated and "look-alike" tumors is exemplified in a study of 265 examples (14 of which were ABC specimens) of metastatic tumors of unknown primary origin by Hammar and associates.[66]

## Large/Polygonal Cell Neoplasms

Neoplasms with large and polygonal cells include anaplastic or poorly differentiated carcinomas (originating in such organs as lung, ovary, pancreas, kidney, liver, and adrenal gland); malignant melanoma; large-cell malignant lymphoma; germ-cell tumor; malignant mesothelioma (epithelial type); and certain sarcomas, such as epithelioid sarcoma, epithelioid leiomyosarcoma, and clear-cell sarcoma. Using an initial

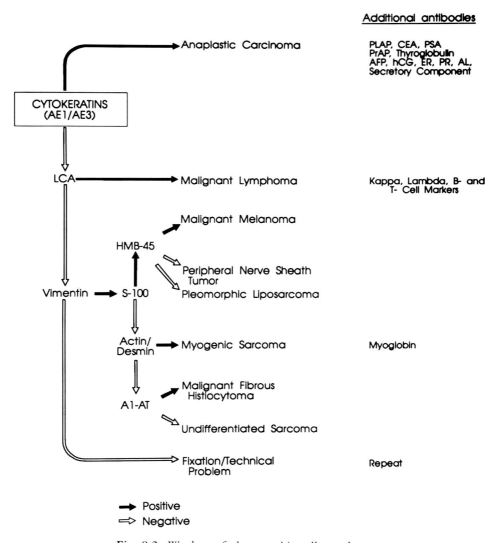

Fig. 8.2. Workup of pleomorphic-cell neoplasms.

panel of cytokeratins, LCA, vimentin, and epithelial membrane antigen, a workup using the algorithm illustrated in Figure 8.1 will help to characterize the tumor. Ultrastructural characteristics of some of the more common large or polygonal cell tumors (these also can be useful in the pleomorphic cell group) posing diagnostic problems are summarized in Table 8.3.

## Pleomorphic-Cell Neoplasms

Pleomorphic neoplasms include pleomorphic variants of certain sarcomas (e.g., malignant fibrous histiocytoma, rhabdomyosarcoma, liposarcoma, and leiomyosar-

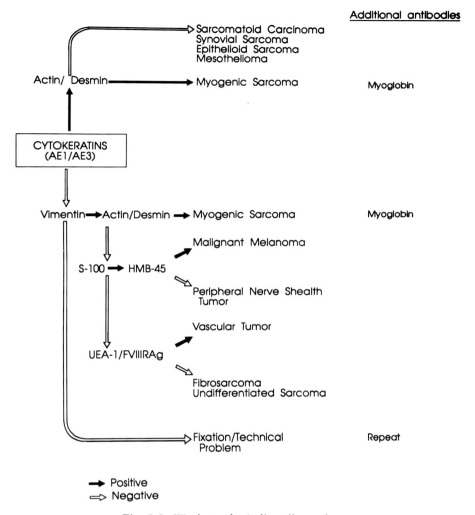

Fig. 8.3. Workup of spindle-cell neoplasms.

coma), malignant melanoma, anaplastic carcinomas, and germ-cell tumors. This type of neoplasm can be worked up using an initial panel of antibodies to cytokeratins, LCA, and vimentin and using the algorithm pathway illustrated in Figure 8.2., as well as the electron microscopic characteristics outlined in Table 8.3.

## Spindle-Cell Neoplasms

Spindle-cell neoplasms include most sarcomas, sarcomatoid carcinomas, sarcomatoid and desmoplastic mesotheliomas, and malignant melanoma. In this group, a panel of antibodies to cytokeratins, vimentin, actin, and desmin is very useful (Fig. 8.3). Electron microscopy (Table 8.4) also has a valuable role in distinguishing the various spindle-cell neoplasms, whether they are sarcomas or epithelial

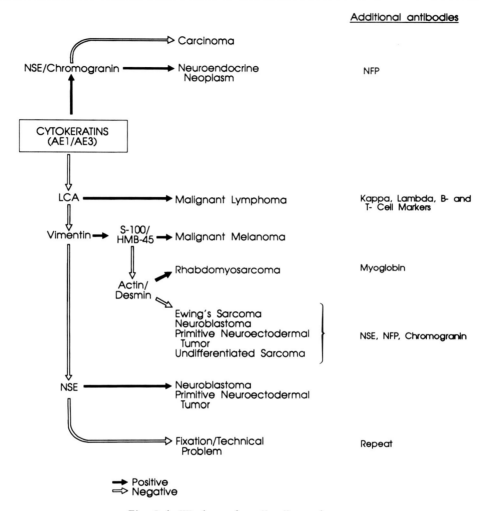

Fig. 8.4. Workup of small-cell neoplasms.

tumors. This distinction is illustrated in Figure 8.5 by the ABC of an anterior mediastinal mass.

A chest radiograph and a computed tomographic (CT) scan showed that this 72-year-old woman had an anterior mediastinal mass. On ABC, this mass proved to be a spindle-cell tumor (Fig. 8.5A), but differentiation between a sarcoma and a thymoma was difficult because a lymphocytic component was not identifiable in the smears. A portion of the ABC was examined by electron microscopy (Fig. 8.5B,C), and the numerous well-developed and intermediate filament-associated desmosomes, scattered tonofilaments, and many long, narrow cellular processes, often outlined by a basal lamina, eliminated a sarcoma and established the diagnosis of a spindle-cell thymoma. The lack of intercellular bridges coupled with widened extracellular spaces, as illustrated in Figures 6.3 and 6.11B, ruled out a spindle-cell

TABLE 8.2. Ultrastructural Features Useful in Distinguishing Undifferentiated Neoplasms

| Feature | Carcinoma | Malignant Lymphoma | Malignant Melanoma | Mesenchymal Neoplasms |
|---|---|---|---|---|
| Cell junctions | + | − | − | ± |
| Basal lamina | + | − | − | ± |
| Tonofilaments | ± | − | − | −* |
| Thin filaments | − | − | − | ± |
| Thick filaments | − | − | − | ± |
| Gland lumens | ± | − | − | − |
| Microvilli | ± | − | − | − |
| Melanosomes | − | − | + | − |
| Mucin granules | ± | − | − | − |
| Organelle-rich | + | − | + | + |
| Organelle-poor | − | + | − | − |

*Except for synovial and epithelioid sarcomas
+ = present in majority of cases; ± = present in some cases; − = usually absent.

squamous cell carcinoma. Immunocytochemistry and electron microscopy of a spindle-cell carcinoid and a metastatic leiomyosarcoma are illustrated in Figures 8.8 and 8.14, respectively.

## Small-Cell Neoplasms

Small-cell neoplasms include small-cell carcinoma, neuroendocrine neoplasms, malignant lymphoma, malignant melanoma, Ewing's sarcoma, neuroblastoma, rhabdomyosarcoma, and malignant peripheral neuroectodermal tumor. A panel comprising antibodies to cytokeratins, LCA, vimentin, neuron-specific enolase, chromogranin, actin, and desmin will often distinguish between the most common tumors of this group (Fig. 8.4; Table 8.5). The principal diagnostic electron microscopic features for small-cell tumors are provided in the "Small Cell Neoplasms" section and in Table 8.6.

# MALIGNANT MELANOMA

Malignant melanomas, particularly the amelanotic type, are great simulators among poorly differentiated neoplasms in cytological[147] and in surgical pathology[119] specimens. When all the cytological features, such as dispersed large-pleomorphic cells with abundant cytoplasm, eccentric nuclei with binucleated/multinucleated forms, and intranuclear cytoplasmic inclusions are present in combination, the diagnosis of malignant melanoma can be suggested.[82] However, none of these features are specific individually. In addition, malignant melanoma cells may appear polygonal, have abundant vacuolated or clear cytoplasm, or become spindled or composed of small cells resembling poorly differentiated carcinomas or malignant lymphomas.

TABLE 8.3. Comparative Diagnostic Ultrastructural Features of Certain Large/Pleomorphic Cell Neoplasms

*Adenocarcinoma*
- Intercellular lumens
- Microvilli, some with filamentous rootlets
- Apical junctional complexes
- Desmosomes
- Tonofilaments
- Basal lamina
- Secretory granules (mucin)
- Glycocalyx
- Prominent rough endoplasmic reticulum

*Lymphoma*
- High nucleus/cytoplasm
- Many polyribosomes, but few other organelles
- Nuclear shape variability
- Nuclear bullae
- Variable filopodia but no true microvilli
- Lack true junctions
- Lack basal lamina
- Lack obvious cytoplasmic filaments
- Cells loosely organized

*Amelanotic melanoma*
- Typical and aberrant melanosomes, some at least partly melanized
- Numerous cytoplasmic organelles (mainly endoplasmic reticulum and Golgi complexes)
- Small membrane-bound vesicles, often numerous
- Lack intercellular junctions
- Loosely arranged cells

*Squamous cell carcinoma*
- Desmosomes with short tufts of intermediate filaments
- Tonofilament bundles
- Focal widened intercellular spaces bridged by cytoplasmic processes bearing junctions parallel or perpendicular to the cell surface
- Limited rough endoplasmic reticulum

*Paraganglioma*
- Neurosecretory-type cytoplasmic granules (approximately 100–250 nm in diameter)
- Occasional microtubules
- Few intercellular junctions
- Focal or continuous basal lamina
- Diffuse intermediate-sized cytoplasmic filaments

*Histiocytic neoplasms*
- Loosely arrange cells
- Lack cell junctions
- Numerous and variably shaped fine-surface projections
- Variable lysosomes
- Prominent endoplasmic reticulum and Golgi units
- No true microvilli
- Lack external lamina
- No obvious filaments
- Birbeck granules in histiocytosis X

The aspirate may have gland-like structures or signet-ring–type cells resembling adenocarcinoma or have a myxoid background simulating mesenchymal neoplasms.

Clinically, most aspirates are obtained from patients with a documented history of malignant melanoma who then develop an enlarged lymph node or a mass in soft tissue, lung, or abdominal organs or sites. The main difficulty is to distinguish a metastatic melanoma from a new primary tumor that is not a melanoma. Of course, the history may not be known to the radiologist or the cytopathologist. The second situation is when a poorly differentiated neoplasm is seen in an ABC specimen with no history of a known primary neoplasm. Amelanotic malignant melanoma should always be included in the differential diagnosis. In both situations,

TABLE 8.4. Comparative Diagnostic Ultrastructural Features of Certain Spindle Cell Neoplasms

*Fibrosarcoma*
    Cells tend to be separated
    Prominent rough endoplasmic reticulum, often dilated by granular material
    No intercellular junctions
    Lack basal lamina
    Intercellular collagen and proteoglycans

*Smooth-muscle tumor*
    Cells closely and often regularly associated
    Rectangular to angular cell outlines forming a mosaic-like growth pattern
    Cytoplasmic filaments (5–8 nm)
    Linear, dense bodies among the filaments
    Cell membrane–associated dense plaques
    Micropinocytotic vesicles
    Basal lamina
    Few tight or gap junctions

*Spindle-cell carcinoma*
    Closely associated cells
    Desmosomes or desmosomelike junctions
    Few to many tonofilaments
    Focal basal lamina
    Minimal endoplasmic reticulum and small Golgi zones

*Nerve sheath tumor*
    Intimately associated cells
    Narrow, markedly elongated cytoplasmic processes often in a parallel alignment
    Complex entwinning processes often circling collagen or processes (mesaxon-like)
    Small intercellular junctions
    Basal lamina, often markedly reduplicated

*Malignant fibrous histiocytoma*
    Mixture of cells, including fibroblasts, myofibroblasts, histiocytes, giant cells, and undifferentiated cells
    Cells separated
    Collagen fibers/proteoglycans forming the stroma
    No basal lamina or junctions
    No cytoplasmic filaments seen

*Osteogenic sarcoma*
    Cells separated by collagen- and proteoglycan-rich stroma
    Tiny matrix vesicles among collagen fibers
    Mineralization of collagen in the form of hydroxyapatite crystals
    Much-dilated endoplasmic reticulum
    Rarely basal lamina

accurate diagnosis may be difficult or impossible on routinely stained smears or cell block sections, and may result in inappropriate treatment or a more invasive surgical procedure. In most cases, the classification will be facilitated by application of immunocytochemical or electron microscopic study, or both.

When the diagnosis of malignant melanoma is suspected, we perform a diagnostic panel consisting of antibodies to cytokeratins, vimentin, LCA, and both S-100 protein and HMB-45. S-100 protein is strongly expressed in almost all malignant melanomas in a majority of the tumor cells (see the "S-100 Protein" section in Chapter 5). Expression of HMB-45 is relatively specific for malignant melanoma, but it is not as sensitive as S-100 protein and the staining may be focal in nature (see the "HMB-45" section in Chapter 5). When both antigens are expressed and the neoplastic cells are negative for cytokeratins and LCA an unequivocal diagnosis of malignant melanoma can be made (Fig. 8.6). Some malignant melanomas express

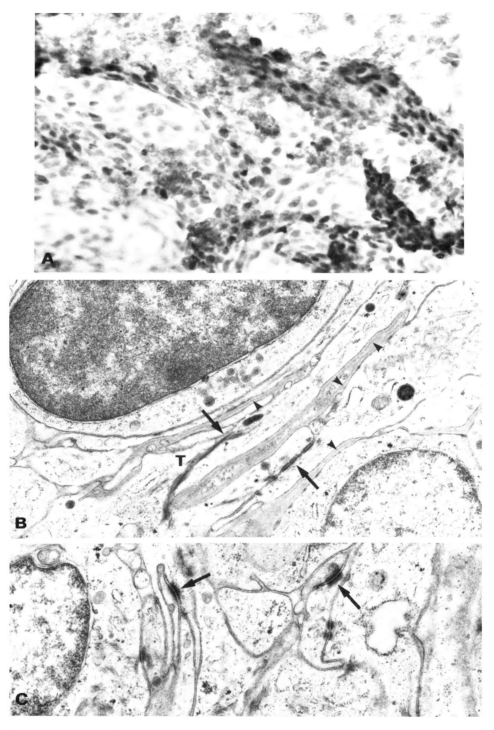

Fig. 8.5. Spindle cell thymoma, ABC of the anterior mediastinum. **A.** Fragments of closely arranged, uniform-appearing spindle cells in the smear preparation. × 160. **B.** The spindle-shaped tumor cells have many processes, are joined by desmosomes (arrows), contain tonofilaments (T), and are lined by basal lamina (arrowheads). × 6,000 **C.** The multiplicity of cellular processes are connected by well-developed desmosomes (arows), from which project intermediate filaments. × 16,000.

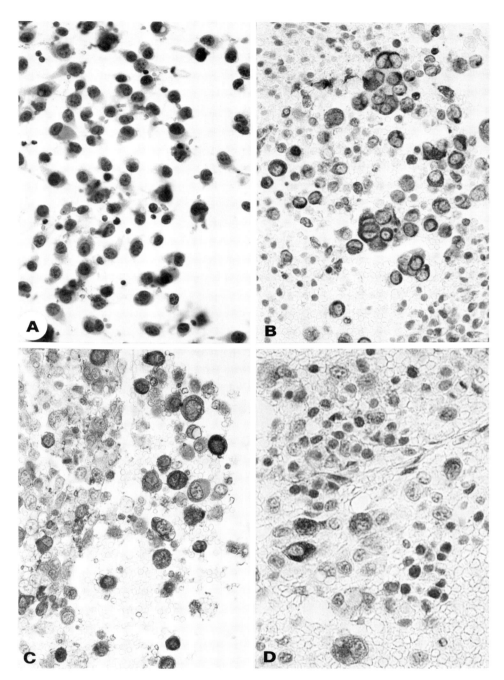

Fig. 8.6. Malignant melanoma, ABC of a retroperitoneal mass. This 63-year-old woman presented with a retroperitoneal periaortic mass, clinically suggestive of a malignant lymphoma. The patient also had a history of malignant melanoma 18 years previously. A. Smear preparation. Note dispersed large cells with abundant cytoplasm and eccentric nuclei suggestive of melanoma. × 260. B.–D. The neoplastic cells strongly expressed vimentin, S-100 protein, and HMB-45, supporting the diagnosis of a melanoma. Stains for LCA and cytokeratins were negative. (B) Vimentin, × 260; (C) S-100 protein, × 260; (D) HMB-45, × 416.

cytokeratins,[10,107,191] which limits the diagnostic value of cytokeratin positivity in an undifferentiated neoplasm and underscores the importance of using a panel of antibodies rather than a single one. Neuron-specific enolase is expressed in a majority of malignant melanomas,[90] but it has a wide distribution in nonmelanomatous tumors and is therefore nonspecific.

In electron microscopy, the hallmark for the diagnosis of malignant melanoma is the identification of the melanosome (see the "Melanosome" section in Chapter 6). This diagnostic criterion is illustrated in Figure 8.7. An 84-year-old man presented with multiple tumor masses in kidney, spleen, liver, and various muscle sites. Lymphoma was suspected, but an ABC specimen obtained from a mass in the buttock showed individualized tumor cells with pigment granules in the cytoplasm (Fig. 8.7A). There was no history provided of a known primary malignant melanoma. Immunostaining with HMB-45 and S-100 protein was positive, and electron microscopy of a portion of the ABC revealed classic stages II, III, and IV melanosomes (Fig. 8.7B).

As outlined in the "Melanosome" section in Chapter 6, the ultrastructural features of melanosomes synthesized by the tumor cells in amelanotic malignant melanomas are not always typical. Referred to as aberrant melanosomes,[45,53,57,102] they are still diagnostically useful. In fact, their recognition is essential for the diagnosis of many amelanotic malignant melanomas (Fig. 6.26). In a proportion of cases of needle aspiration and surgical pathology biopsies few if any classic melanosomes can be identified, and no history of a previous excision of a malignant melanoma may be provided. It also should be appreciated that in a small number of metastatic amelanotic malignant melanomas, typical or atypical by light microscopy and even positive with S-100 protein or HMB-45, or both, no melanosomes (either usual or aberrant) may be synthesized and therefore cannot be found ultrastructurally. It is essential to also be aware of the spectrum of nonmelanocytic tumors in which melanasomes have occasionally been identified; these are effectively tabulated by Szpak and associates.[161]

# NEUROENDOCRINE NEOPLASMS

Neuroendocrine neoplasms comprise a heterogeneous group of relatively common tumors that arise from many organs or sites throughout the body. Gould and De-Lellis[60] defined a neuroendocrine neoplasm as "one whose predominant cells display convincing features of neuroendocrine differentiation either by electron microscopy or by immunohistochemical and/or biochemical techniques." A neuroendocrine neoplasm such as carcinoid tumor may be predominantly or exclusively composed of neuroendocrine cells (Fig. 8.8). Many nonneuroendocrine neoplasms, however, such as carcinomas of colon, pancreas, breast, and endometrium, may have isolated or clusters of neuroendocrine cells.[9,25,68,71,94,121] The presence of a neuroendocrine subpopulation in a predominantly exocrine carcinoma may be associated with a more aggressive clinical behavior.[71]

With the present understanding of neuroendocrine neoplasms, they can be divided into three groups: group A—neural-type, such as neuroblastomas, paragangli-

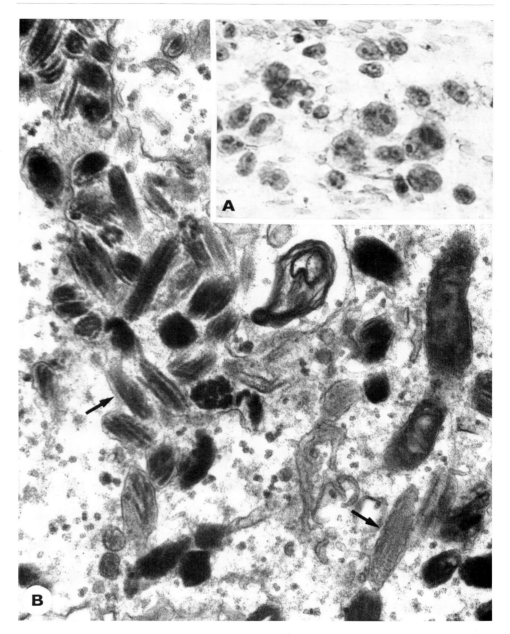

Fig. 8.7. Malignant melanoma, ABC of a mass in the buttock. **A.** Smear preparation. Separated, generally round cells with an obvious nucleolus and a cytoplasm containing pigment granules indicated a diagnosis of malignant melanoma. × 416. **B.** The presence of numerous, typical stage II (arows) to IV melanosomes confirmed the diagnosis. × 58,000.

omas, and pheochromocytomas; group B—epithelial-type, such as carcinoid tumors and neuroendocrine adenomas and carcinomas; and group C—miscellaneous neuroendocrine neoplasms that show a predominantly mesenchymal differentiation and vimentin expression.[59,60] Neuroendocrine neoplasms of epithelial lineage are more commonly encountered in ABC specimens, and by definition they express cytokeratin polypeptides (usually low molecular weight) and desmoplakin. Coexpression of cytokeratins and neuroendocrine markers such as synaptophysin or chromogranin (and even calcitonin in the case of medullary carcinoma of the thyroid[141,149]) highly suggests this class of neuroendocrine neoplasms. Neuroendocrine neoplasms of neural lineage express neurofilament proteins. Some epithelial neuroendocrine neoplasms, however, such as islet-cell tumors, bronchial carcinoids, and neuroendocrine carcinomas of skin, frequently express neurofilament proteins.[59,60] Expression of neurofilament proteins and neuroendocrine markers such as chromogranin and the absence of cytokeratins support the neural-type of neuroendocrine neoplasms.[61,73] Group C tumors, such as medulloblastomas, some malignant melanomas, and primitive neuroectodermal tumors, express vimentin predominantly but not exclusively.[60]

Cytological diagnosis and precise classification of neuroendocrine neoplasms are mandatory for optimal management and therapy as well as prognosis, especially when histological material is not available. Depending on the cytomorphology, the differential diagnoses may include malignant lymphoma, Ewing's sarcoma, embryonal and alveolar rhabdomyosarcoma, and small- and large-cell nonneuroendocrine carcinoma. Demonstration of neurosecretory granules, the key ultrastructural feature indicating neuroendocrine differentiation, is considered a good criterion (see the "Neuroendocrine Granule" section in Chapter 6). The number of neurosecretory granules usually reflects the degree of tumor-cell differentiation, and they may be difficult to find in poorly differentiated lesions such as small-cell neuroendocrine carcinoma of lung. In this circumstance, the granules are best looked for in the cytoplasmic extensions wedged between adjacent tumor cells (Figs. 6.22, 6.23B, 8.8G).

There has been a continuous search to find a "generic," sensitive, and specific marker for neuroendocrine differentiation; this search has resulted in numerous commercially available antibodies. Neuron-specific enolase is an extremely sensitive marker for neuroendocrine differentiation; however, it is not specific and may be expressed in a number of different nonneuroendocrine neoplasms;[15,59,139,174,187] therefore, it cannot be used in isolation.[30] Chromogranin is specific for neuroendocrine neoplasms,[25,78,96,97,126,139,181] but the intensity of staining and the percentage of positive tumor cells are highly dependent on the number of neurosecretory granules.[188] In this situation, chromogranin is not a sensitive marker for poorly differentiated neuroendocrine neoplasms. This is the group in which special studies are particularly needed and where electron microscopy can be especially valuable.

Synaptophysin is a relatively recently introduced neuroendocrine marker that is reported to be present in most neuroendocrine neoplasms that have been examined.[59,104] It has been consistently absent, so far, in nonneuroendocrine neoplasms.[59,104] According to the literature, synaptophysin appears to be a specific and sensitive marker for neuroendocrine differentiation,[59,61] but in our experience it is not a sensitive marker for neuroendocrine neoplasms of the lung, and in 50% of the positive cases the staining is weak or focal. Neurofilament protein expression is seen in neural and some epithelial neuroendocrine neoplasms.[67,88,91,108,113,115,118,172] Neurofilament proteins may be absent in many neuroendocrine neoplasms; therefore, as a neuroendocrine marker it is of limited value in diagnosis by ABC. Other significant neuroendocrine markers include bombesin, which may not be specific,[139] and Leu-7, which may be expressed by a variety of neuroendocrine neoplasms,[23] but it is also not specific.

Neuroendocrine neoplasms may express a variety of polypeptides, such as serotonin, calcitonin, insulin, glucagon, somatostatin, gastrin, adrenocorticotropic hormone, vasoactive intestinal polypeptide, leu-enkephalin, alpha-melanocyte stimulating hormone, and substance P.[17,60] Many neuroendocrine neoplasms synthesize more than one of these polypeptides, which may be very useful in particular clinical contexts.[59,60,181] However, their expression is variable. Some clinically silent or nonsecreting neoplasms may show focal or no staining for any of these antigens,[95] and none can be used as panneuroendocrine marker. These markers are therefore of limited value in diagnostic cytopathology.

Although the neuroendocrine neoplasm is a major differential diagnosis in ABC of many organs or sites, this problem is encountered more frequently in ABC specimens of certain organs, such as lung, pancreas, and liver. Lung neoplasms are a complex group that may display multiple pathways of differentiation.[139] Clinicians have been mainly interested to know if the tumor is composed of small or large cells. Not all small-cell carcinomas of lung, however, are demonstrably neuroendocrine in nature,[163] and some represent small-cell squamous or adenocarcinomas. In contrast, a significant proportion of pulmonary large-cell carcinomas have neuroendocrine features demonstrated immunocytochemically,[92,93,175] which might have therapeutic and prognostic implications.[92,117] Therefore, precise classification of pulmonary neuroendocrine neoplasms (carcinoid tumor, atypical carcinoid tumor, small-cell neuroendocrine carcinoma, and large-cell carcinoma with neuroendocrine features), and distinction from nonneuroendocrine neoplasms is extremely important. Both immunocytochemistry and electron microscopy are useful in establishing neuroendocrine differentiation (Figs. 8.8, 8.9). In our experience with ABC specimens, all neuroendocrine neoplasms express neuron-specific enolase, whereas chromogranin is expressed in 76% of cases. In approximately half of the chromogranin-positive cases, the staining is weak or focal in nature. Positive staining with both antibodies or convincing staining with chromogranin alone is required to support neuroendocrine differentiation. Our results with synaptophysin are disappointing. Only 40% of neuroendocrine neoplasms express synaptophysin, which is weak or focal in half of the cases. We have noted that neuroendocrine differentiation was confirmed ultrastructurally in 89% of pulmonary neuroendocrine neoplasms. Electron microscopy is particularly sensitive in small-cell neuroendocrine carcinomas. Absence of neurosecretory granules or lack of expression of neuroendocrine markers in ABC specimens does not totally rule out the possibility of neuroendocrine differentiation. Furthermore, these techniques are of limited use in discriminating atypical carcinoid tumor from combined small-cell and large-cell carcinoma or in differentiating a combined tumor from small-cell carcinoma. In such cases, one should rely on light microscopic cytomorphological findings.

Pancreatic endocrine neoplasms are relatively uncommon and are characterized in ABC specimens by single cells and small fragments of cells with granular cytoplasm and round-to-oval uniform nuclei. The neoplastic cells generally express

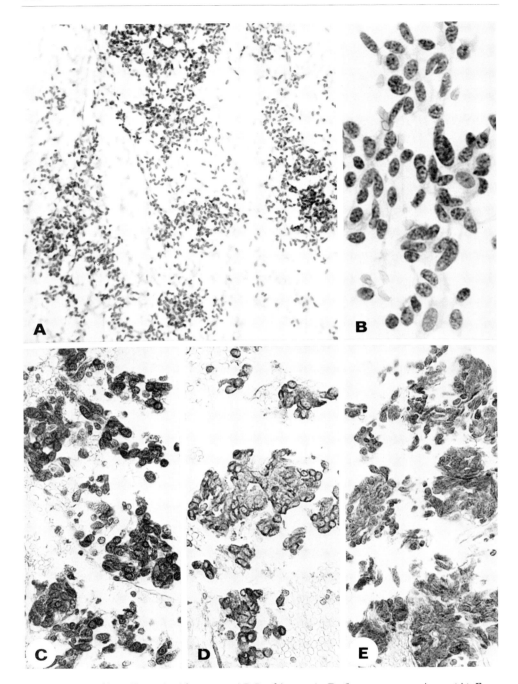

Fig. 8.8. Spindle-cell carcinoid tumor, ABC of lung. A.,B. Smear preparations, (A) Fragments of compactly organized uniform cells suggestive of carcinoid tumor. × 104. (B) Note elongated nuclei. × 416. C.–E. the neoplastic cells express neuron-specific enolase (NSE), chromogranin, and synaptophysin, supporting the diagnosis of carcinoid tumor. (C) NSE, × 260; (D) Chromogranin, × 260. (E) Synaptophysin, × 260.

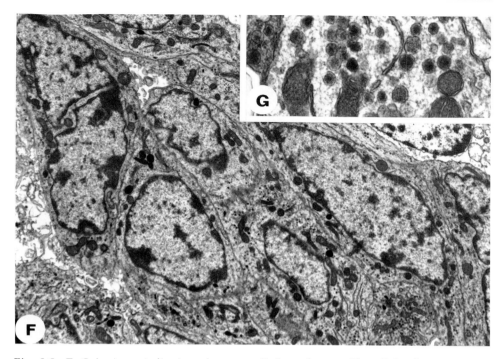

Fig. 8.8. F. Cohesive spindle-shaped tumor cells have few specific cellular features except for focal accumulations of small, secretory granules (arrows). × 6,000. G. Round, regular, and uniformly sized secretory granules have a central darkly staining region and a narrow peripheral lucent zone. × 31,500. (F,G) from Dardick I, Yazdi HM, Brosko C, Rippstein P, Hickey NM: A quantitative comparison of light and electron microscopic diagnoses in specimens obtained by fine-needle aspiration biopsy: *Ultrastruct Pathol* 15:105–129, 1991. Used with permission.

neuron-specific enolase, but chromogranin or synaptophysin may not be expressed. Shaw and colleagues[146] demonstrated chromogranin expression in 5 of the 10 islet-cell neoplasms examined. The neuroendocrine nature of the tumor can also be confirmed by the demonstration of neurosecretory granules ultrastructurally, or by expression of serotonin, insulin, glucagon, or a variety of other polypeptides immunocytochemically.[77,154]

In Chapter 6, the "Neurosecretory Granule" section detailed the electron microscopic characteristics of neurosecretory granules, the various types of tumor in which the ultrastructural identification of these granules can be diagnostically useful (Table 6.4), and the considerable frequency of neuroendocrine neoplasms in ABC specimens of the lung and liver (Table 6.5).

Diagnostic light and electron microscopic features of carcinoids, atypical carcinoids, and small-cell carcinomas of the lung have been studied in ABC specimens,[162] as have pancreatic endocrine[154] and Merkel-cell tumors.[133] The diagnostic usefulness of identifying neurosecretory granules in primary spindle-cell tumors, such as a spindle cell form of carcinoid, is shown by the ABC of a 2-cm discrete nodule

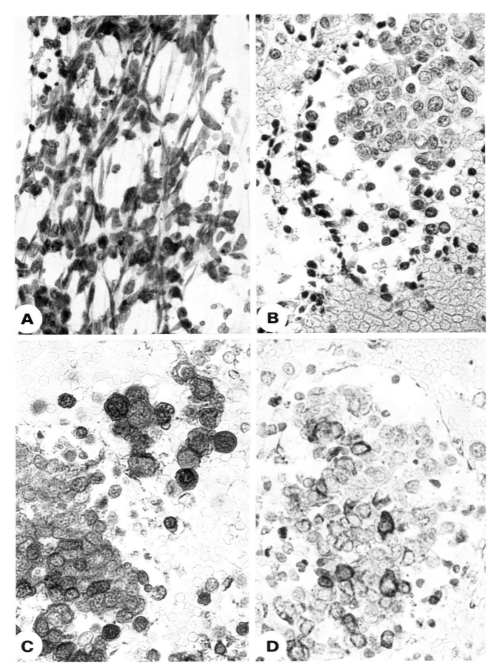

Fig. 8.9. Small-cell neuroendocrine carcinoma, ABC of lung. **A.** Smear preparation. Fragments of tumor cells with hyperchromatic pleomorphic nuclei and nuclear streaks. Occasional pyknotic nuclei are seen. × 260. **B.** Cell block section. A fragment of better preserved larger cells with a moderate amount of cytoplasm. Several tumor cells with pyknotic nuclei are again evident. × 416. **C.** Strong immunocytochemical positivity for neuron-specific enolase (NSE). × 416. **D.** A few tumor cells express chromogranin. × 416. Expression of NSE and chromogranin supports the diagnosis of neuroendocrine carcinoma.

in the upper lobe of the left lung in a 52-year-old man. Tissue fragments were composed of compactly organized, uniform spindle cells (Fig. 8.8A,B), suggestive of a spindle-cell carcinoid tumor. The main portions of cytoplasm of the spindle cells contained a few, small secretory granules (Fig. 8.8F), but more numerous typical neurosecretory granules were present in some of the cytoplasmic extensions (Fig. 8.8G) confirming the light microscopic impression of the ABC smears. The tumor was subsequently resected and confirmed as a spindle-cell carcinoid.

The neuroendocrine nature of metastatic lesions can also be detected ultrastructurally. A 59-year-old woman complained of abdominal bloating, and ultrasound showed multiple masses in the liver, one of which underwent needle aspiration biopsy (NAB). The smear preparation showed relatively small neoplastic cells that were cytologically compatible with a neuroendocrine carcinoma (Fig. 8.10A). By electron microscopy, the tumor cells were closely associated and had a moderate amount of cytoplasm (Fig. 8.10B). The principal feature was the many small but uniformly sized neurosecretory-type granules tending to aggregate in one part of the cytoplasm (Fig. 8.10C). Although these granules had no specific characteristics, their numbers and the general cellular features suggested that either the pancreas or the gastrointestinal tract might be the primary source. Five months later, the patient died of metastatic disease, and autopsy revealed a 2-cm diameter islet-cell tumor in the tail of the pancreas. There were multiple metastases to the liver and peripancreatic lymph nodes.

In certain types of tumor, the combination of neuroendocrine granules and intermediate filaments may have specific diagnostic relevance. In both carcinoids and Merkel-cell tumors (particularly the latter), but only very infrequently in small-cell anaplastic carcinomas of lung, round, discrete accumulations of intermediate filaments, referred to as globular filamentous bodies,[57] may be seen adjacent to the nucleus in electron micrographs.[14,41,57] This feature produces a punctate spot of intense cytoplasmic positivity with antibodies to low-molecular-weight cytokeratins, and in a smaller proportion of Merkel-cell tumors with antibodies to neurofilament proteins.[14,41,134,148,153] Such filamentous aggregates have been observed in ABC of primary, recurrent, and metastatic neuroendocrine carcinomas of the skin (Merkel-cell tumors), both ultrastructurally and immunocytochemically.[41] This is a useful marker to distinguish a primary small-cell carcinoma of skin from non-Hodgkin's lymphoma and metastatic small-cell anaplastic carcinoma of lung.

One of the principal values of electron microscopy in the diagnosis of neuroendocrine lesions in ABC is its exquisite sensitivity. The usual panel of immunocytochemical markers of neuroendocrine differentiation may be negative because the quantity of these peptides may be insufficient to produce a detectable signal at the light microscopic level. Ultrastructurally, however, even small numbers of neurosecretory granules are apparent. In fact, there appears to be a close relationship between the number of granules and the degree of detection of chromogranin immunohistochemically.[188] The location of these granules, particularly in poorly differentiated neuroendocrine neoplasms, has practical value diagnostically. These granules may be found very infrequently in the main portion of tumor-cell cytoplasm, and a search should be made for the sections of cytoplasmic extensions wedged between adjacent tumor cells because these extensions may contain a group of small, uniformly sized, and typical-appearing secretory granules (Fig. 6.22, 6.23B, 8.18G).

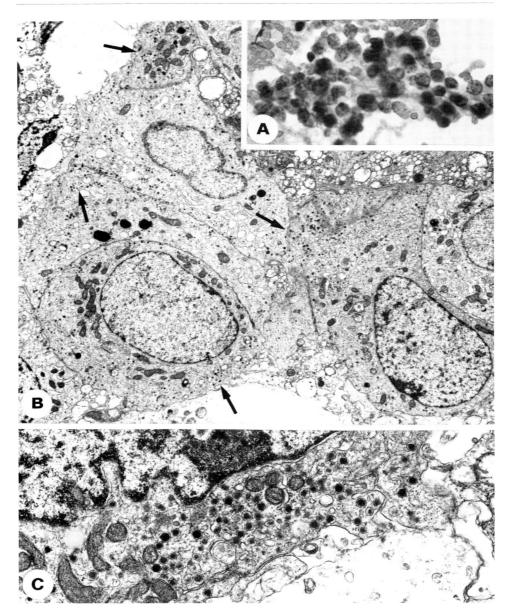

Fig. 8.10. Metastatic neuroendocrine carcinoma, ABC of liver. **A.** Fragment of cohesive, relatively small uniform-appearing tumor cells suggestive of a neuroendocrine neoplasm. × 416. **B.** At low magnification, the compactly arranged tumor cells have many darkly staining granules (arrows) that tend to aggregate in certain regions of the cytoplasm. × 3,400. **C.** The neuroendocrine nature of the granules, at higher magnification, confirms the light microscopic impression of a metastatic neuroendocrine carcinoma. × 14,300.

# SMALL CELL NEOPLASMS

Classification of small cell neoplasms in ABC specimens presents a challenge to cytopathologists. Accurate typing and staging of these neoplasms is essential for selection of the appropriate treatment and for determining prognosis.[29,89] In some cases, such as small-cell carcinoma of the lung, NAB is usually the final diagnostic procedure before treatment. The group of small-cell neoplasms that cause difficulties in the differential diagnosis includes neuroblastoma, malignant lymphoma, embryonal rhabdomyosarcoma, Ewing's sarcoma, and metastatic small-cell carcinoma. The latter is mainly a tumor of adulthood and is discussed in the preceeding "Neuroendocrine Neoplasms" section. The other four entities are more common in children and account for approximately 25% of childhood malignancies.[167,190] Other less commonly encountered neoplasms include primitive neuroectodermal tumor, medulloblastoma, and retinoblastoma.

Cytomorphology alone is inadequate to distinguish different types of small-cell neoplasms. The only exceptions are the better differentiatied tumors and the finding of diagnostic cells such as rhabdomyoblasts. Even in these cases, some form of confirmation is mandatory. Light microscopy of ABC coupled with immunocytochemical or ultrastructural studies, or both, can be effective in diagnosis of small cell tumors.[4,6,29,89,114,151,167,169] A diagnosis should be made in the context of the clinical history, including the age of the patient, physical examination, and laboratory and radiological findings. Several antibodies are available commercially that may be useful. A panel of antibodies to cytokeratins, vimentin, LCA, actin, desmin, neuron-specific enolase, and chromogranin can be a useful approach for the accurate diagnosis of these neoplasms (Table 8.5). When the results are equivocal, other antibodies such as neurofilament proteins, synaptophysin, myoglobin, and lymphoid markers may prove helpful (Fig. 8.4).

The value of imunocytochemical studies is demonstrated in the case illustrated in Figure 8.11. This 20-year-old man presented with left inguinal pain and a large mass in the left flank and lower abdomen. A CT scan of the pelvis showed a large soft-tissue mass attached to the ilium. An NAB was performed. The aspirate consisted of dispersed, relatively small uniform cells with scanty cytoplasm (Figs. 8.11A,B). Rosette-like structures were occasionally seen. The differential diagnoses

TABLE 8.5. Antigen Expression in the Most Common Small Round-Cell Tumors of Children

| Neoplasm | Neuron-Specific Enolase | Leucocyte Common Antigen | Actin/Desmin | Chromogranin | Cytokeratin |
|---|---|---|---|---|---|
| Neuroblastoma* | + | − | − | + | − |
| Malignant lymphoma | − | + | − | − | − |
| Rhabdomyosarcoma | ± | − | + | − | − |
| Ewing's sarcoma* | ± | − | − | − | ± |

*Neuroblastoma and Ewing's sarcoma may have the same antigenic profile.

included Ewing's sarcoma, primitive neuroectodermal tumor, neuroblastoma, malignant lymphoma, and, less likely, rhabdomyosarcoma. A periodic acid-Schiff stain showed no glycogen in the tumor cells. The neoplastic cells strongly expressed neuron-specific enolase (Fig. 8.11C) and vimentin (focally) (Fig. 8.11D), and S-100 protein. Staining with antibodies to actin, desmin, LCA, cytokeratin, chromogranin, and neurofilament proteins was negative. On the basis of light microscopy, immunocytochemical findings, and the age of the patient, a diagnosis of primitive neuroectodermal tumor was made on the ABC. This diagnosis was subsequently confirmed by open biopsy and the patient was treated with chemotherapy and radiotherapy. When last seen 15 months later, the patient had pain and locally progressive disease.

In the ultrastructural study of ABC specimens of small-cell neoplasms in pediatric patients, the largest and most varied series is that by Akhtar and associates,[4] in which 62 ABC specimens, including cases of lymphoma, neuroblastoma, Ewing's sarcoma, metastatic retinoblastoma, Wilms' tumor, rhabdomyosarcoma, and unclassifiable lesions were included; the distinctive cytological and ultrastructural features of each of these classes of tumor are succinctly detailed by these authors. In an article on the diagnostic application of electron microscopy to ABC specimens, some aspects of the differential diagnosis of pediatric small-cell neoplasms are also discussed by Mackay and associates.[99] The principal diagnostic ultrastructural criteria for the various types of small round-cell tumors of children (Table 8.6) include pleomorphic cells with thick and thin filaments, Z-band material, and basal lamina in rhabdomyosarcoma; relatively uniform and separated cells with sparse cytoplasmic organelles in lymphoma; primitive-appearing tumor cells with disaggregated chromatin, glycogen aggregates, and a few primitive junctions in Ewing's sarcoma; neuritic cellular processes, neurosecretory granules, microtubules, intermediate filaments, and simple intercellular junctions in neuroblastomas; and epithelial-type lumens with microvilli, basal lamina, many well-formed intercellular junctions, and at times aggregates of intermediate filaments in Wilms' tumor.

The diagnostic differentiation of the small-cell tumors of children is a classic example of the successful application of electron microscopy to surgical specimens[38,56,157,166,167] and, indeed, to ABC specimens.[4-7,122,151] On the basis of the distinct advantages of ABC in the pediatric age group,[20] it is somewhat surprising that reports of electron microscopy in this situation are not more prevalent. In fact, no pediatric studies have systematically compared the light microscopic diagnosis of small-cell tumors established on the basis of the smear cytology with subsequent ultrastructural studies. Despite ultrastructural study, a small number of ABC specimens from children remain unclassifiable.[4] A limiting factor for successful interpretation of ultrastructural features in ABC specimens may be the degree of differentiation.[4] In poorly differentiated tumors, there may be insufficient structural evidence in the small amount of tissue procured to establish a specific diagnosis. Such aspects need to be addressed to establish the true role for electron microscopy in ABC of childhood tumors.

## Neuroblastoma

Neuroblastoma constitutes the third most common malignant childhood neoplasm.[111,151] In ABC specimens, the neoplasm is characterized by small primitive

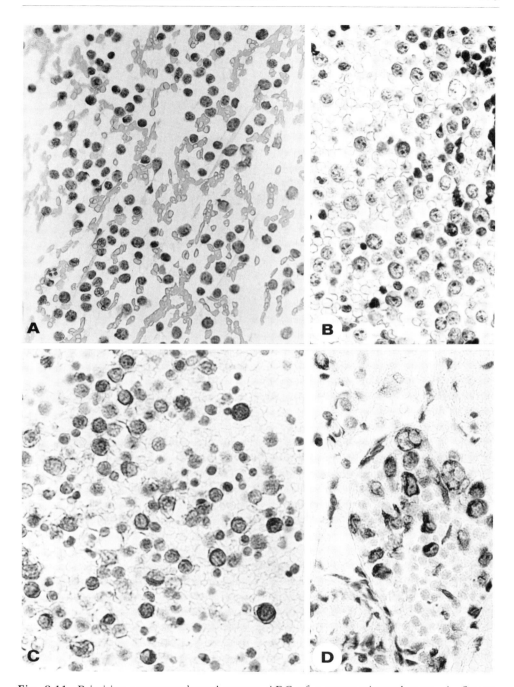

Fig. 8.11. Primitive neuroectodermal tumor, ABC of a retroperitoneal mass. A. Smear preparation. Dispersed, relatively small uniform cells with scanty cytoplasm. × 260. B. Cell block section. Note the small rim of cytoplasm. × 416. C. Most neoplastic cells express neuron-specific enolase. × 260. D. A few cells stained positively with antibody to vimentin. × 416.

TABLE 8.6. Practical Ultrastructural Diagnostic Features in Small Round-Cell Tumors of Children

Neuroblastoma and primitive neuroectodermal tumor
  Elongated cellular processes often in "tangles"
  Neurosecretory granules (best seen in processes)
  Microtubules and intermediate filaments
  Primitive cell junctions
Ewing's sarcoma
  Primitive-appearing cells
  Regular nuclear contours and disaggregated chromatin
  Excessive glycogen deposits
  Primitive cell junctions
  Occasional lipid droplets
Rhabdomyosarcoma
  Thick and thin filaments
  Z-band material
  Pleomorphic cells and nuclear profiles
  Prominent rough endoplasmic reticulum
  Occasionally basal lamina synthesis
Leukemia/lymphoma
  Characteristics of lymphocytes
  May have nuclear irregularities or "blebs" (bullae)
  Absent junctions and neurosecretory granules
  Few cytoplasmic organelles other than ribosomes
Retinoblastoma
  Cell junctions
  No cytoplasmic processes
  No neurosecretory granules or microtubules
  Gland-like spaces and tight junctions
  A few microvilli projecting into lumens (rosettes)
Wilms' tumor
  Frequent cell junctions
  Basal lamina at margins of cell clusters
  Occasional tubular lumens with microvilli
  Prominent clusters of intermediate filaments in the rhabdoid form

cells, slightly larger than lymphocytes, with scanty cytoplasm and a delicate fibrillary matrix. It is usually difficult to distinguish neuroblastoma from other small-cell tumors of childhood.[111] Better-differentiated neoplasms may exhibit pseudorosettes with peripherally located nuclei and a central syncytium of cytoplasm.[111] Multinucleated ganglion cells may also be seen. Silverman and colleagues[151] applied electron microscopic and immunocytochemical studies to five ABC specimens of neuroblastoma; all case were confirmed by electron microscopy. Four of the five neoplasms tested expressed neuron-specific enolase and two of the five expressed neurofilament proteins. Studies for cytokeratins, lymphoid markers, and myoglobin were negative. Although neuron-specific enolase is a sensitive marker for neuroblastoma, it is not specific (Table 8.5) and may be expressed by other small-cell tumors such as embryonal rhabdomyosarcoma and Ewing's sarcoma.[89,169] Most neuroblastomas also express chromogranin and synaptophysin.[29,62] The latter is coexpressed with neurofilament protein.[62] The diagnosis of neuroblastoma should be confirmed with antibodies to neuron-specific enolase and chromogranin or synaptophysin.

## Malignant Lymphoma

Non-Hodgkin's lymphoma comprises 6% of all childhood cancers and is a major differential diagnosis of small-cell neoplasms.[167] ABC specimens of malignant lymphoma are cellular and composed of dispersed cells with small amounts of cytoplasm. LCA is a reliable marker for confirmation. Lymphoma cells also show positivity with most of the commercially available antibodies to vimentin. Most other markers, such as cytokeratins and neuroendocrine markers, are negative.

## Rhabdomyosarcoma

Embryonal and alveolar rhabdomyosasrcomas are the most common soft-tissue sarcomas of childhood. The aspirates are generally cellular and composed of small cells with scanty cytoplasm. Nuclear pleomorphism is usually present. Finding rhabdomyoblasts (somewhat larger cells with eosinophilic cytoplasm, and a well-defined cell border) is a useful clue to the diagnosis.[82] In most cases ancillary techniques are necessary for differentiation from other small-cell neoplasms.

A large number of muscle markers are available commercially (see Chapter 5 and the "Mesenchymal Neoplasms" section in this chapter). A combination of antibodies to muscle-specific actin and desmin is useful to confirm the muscle lineage of tumors. These antibodies do not differentiate between tumors of skeletal and smooth-muscle origin. Smooth-muscle tumors, however, are uncommon in children. Myoglobin is the only muscle marker that is expressed in the better-differentiated rhabdomyosarcoma and is absent in smooth-muscle neoplasms.[89] Myoglobin, however, can be expressed in nonmyogenic neoplasms such as carcinoma, malignant lymphoma, and malignant melanoma.[47] Because neuron-specific enolase may be expressed in rhabdomyosarcoma (Table 8.5), reliance for differential diagnosis cannot be placed on staining with a single antibody.

## Ewing's Sarcoma

Ewing's sarcoma is a highly malignant tumor occuring in children and adults under the age of 30 years.[137] It is usually a tumor originating in bone; however, some cases do not have any bone involvement and are regarded as primary soft-tissue neoplasms. It is a tumor derived from primitive pluripotential cells that may differentiate into tumor cells with mesenchymal, epithelial, and, rarely, neural features.[114] The latter aspect is supportive of the hypothesis that at least some of these tumors represent primitive neuroectodermal tumors of bone and soft tissues.[103,137,157] In ABC specimens, Ewing's sarcoma is characterized by small- to medium-sized, round cells with scanty cytoplasm. The cells are monotonous in size and some have short, thick cytoplasmic processes.[82] Occasionally, rosette-like structures are seen. Most neoplastic cells contain large amounts of cytoplasmic glycogen. Although this is an important finding, it is not specific and may be seen in other small cell neoplasms (e.g., rhabdomyosarcoma and, to a lesser extent, in neuroblastoma).[137,151] In addition, some Ewing's sarcomas lack glycogen. Immunocytochemically, vimentin is the main intermediate filament protein expressed by the tumor.[114] Desmoplakin, a protein associated with the plaques of the small intercellular junctions commonly seen ultrastructurally in Ewing's sarcomas, and sometimes cytokeratins, can be detected using appropriate antibodies.[114,137] Some Ewing's sarcomas also express neuron-specific enolase and neurofilament proteins, further evidence for the possibility of neuroectodermal derivation or differentiation in certain of these tumors.[89,114] Again, the results of immunocytochemistry (Table 8.5) and electron microscopy (Table 8.6) should be used in conjunction with light microscopy and clinical findings.

## Primitive Neuroectodermal Tumor

Primitive neuroectodermal tumor (PNET) is a highly malignant neoplasm and occurs in children and young adults as large soft-tissue masses; at times, bone is also involved. The chest wall is the most common location for this tumor; however, it may arise in the extremities and the abdomen.[74] The tumor exhibits immunocytochemical and electron microscopic features of neural differentiation and is thought to represent a primitive neoplasm of neural crest derivation.[137] In ABC specimens, PNET is composed of small round cells with scanty cytoplasm similar to the other small-cell tumors (Fig. 8.11). They share some features with neuroblastoma cytologically, ultrastructurally (presence of dense core granules and neurites), and immunocytochemically.[151] They also share some features with Ewing's sarcoma cytologically, immunocytochemically, and cytogenetically, which favors a close relationship between these two entities,[137] and the possibility that perhaps some extraskeletal Ewing's sarcomas are PNETs.[103,157] Cytoplasmic glycogen is usually absent in PNET. These neoplasms have been shown to express vimentin, neuron-specific enolase, synaptophysin, Leu-7, S-100 protein, and neurofilament proteins.[122,137]

# MESENCHYMAL NEOPLASMS

ABC can be used for the preoperative diagnosis of primary and recurrent, as well as metastatic, soft-tissue sarcomas. These tumors form a large heterogeneous group with a diverse cytomorphology. In high-grade sarcomas, primary or secondary, a general diagnosis of malignancy can be made with ABC,[82] but precise classification and differentiation from sarcomatoid carcinoma and malignant melanoma may be difficult or impossible. Difficulty persists in assigning a correct histogenetic diagnosis in many cases of ABC of sarcomas.[3,31,79-81] Selective cases in which electron microscopy or immunocytochemistry have been applied illustrate that a specific diagnosis is always a possibility. These include a spectrum of soft-tissue and bone tumors,[31,65,79] cartilaginous tumors,[180] osteosarcomas and osteoblastomas,[177,185] synovial sarcoma,[81,140] rhabdomyoma,[16] chordoma,[136,178] alveolar soft-part sarcoma,[132] and stromal sarcoma of breast.[138] A useful taxonomic approach to the cytological classification of soft-tissue tumors and the general advantages of ancillary studies of such cases is provided by Nguyen.[123]

Some sarcomas, such as epithelioid sarcoma and leiomyosarcoma, may have an epithelioid appearance on ABC (Fig. 8.12). Neoplasms of smooth muscle, synovial, fibroblast, vascular, and nerve sheath origin may all be composed chiefly of spindle cells. In this group, only a general diagnosis of spindle cell neoplasm or sarcoma can be achieved by light microscopy.[156] Likewise, pleomorphic variants of liposarcoma, rhabdomyosarcoma, and malignant fibrous histiocytoma may have similar cytological appearances on ABC, indistinguishable from each other and from anaplastic carcinoma (Fig. 8.13) and malignant melanoma. Myxoid mesenchymal lesions also present problems in differential diagnosis in aspiration biopsies (Fig. 8.17); a correlative cytological, histological, and electron microscopic study of ABCs of myxoid soft-tissue tumors[58] and a comparative ultrastructural study of such lesions[176] may assist with this problem. Embryonal and alveolar rhabdomyosarcoma are composed of small- to medium-sized cells and many resemble those of malignant lymphoma, neuroblastoma, Ewing's sarcoma, or Wilms' tumor. These difficulties may be further complicated by the potential for sampling error.

Immunocytochemical and ultrastructural studies may resolve some of these problems.[37,79,145] These techniques significantly contribute to (1) separation of sarcomas from nonsarcomatous neoplasms, and (2) further characterization of mesenchymal tumors.[51] There are increasing reports of the lack of immunocytochemical specificity in sarcomas, such as "expression" of cytokeratins in smooth-muscle tumors, rhabdomyosarcomas, malignant fibrous histiocytomas, malignant peripheral nerve sheath tumors, epithelioid vascular tumors, epithelioid sarcomas, synovial sarcoma, and Ewing's sarcomas; epithelial membrane antigen expression in leiomyosarcomas, chondrosarcomas, malignant peripheral nerve sheath tumors, synovial, and epithelioid sarcomas; and expression of desmin in malignant fibrous histiocytomas and malignant peripheral nerve sheath tumors.[11,21,32,36,51,64,85,105,106,109,114,125,142,160,182,186] The validity and significance of reported unexpected "expression" of these antigens, such as cytokeratins, in soft-tissue sarcomas remains to be determined.[158] This unexpected expression is not usually an obstacle, however, provided that a panel of antibodies is used and the staining results are interpreted in the context of cytomorphological and clinical findings. Also, the vagaries of intermediate filament expression in sarcomas does not effect the utility of electron microscopy in further classifying these lesions in ABC specimens, where many additional characteristics of mesenchymal tumors (see Chapter 6) become available.

The first step in the workup of an undifferentiated neoplasm or a tumor with probable sarcomatous differentiation is to confirm it as a sarcoma and to rule out other probabilities. A simple diagnostic panel consisting of antibodies to cytokeratins, vimentin, LCA, and HMB-45 is a practical beginning. With some exceptions, strong vimentin expression and negative staining with antibodies to cytokeratin, LCA, and HMB-45 support a diagnosis of sarcoma. Unfortunately, not all malignant melanomas are HMB-45–positive, and smooth-muscle tumors may express cytokeratin. Despite vimentin expression in the majority of sarcomas, it is detectable as well in some carcinomas and therefore is not a specific marker for mesenchymal neoplasms. It is, however, useful indirectly in showing whether the antigenicity of other markers, (e.g., desmin and actin) have been preserved, especially when internal control specimens for these antigens are absent in a given smear or section.[12]

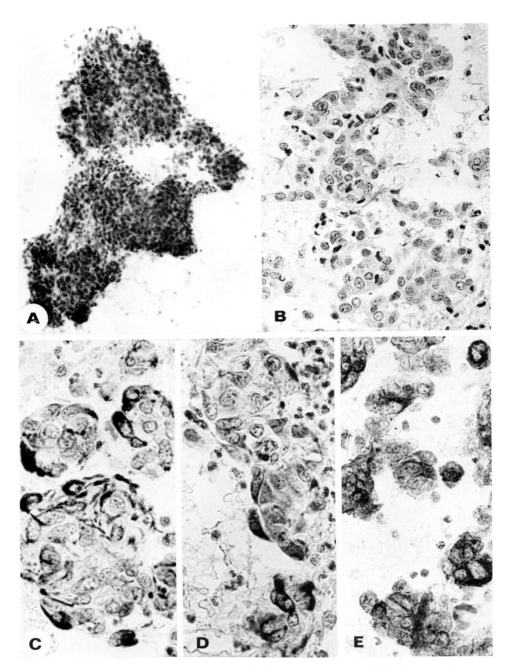

Fig. 8.12. Epithelioid leiomyosarcoma, ABC of an epigastric mass. This 77-year-old woman had a remote history of breast carcinoma. **A.** Smear preparation. Cohesive fragment of neoplastic cells with spindling of some nuclei. × 104. **B.** Cell block section. Note the epithelioid appearance of the tumor cells. × 260. The tumor was initially interpreted as carcinoma. **C.–E.** The neoplastic cells strongly expressed vimentin, desmin, and muscle-specific actin, supporting diagnosis of a myogenic sarcoma rather than a carcinoma. Stains for cytokeratin were negative. (C) Vimentin, × 416; (D) Desmin, × 416. (E) Actin, × 416. Subsequent surgery demonstrated an epithelioid leiomyosarcoma involving the wall of the stomach, mesentery, and omentum.

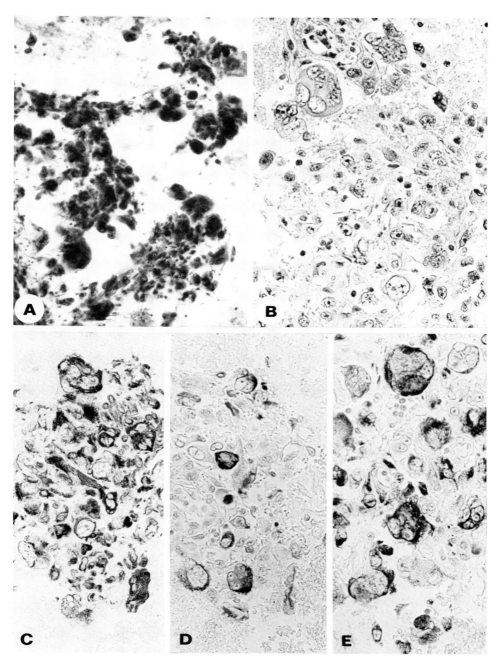

Fig. 8.13. Anaplastic carcinoma, ABC of a pelvic mass of a 78-year-old woman, involving the anterior wall of the urinary bladder. **A.** Smear preparation. Fragments of spindled and pleomorphic bizarre giant cells. × 104. **B.** Cell block section. Note multinucleated giant cells. × 260. Although the cytological features were suggestive of sarcoma, particularly malignant fibrous histiocytoma, strong expression of cytokeratins and vimentin, along with negative staining with antibodies to muscle-specific actin, desmin, and alpha-1-antitrypsin supported the diagnosis of anaplastic carcinoma. **C.** Vimentin, × 260; **D.** Anti-cytokeratin AE1, × 260; **E.** Anti-cytokeratin AE3, × 260. Subsequent surgery demonstrated an anaplastic carcinoma of the urinary bladder.

In practical terms, a simple checklist of electron microscopic features will identify the common epithelioid (Table 8.3) and spindle (Table 8.4) tumors encountered in ABCs. In fibrosarcomas, the fairly uniform-appearing tumor cells are separated by collagen and stromal materials, infrequently contact adjacent cells, lack intercellular junctions and basal lamina, and have a prominent complement of rough endoplasmic reticulum. Similar tumor cells and organization are seen in malignant fibrous histiocytomas, but there is more tumor cell heterogeneity due to the presence of lipid- and lysosomal-rich histiocytes, multinucleated giant cells, and undifferentiated cells. In myofilament-containing smooth-muscle tumors, tumor cells tend to be rectangular, or at least angular, in shape and have a regular arrangement with a fairly uniform spacing, producing a "mosaic" growth pattern ultrastructurally. Also, one or more features, such as micropinocytotic vesicles, basal lamina, dense plaques associated with the inner surface of the cell membrane, and linear dense bodies among the myofilaments, may be seen. Peripheral nerve-sheath tumors are distinguished by the many narrow, markedly elongated cell processes often aligned in a parallel fashion; these cell processes are often complexely intertwined, and may encircle collagen fibers or other cellular processes. Basal lamina, often considerably redundant, and intercellular junctions are other useful diagnostic markers in tumors with neurogenic differentiation. Identification of thick and thin cytoplasmic filaments, with or without Z bands, are the cardinal diagnostic criteria for rhabdomyosarcoma. Nuclear and cytoplasmic features are generally more variable than in the other sarcomas described, and ribosomal or filament complexes (i.e., rows of ribosomes aligned along rigid cytoplasmic filaments), basal lamina, junctions, and glycogen accumulations may also be present. A general review of the value of electron microscopy and immunohistochemistry in the diagnosis of soft-tissue sarcomas is provided by Fisher.[51]

Low-grade sarcomas are difficult to differentiate from reactive, nonneoplastic lesions and benign mesenchymal neoplasms using ABC.[112] There is no general role, however, for immunocytochemistry and electron microscopy in nonneoplasic and benign mesenchymal neoplasms except for possible determination of the histogenesis of the cells. An example of the latter is provided in Figure 8.16. With this background, the value of applying immunocytochemistry and electron microscopy to ABC specimens in some of the sarcomas will be discussed.

## Myogenic Sarcoma

The diagnosis of myogenic sarcomas as malignant tumors is relatively easy on ABC specimens. Precise classification, however, is difficult when no evidence of muscle differentiation, in the form of cross striations, can be identified. As previously discussed (see the "Small Cell Neoplasms" section in this chapter), the embryonal and alveolar forms are difficult to separate from other small-cell neoplasms of childhood, whereas pleomorphic rhabdomyosarcoma of either adults or children are difficult to differentiate from other pleomorphic sarcomas and pleomorphic types of carcinoma or melanoma. Whether in adults or children, the ultrastructural identification of the minimal criteria for skeletal muscle differentiation (i.e., rigid, thick filaments with associated ribosomal arrays or groups of thick and thin filaments with or without Z bands[46]) confirms this diagnosis (see the "Cytoplasmic Filaments" section in Chapter 6). In studying the cytology of embryonal rhabdomyosarcoma,

Seidal and colleagues[145] demonstrated abundant myofilaments and Z bands ultrastructurally in tumor cells in one example obtained by ABC.

Leiomyosarcoma is relatively uncommon and is composed of spindle cells with elongated, blunt-ended nuclei and eosinophilic cytoplasm. Differentiation from other spindle-cell sarcomas, such as fibrosarcoma, synovial sarcoma, angiosarcoma, and nerve-sheath tumors, can be very difficult in ABC specimens (Fig. 8.14). Useful guidelines to the ultrastructural diagnosis of a smooth-muscle tumor were presented in the introduction to the "Mesenchymal Tumors" section in this chapter. The placement of myofilaments and their associated dense bodies have been described and illustrated in the "Cytoplasmic Filaments" section in Chapter 6, and some of these features can be seen in Figure 7.2. Cytoplasmic myofilaments may be quite limited in some smooth-muscle tumors, particularly in poorly differentiated lesions. If the filaments are profuse, contraction may result in a wrinkled appearance of the nuclear and cell surfaces.[101] The role for the identification of intracytoplasmic filaments is illustrated in the following case. A 56-year-old woman with a history of a malignant peripheral nerve-sheath tumor of the thigh resected 7 years previously presented with multiple liver nodules. NAB produced large fragments of cohesive pleomorphic spindle cells consistent with a sarcoma (Fig. 8.14A,B). The neoplastic cells expressed vimentin, desmin, and muscle-specific actin (Figs. 8.14C–E) indicating a myogenic sarcoma. By electron microscopy, the somewhat rectangular shape of the tumor cells and the presence of accumulations of cytoplasmic filaments with dense bodies (Fig. 8.14F) indicated that the lesion was a metastatic leiomyosarcoma and that the original diagnosis of a malignant peripheral nerve-sheath tumor was incorrect.

A large number of muscle-associated markers have been described, including actin, desmin, myoglobin, myosin, titin, Z protein, tropomyosin, and creatine phosphokinase.[22,28,35,36,47-49,55,72,86,98,127,128,130,131,145,152,170,173,184] Most of these markers lack specificity or have relatively low sensitivity, or both, which limits their usefulness in diagnostic cytopathology. We use antibodies to actin and desmin in combination to confirm the muscle lineage of myogenic sarcomas (Figs. 8.14, 8.15). These antibodies do not differentiate leiomyosarcoma from rhabdomyosarcoma. Myoglobin is expressed in better-differentiated rhabdomyosarcomas and not in smooth-muscle tumors (Fig. 8.15). Antibodies to myoglobin, however, can also stain certain nonmyogenic lesions, such as carcinomas, malignant lymphomas, and malignant melanomas.[47] Antibodies to actin and desmin react with most myogenic neoplasms, and many cells in a given tumor express these markers, especially in alcohol-fixed material. In our experience, as well as that of others, muscle-specific actin appears to be a more specific and sensitive marker than desmin in myogenic sarcomas.[12] In contrast, Desai and colleagues,[37] in studying 9 cytological specimens (6 were ABC samples) from patients with rhabdomyosarcoma, found desmin to be a more sensitive marker than muscle-specific actin. Desmin is also known to be expressed in nonmyogenic neoplasms.[85,168,186] We recommend the use of both markers until further experience is obtained with anti-actin antibodies. Both actin and desmin can be expressed in myofibroblasts, whereas only actin is located in myoepithelial cells.

As noted previously, smooth-muscle tumors may express cytokeratins and further complicate the immunophenotyping of myogenic sarcomas. A number of other mesenchymal lesions, neoplastic and nonneoplastic, may have a stromal component

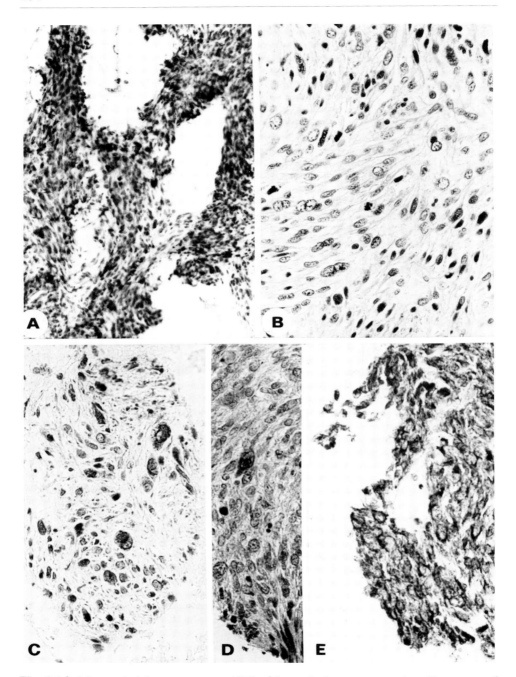

Fig. 8.14. Metastatic leiomyosarcoma, ABC of liver. A. Smear preparation. Fragments of spindle-shaped tumor cells with elongated, hyperchromatic, and pleomorphic nuclei consistent with sarcoma. × 104. B. Cell block section. The compactly organized spindle cells cannot be further classified with certainty. × 260. C.–E. The neoplastic cells expressed vimentin, desmin, and muscle-specific actin, supporting the diagnosis of myogenic sarcoma. (C) Vimentin, × 260; (D) Desmin, × 260; (E) Actin, × 260.

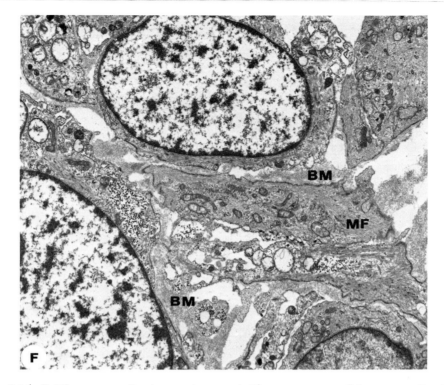

Fig. 8.14. F. Ultrastructurally, the angular, mosaic-like arrangement of the tumor cells, their content of microfilaments (MF) with dense bodies, cell membrane–associated dense staining plaques, and intercellular basal lamina–type materials (BM) confirm the diagnosis of metastatic leiomyosarcoma. × 4700. (From Dardick I, Yazdi HM, Brosko C, Rippstein P, Hickey NM: A quantitative comparison of light and electron microscopic diagnoses in specimens obtained by fine-needle aspiration biopsy. *Ultrastruct Pathol* 15:105–129, 1991.)

with myofibroblastic differentiation. In these tumors, staining for actin and desmin is confined to cells in the stromal tissue and is thus usually focal in nature. Sarcomas and epithelial neoplasms may infiltrate muscle. Because normal, reactive, and degenerating muscle cells (some multinucleated) express muscle markers, care must be taken not to misinterpret them as tumor cells. Most of these cells have benign-appearing nuclei, and the staining is again usually focal.

Other sarcomas, histogenetically of myogenic differentiation, can have insufficient expression of cytoplasmic filaments for ultrastructural diagnosis. A 62-year-old man presented with a large soft-tissue mass extending from the right thigh up to the abdomen. An NAB of the abdominal mass was performed. The cellular aspirate was composed of single cells and large fragments of cohesive cells (Fig. 8.15A,B). The nuclei were round or oval and contained nucleoli. Some of the cells had a moderate amount of cytoplasm. The overall cytological appearance was suggestive of sarcoma. The neoplastic cells expressed vimentin, muscle-specific actin, and, focally, myoglobin (Fig. 8.15C–E) and were negative with antibodies to cytokera-

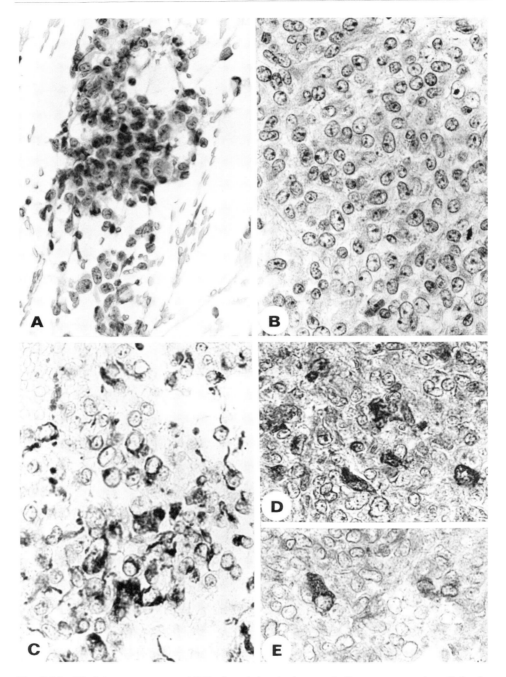

Fig. 8.15. Rhabdomyosarcoma, ABC of an abdominal mass. **A.** Smear preparation. Cohesive fragments and single cells. Some cells have a moderate amount of cytoplasm. × 416. **B.** Cell block section. Round or oval nuclei containing nucleoli. × 416. **C.–E.** The neoplastic cells expressed vimentin, muscle-specific actin, and myoglobin (focally), supporting the diagnosis of rhabdomyosarcoma. Staining for cytokeratins was negative. (C) Vimentin, × 416; (D) Actin, × 416; (E) Myoglobin, × 416.

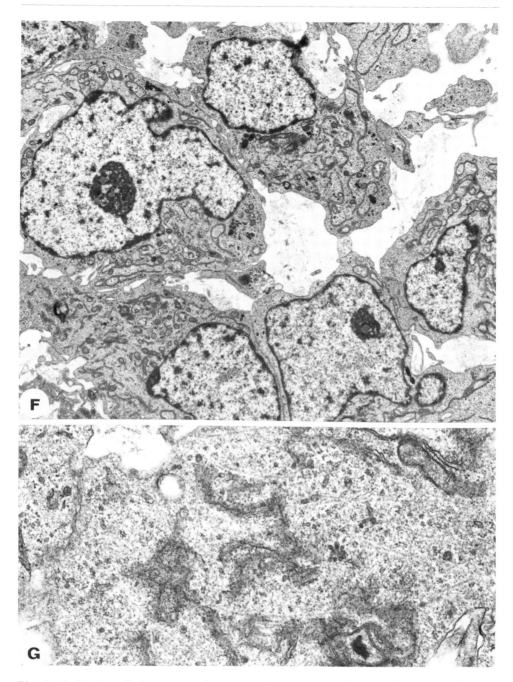

Fig. 8.15. F. Irregularly contoured tumor cells are separated by focal accumulations of glycosaminoglycans. There are no specific features of the cytoplasm and no development of junctions, basal lamina, or glandular lumens. × 2,500. G. Higher magnification reveals the absence of diagnostic organelles, particularly myofilaments; a few microtubules are present. × 9,500.

tins. The immunocytochemical findings supported the diagnosis of rhabdomyosarcoma. Electron microscopy of a portion of the ABC specimen showed that the tumor cells lacked both thin (actin-type) and thick (myosin-type) cytoplasmic filaments and that the cells had the primitive appearance of an undifferentiated sarcoma (Fig. 8.15F,G). This case illustrates that skeletal muscle-specific polypeptides may be detectable in the absence of the structural differentiation indicative of rhabdomyoblasts.

## Neurogenic Neoplasms

Neurogenic tumors that may be aspirated include benign (schwannoma, neurofibroma) and malignant (malignant schwannoma) peripheral nerve-sheath tumors. Aspirates from both the benign and malignant forms are composed of elongated cells with spindle-shaped, rather pointed nuclei and eosinophilic fibrillary cytoplasm (Fig. 8.16). Palisading of nuclei may be seen. Nuclear atypia and cellular pleomorphism may be seen in both benign and malignant schwannomas. Malignant schwannomas may resemble other spindle-cell or pleomorphic sarcomas on ABC and may even consist of epithelioid cells. Neurofibromas are difficult to aspirate and they are generally less cellular. The ABC of these tumors consists of Schwann cells and fibroblastic cells, and these smears can be difficult to distinguish from tumors of fibroblastic origin.[82]

Virtually all benign schwannomas and neurofibromas and one-third to one-half of malignant schwannomas express S-100 protein.[63,75,156,183,186,192] S-100 protein, however, is not specific for neurogenic tumors and may be expressed by other sarcomas. Wick and colleagues[159,186] recommended a panel of antibodies to S-100 protein, Leu-7, and myelin basic protein for the workup of malignant schwannomas. These tumors may also express epithelial markers and desmin, which may lead to further confusion. Ultrastructural studies, therefore, have a definite role for the precise classification of neurogenic tumors.

A number of electron microscopic features assist in the diagnosis of neurogenic tumors.[39,192] These features include a profusion of elongated, very narrow and often aligned cytoplasmic processes (Fig. 8.16); scanty organelles in both the processes and the main portion of cytoplasm; variable numbers of the processes endowed with discrete basal lamina (see the "Basal Lamina" section in Chapter 6); and attempts at mesaxon formation due to the encompassing of collagen bundles or other tumor-cell extensions by multiple, thin layers of cell cytoplasm.[44] Some of these features are illustrated by the following patient with a peripheral nerve-sheath tumor with an unusual clinical presentation, in which electron microscopy was instrumental in establishing the diagnosis from an ABC specimen.

Due to chest-wall pain, a 47 year-old man had a chest radiograph and CT scan (Fig. 8.16A,B) that were interpreted as showing a pleural-based tumor. NAB produced fragments of a cellular, spindle-cell tumor with a differential diagnosis of fibroma, mesothelioma, and sarcoma (Fig. 8.16C,D). Ultrastructurally (Fig. 8.16G,H), the correct diagnosis was readily established by the innumerable narrow cytoplasmic processes often having a parallel alignment and an intricately entangled organization, features that are characteristic of a schwannoma. Some processes were lined by a discrete external lamina (Fig. 8.16H). A subsequent S-100 protein immu-

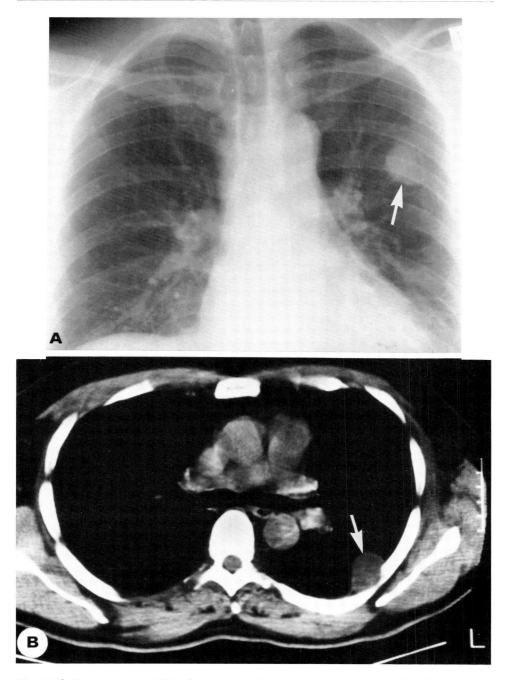

**Fig. 8.16.** Schwannoma, ABC of the chest wall. **A.** A left pleural-based density (arrow) is evident in a routine chest radiograph. **B.** The neoplasm (arrow) as seen in the computed tomographic scan.

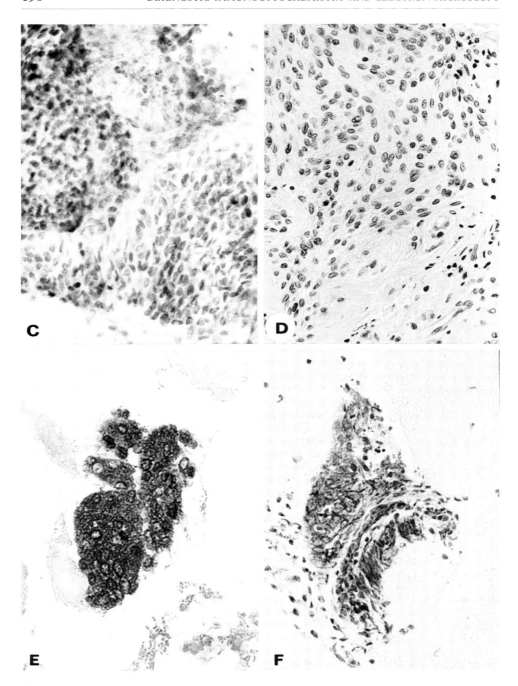

Fig. 8.16. C. Smear preparation. Compactly organized tumor cells with oval or spindle nuclei. × 260. D. Cell block section. Note the spindle-cell nature of the tumor and its fibrillar background. × 260. The neoplastic cells strongly expressed S-100 protein and vimentin, corroborating the diagnosis established by electron microscopy. Stains for cytokeratins and desmin were negative. E. S-100 protein, × 416; F. Vimentin, × 260.

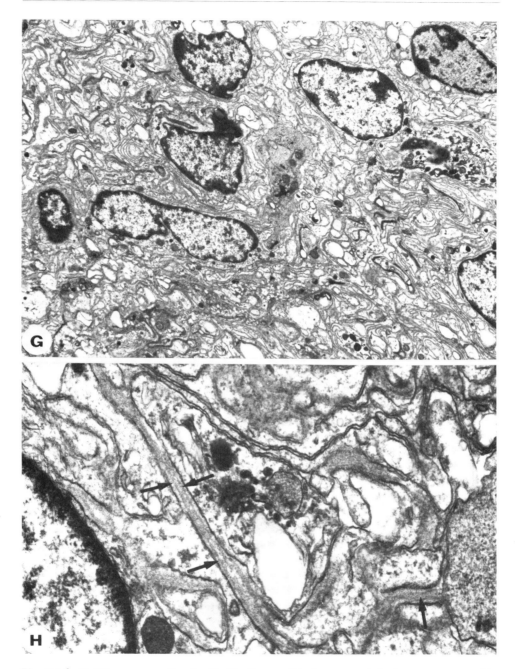

**Fig. 8.16. G.** Ultrastructurally, the fibrillar "stroma" is due to myriad of narrow, reasonably aligned cytoplasmic processes, a feature diagnostic of schwannomas. × 5,500. **H.** At higher magnification, some interlocking cell processes form basal lamina (arrows). × 30,000. (From Dardick I, Yazdi HM, Brosko C, Rippstein P, Hickey NM: A quantitative comparison of light and electron microscopic diagnoses in specimens obtained by fine-needle aspiration biopsy. *Ultrastruct Pathol* 15:105–129, 1991.)

nostain (Fig. 8.16E), not done initially because the diagnosis was unsuspected, was strongly positive. The neoplastic cells were also vimentin-positive (Fig. 8.16F) but were negative with anti-desmin and anti-cytokeratin antibodies.

## Malignant Fibrous Histiocytoma

Malignant fibrous histiocytoma (MFH) is now the most common soft-tissue sarcoma of adults, and often is located in the deep soft tissue. The smears in ABC specimens are usually cellular, and a typical aspirate consists of two components: (1) large, bizarre, often multinucleated cells presumed to be histiocytic in origin, and (2) a fibroblastic component consisting of spindle cells.[82] The absence of most markers of macrophage or monocyte lineage in MFH is supportive of a fibroblastic rather than histiocytic phenotype.[19,189] Vimentin, alpha-$_1$-antichymotrypsin and alpha-$_1$-antitrypsin are usually expressed in the tumor cells of MFH.[43] A lack of reactivity for alpha-$_1$-antichymotrypsin in a suspected MFH should make one reconsider the diagnosis. Alpha-$_1$-antichymotrypsin expression in isolation, however, is not specific for MFH and is seen in most sarcomas, many carcinomas, and malignant melanoma.[42,51,87] Occasional cells may express cytokeratins, which may be indicative of epithelial differentiation, cross-reactivity, or aberrant reaction.[85,182] Sarcomatoid or pleomorphic carcinomas (Fig. 8.13) or other sarcomas may simulate MFH. Using a diagnostic panel of antibodies to cytokeratins, vimentin, desmin, muscle-specific actin, and S-100 protein, in addition to alpha-$_1$-antichymotrypsin or alpha-$_1$-antitrypsin usually resolves the problem (Fig. 8.2). According to Nemes and Thomázy,[120] the ubiquitous expression of factor XIIIa in MFH distinguishes these tumors from morphologically similar soft-tissue tumors.

On the basis of personal experience with surgically derived tumor tissue, MFH exhibits two main patterns of cellular differentiation ultrastructurally. In one case, the spindle cells have all of the electron microscopic attributes of well-differentiated fibroblasts, usually with prominent, dilated segments of rough endoplasmic reticulum. This form of MFH is nothing more than a high-grade fibrosarcoma, a situation that accounts for the apparent decline of soft-tissue fibrosarcomas. In the second category are those MFHs in which the tumor-cell populations are more variable; present are cells with features of histiocytes, fibroblasts, myofibroblasts, intermediate cells (mixed features of histiocytes and fibroblasts), multinucleated giant cells, and undifferentiated cells.[51,54,70,84] Electron microscopy of ABC-derived specimens can provide a rapid means of segregating MFH in spindle-cell and pleomorphic neoplasms.[58,79,124,179]

## Liposarcoma

Liposarcoma is the second most frequent soft-tissue tumor in adults, is often located in the deep soft tissues and retroperitoneum, and may be the subject of ABC. The key cytological marker is the lipoblast; however, it is not always present. The myxoid variant presents a characteristic appearance in ABC specimens. The other types of liposarcomas, however, are either difficult to diagnose or type. The well-differentiated variant is difficult if not impossible to differentiate from lipoma. Although a general diagnosis of malignancy can be made in aspirates of round-cell and pleomorphic liposarcomas, precise classification is difficult, if not impossible. Vimentin and S-100 protein are consistently expressed in lipomatous tumors; however, they are not specific.[183]

Ultrastructurally, four features can assist in identifying a liposarcoma in ABC fragments:[18,38,176] (1) the generally undifferentiated or fibroblast-like appearance of the tumor cells, (2) the association of the tumor cells with capillaries or small groups of endothelial cells, (3) a relatively thick "fuzzy" coating on the surface of the tumor cells (Fig. 8.17B), and (4) the presence of variably sized lipid droplets (one or more may be particularly large) in the cytoplasm (see the "Secretory Products" section in Chapter 6). ABC specimens of a few liposarcomas and intramuscular lipomas have been examined by electron microscopy,[58,79,124] and a myxoid liposarcoma has been compared with other myxoid neoplasms of soft tissues.[58] Figure 8.17 illustrates the application of these characteristics in identifying an unsuspected liposarcoma in an unusual location.

An ABC was performed on a solitary expansile rib lesion in a 30-year-old man who presented with chest-wall pain. Clinically, the lesion was felt to be a metastatic tumor or Ewing's sarcoma, but the ABC smear showed a myxoid spindle-cell tumor (Fig. 8.17A). Ultrastructurally, most tumor cells were considerably separated by a proteoglycan-rich stroma, were irregular in shape, were bordered by a relatively thick coat of granular material, and contained little or no lipid (Fig. 8.17B). Despite damage to some of the tumor cells during aspiration, it was still evident that, occasionally cells contained 2 or 3 relatively large droplets and a number of smaller-sized ones, which along with the other features was diagnostic of a myxoid liposarcoma (Fig. 8.17C). Subsequent surgical resection of the rib lesion confirmed this diagnosis.

## Synovial Sarcoma

The aspirates of synovial sarcoma are usually cellular and consist of fragments of cohesive small spindle cells with scanty cytoplasm and, occasionally, small groups of larger epithelial-like cells with clear cytoplasm.[82] The latter component may be difficult to find or may be completely absent in the monophasic variant. Although the spindle-cell portion expresses vimentin, the epithelial component demonstrates both cytokeratins and epithelial membrane antigen; the latter seems particularly useful in accentuating the epithelial-type tumor cells.[1,34,129] In contrast to other spindle-cell sarcomas, synovial sarcomas are distinctive in expressing cytokeratins 7 and 19, along with desmoplakins.[106]

Only a limited number of ABC specimens of synovial sarcoma have been studied by electron microscopy.[73,124,140] In biphasic synovial sarcomas, electron microscopy will reveal the formation of glandular lumens (complete with apical tight junctions and surface microvilli) by cuboidal-to-columnar epithelial cells that are separated from the surrounding fibroblast-like stromal component by a distinct basal lamina.[34,51,129] In the monophasic variant, the tumor cells are plumper, more closely associated, and produce less collagen and other intercellular materials than the tumor cells in fibrosarcomas.[34,51,129] Monophasic synovial sarcomas also differ from fibrosarcomas in that they show, at least in some cases, focal basal lamina formation and intercellular junctions.[34,129]

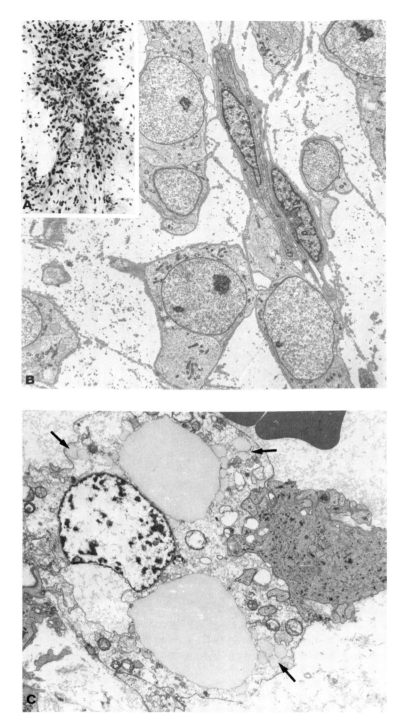

Fig. 8.17. Myxoid liposarcoma, ABC of a rib lesion. A. Smear preparation. Numerous spindle cells in a myxoid stroma. × 200. B. Spindle cells separated by a proteoglycan containing stroma and bordered by a moderately electron-dense staining granular material are features suggesting a myxoid liposarcoma. × 1,600. C. This diagnosis was confirmed by the occasional cell with multiple small lipid droplets (arrows) and one or two large droplets. × 3,000.

## Angiosarcoma

Sarcomas of vascular origin are rarely encountered in ABC specimens and may be a source of confusion. The vascular nature of the tumor may be confirmed with antibodies to Factor VIII—related antigen and *Ulex europaeus* agglutinin-1.[26,51,150] Factor VIII—related antigen is a more reliable and specific endothelial marker than *Ulex europaeus* agglutinin-1, but the former is less sensitive.[100]

Ultrastructurally, recognition of vascular lesions relies on observing attempts at lumen formation by the tumor cells and association of these cells with erythrocytes, basal lamina formation (often prominent), and primitive junctions in the case of angiosarcomas.[84,100] Weibel-Palade bodies,[27] although identifiable in many benign vascular lesions, are usually not detected in angiosarcomas. Hemangiopericytomas may also have distinctive features by electron microscopy, particularly the relationship between neoplastic pericytes (with their numerous cytoplasmic processes) and ramifying capillary-like structures, as well as the excessive formation of basal lamina in relation to both of these cellular components.[33]

## Osteogenic Neoplasms

Osteosarcoma is a highly malignant tumor that may occur in bone and rarely in soft tissue. Although a general diagnosis of malignancy is easily made in ABC specimens, in most cases differentiation from metastatic carcinoma, other sarcomas of bone, and some benign lesions may be very difficult. The osteoblastic and chondroblastic types of osteosarcoma may be very epithelioid on ABC,[82,177] and may be interpreted as metastatic carcinoma. Antibodies to epithelial markers may be useful in eliminating the latter diagnosis.[135] Osteosarcomas express vimentin but not cytokeratins, which in conjunction with ultrastructural study would help to exclude the possibility of carcinoma. Sometimes osteoid-like material can be identified in smears, a feature that can be confirmed ultrastructurally[177] or by using an antibody to osteonectin.[143] Osteonectin is a 32-kd noncollagenous bone protein that has been demonstrated even in osteosarcomas with small amounts of matrix, such as the anaplastic and fibroblastic variants.[143]

Ultrastructurally, the main cohort of tumor cells resemble osteoblasts on the basis of cell shape and some fine cytoplasmic processes radiating from the surface, extensive rough endoplasmic reticulum, and a prominent Golgi zone.[177] Osteoclast-like, multinucleated giant cells, if present in ABC specimens, provide additional support for a diagnosis of a primary bone tumor. An extracellular collagen matrix, in which matrix vesicles and the microcrystals of hydroxyapatite can be identified, remains the key diagnostic feature for osteogenic neoplasms in ABC.[177] That such features may be difficult to apply in all cases is evident in the ultrastructural details of small-cell osteosarcomas.[40]

The following case illustrates that electron microscopy of aspiration biopsies may have sufficient diagnostic features to allow detection of certain primary bone tumors, as well as elimination of metastatic neoplasms. A 21-year-old man complained of vague right chest-wall pain, and physical examination disclosed a swelling that on a chest radiograph proved to be a localized defect in the lateral portion of the right sixth rib. Clinically, this defect was suspected of being a Ewing's sarcoma or metastatic tumor. Smears showed primarily spindle cells (Fig. 8.18A), reported as "favor a spindle-cell mesenchymal tumor." Review of the smears following electron

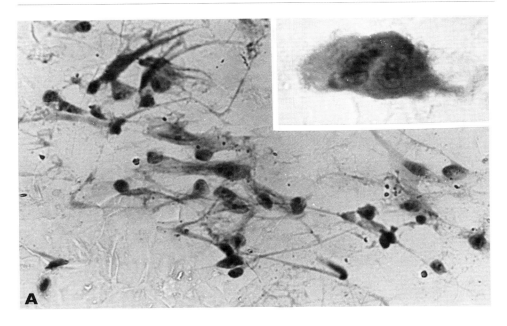
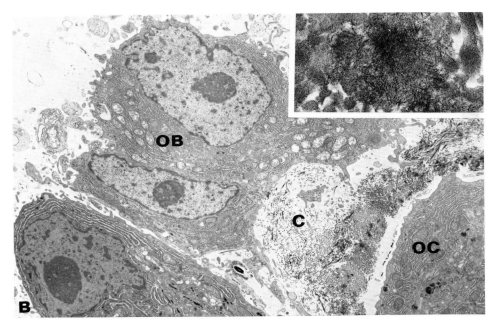

Fig. 8.18. Osteoblastoma, ABC of a rib tumor. **A.** Smear preparation. Noncohesive spindle cells have no specific diagnostic features. × 340. (Inset) An occasional multinucleated giant cell was present. × 340. **B.** Ultrastructurally, the organization of rough endoplasmic-rich cells separated by a band of collagen (C) in which hydroxyapatite crystals are being deposited (inset) indicated the osteoblastic (OB) and osteocytic (OC) nature of these cells. × 3,500; inset, × 62,000.

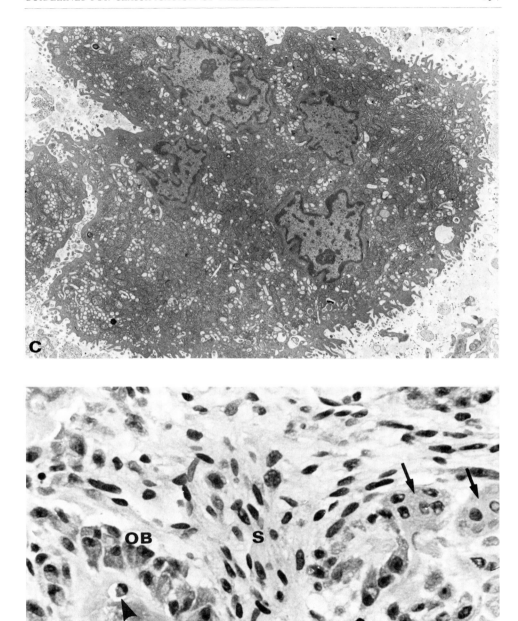

Fig. 8.18. C. Osteoclast-type giant cells, with many surface filopodia, smooth and rough endoplasmic reticulum, and multiple nuclei were initially identified by electron microscopy of this ABC specimen. × 3,500. D. The surgically resected tumor displays the spindle cells (S), osteoblasts (OB), osteoclasts (arrows), and osteocytes (arrowhead) seen in the ABC specimen. Hematoxylin and eosin, × 170.

microscopy of this biopsy sample identified a few multinucleated giant cells (Fig. 8.18A, inset). Electron microscopy of the ABC fragments suggested a primary bone tumor on the basis of mineralizing osteoid associated with both osteoblasts and osteocytes (Fig. 8.18B) and osteoclast-type giant cells (Fig. 8.18C). Subsequent excision of the rib tumor showed that the features seen ultrastructurally represented an osteoblastoma (Fig. 8.18D). The architectural features that can be represented in an ABC specimen and their diagnostic importance are emphasized by comparing the electron micrographs with the histology of the surgically resected tumor.

## Cartilaginous Neoplasms

The cytomorphology of chondrosarcoma varies in ABC, depending on the degree of differentiation.[82] Low-grade tumors are difficult to distinguish from cellular chondromas. High-grade chondrosarcomas are usually cellular in ABC, and the aspirate is composed mostly of large round-to-oval single cells with abundant foamy or vacuolated cytoplasm.[82] When signet-ring–like cells or polygonal cells are present, they may mimic epithelial tumors. S-100 protein may be expressed in chondrosarcomas (Fig. 8.19), but it is not specific. Some low-grade chondrosarcomas may have an anaplastic sarcoma component (chondrosarcoma with additional mesenchymal component or dedifferentiated chondrosarcoma). The aspirate may consist only of

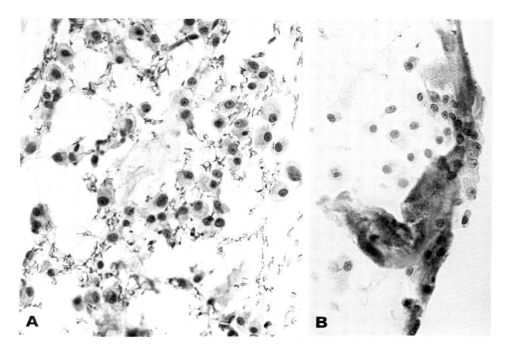

Fig. 8.19. Chondrosarcoma, ABC of a rib mass. A. Smear preparation. Numerous isolated tumor cells, some binucleated, and fragments of chondroid substance. × 260. B. Strong immunoreactivity of chondroid substance with antibody to S-100 protein. × 260.

the anaplastic component, which is consistently negative with antibody to S-100 protein.[165]

Kindblom,[79] in studying the ABC specimens of three chondrosarcomas involving ilium and sacrum, readily determined the typical ultrastructural features (extensive cytoplasmic rough endoplasmic reticulum, glycogen pools, lipid droplets, and many narrow, cytoplasmic extensions from the cell surface) and lacunar arrangement of chondroblasts and chondrocytes. This finding has been confirmed in a larger series with a wider spectrum of chondrocytic neoplasms, some of which were examined by electron microscopy.[177] Compared with osteocytic tumors, the stromal materials have less collagen and larger amounts of fibrillar and granular proteoglycans in chondrogenic neoplasms.

## Chordoma

Chordoma is an uncommon neoplasm that originates from notochordal remnants. In ABC specimens, many small fragments and isolated cells with large amounts of cytoplasm are seen in a background of abundant metachromatic material.[50] Anisonucleosis, intranuclear inclusion and varying-sized intracytoplasmic vacuoles, including classic physalirous cells, may be seen.[50,136] The differential diagnosis includes chondrosarcoma, myxoid liposarcoma, myxopapillary ependymoma, and metastatic clear-cell and mucus-secreting carcinomas. A precise classification can be made by considering the characteristic cytomorphology of the cells in the aspirate; expression of cytokeratins (both low and high molecular weight), vimentin and epithelial membrane antigen, and ultrastructural features.[50,79,136,178] These aspects are well summarized in an assessment of 17 cases of chordoma by Walaas and Kindblom.[178] The differential diagnosis of chordoma by immunocytochemistry is provided in Table 8.7.

## Ewing's Sarcoma

See "Small Cell Neoplasms."

TABLE 8.7. Immunocytochemical Differential Diagnosis of Chordoma

| Neoplasm | S-100 Protein | Neuron-specific Enolase | Cytokeratin | Carcino-embryonic Antigen | Epithelial Membrane Antigen | Vimentin | Glial Fibrillary Acidic Protein |
|---|---|---|---|---|---|---|---|
| Chordoma | + | + | + | − | + | + | − |
| Chondrosarcoma | + | − | − | − | ± | ± | − |
| Myxoid liposarcoma | ± | − | − | − | − | + | − |
| Metastatic mucus-secreting carcinoma | ± | − | + | ± | + | − | − |
| Myxopapillary ependymoma | + | + | − | | | | + |

From Plaza JA, Ballestín C, Pérez-Barrios A, Martínez MA, de Agustín P: Cytologic, cytochemical, immunocytochemical and ultrastructural diagnosis of a sacrococcygeal chordoma in a fine needle aspiration biopsy specimen. *Acta Cytol* 33:89–92, 1989. Used with permission.

## Other Sarcomas

Epithelioid sarcoma is a slow growing soft-tissue sarcoma that is usually located in the extremities. The aspirate consists of dispersed polygonal, triangular, or spindle cells with abundant cytoplasm.[82] The tumor may be misinterpreted as carcinoma or malignant melanoma because of the epithelioid appearance of the cells. The neoplastic cells consistently express cytokeratins, epithelial membrane antigen, and vimentin,[52] a phenotype that is seen in many carcinomas. The diagnosis should be made in the clinical context. Malignant melanomas of soft tissue may be diagnosed on rare occasions in cytological material supported by S-100 protein and HMB-45 expression and ultrastructural studies.[76,144] Fibrosarcoma and other sarcomas are extremely rare and may only express vimentin. Some sarcomas show no evidence of differentiation by either light or electron microscopy and do not express any specific marker and should be regarded as undifferentiated sarcoma.

# GERM-CELL NEOPLASMS

Extragonadal germ-cell tumors may be encountered in NABs of mediastinal and abdominal masses. These neoplasms are not infrequent. Approximately 20% of mediastinal tumors and cysts are germ cell in origin.[137] In ABC specimens, seminomas are usually cellular and composed of loosely arranged large cells with large amounts of cytoplasm and well-defined cell borders. The nuclei are round and usually uniform in size and shape. In a typical case, mature lymphocytes are also seen. The aspirate of embryonal carcinoma consists of isolated cells as well as cohesive papillary-like fragments. The nuclei are pleomorphic. The main differential diagnoses of germ-cell neoplasms include carcinomas and malignant lymphomas. It is very important to consider the diagnosis of germ-cell tumor when dealing with a poorly differentiated neoplasm, especially in young patients. The extragonadal tumors may be primary or metastatic in nature. In patients with a history of previous germ-cell neoplasm, the classification is relatively simple; however, the tumor may be the first presentation of an occult testicular neoplasm. The possibility of a testicular or ovarian primary site, therefore, should always be investigated. The classification may be further compounded by the fact that germ-cell tumors often consist of two or more elements, as illustrated in Figures 8.20 and 8.22.

Except for pure seminoma, most other germ-cell neoplasms are associated with elevated serum levels of human chorionic gonadotrophin (hCG) and alpha-fetoprotein (AFP).[69] Placental alkaline phosphastase (PLAP) is an excellent screening marker for possible germ-cell differentiation in aspirates. Most seminomas, embryonal carcinomas, and endodermal-sinus (yolk sac) tumors express PLAP. Antibody to PLAP is not specific, and some nongerm-cell neoplasms show positivity. Therefore, PLAP-positive neoplasms should be stained further with antibodies to AFP, beta-hCG, alpha-$_1$-antitrypsin, and cytokeratin if adequate smears or a cell block are available. Most endodermal sinus tumors, embryonal carcinomas, and, occasionally immature teratomas express AFP.[69,83] Beta-hCG is secreted by choriocarcinoma and embryonal carcinoma. Alpha-$_1$-antitrypsin may be expressed in embryonal carcinoma and endodermal sinus tumor.[83] Embryonal carcinoma, choriocarcinoma, and endodermal-sinus tumor show cytokeratin positivity.[13,110] Seminomas are either negative or show the occasional positive cell.[13,110] The immature and mature glands of teratomas and teratocarcinomas express cytokeratins as well as carcinoembryonic antigen.[13,69,110] Figure 8.20 illustrates the value of immunocytochemical study in ABC of an anterior mediastinal mass in a 20-year-old man. The aspirate was very cellular and consisted of cohesive fragments of large tumor cells as well as isolated cells. Acinus and papillary-like structures were noted. The neoplastic cells expressed cytokeratin, vimentin (weak), alpha-$_1$-antitrypsin (focal), placental alkaline phosphatase, AFP, and carcinoembryonic antigen (focal), and were negative for beta-hCG. The combination of cytological features and the immunophenotype of the tumor supported the diagnosis of a mixed germ-cell tumor with suggestion of embryonal carcinoma, endodermal-sinus tumor, and teratocarcinoma components. This diagnosis was subsequently confirmed by surgical biopsy.

The cytological features in smear preparations may be suggestive of a germ-cell tumor when they initially present as metastases or occur as primary lesions in sites such as the retroperitoneum or anterior mediastinum.[2,8] Akhtar and associates[2] reported an integrated cytological and electron microscopic study of ABC specimens of seminomas and dysgerminomas. Due to certain distinctive nuclear and other cellular characteristics of germ-cell tumors,[116,155,171] electron microscopy can be a rapid method for confirming this kind of diagnostic impression in ABC. In the seminoma or dysgerminoma category, the regular nuclear profiles, markedly disaggregated chromatin, and the nucleolonemal form of nucleoli, along with scanty cytoplasmic organelles, aggregated glycogen, and small desmosomes, are diagnostic (Fig. 8.21B). A 30-year-old male presented with a 3-month history of intermittent discomfort in his left arm and shoulder. A chest radiograph suggested bilateral hilar lymphadenopathy, but a CT scan showed a large anterior mediastinal mass and an NAB was performed. The ABC smear (Fig. 8.21A) was suggestive of a germ-cell tumor, which was confirmed by electron microscopy on the basis of characteristic nuclear and cytoplasmic features (Fig. 8.21B). Serum AFP and beta-hCG levels were normal.

Embryonal carcinomas are poorly differentiated adenocarcinomas ultrastructurally, with variably developed intercellular lumens and microvilli, but they usually exhibit many aberrant tight junctions. In addition, the tumor cells have a primitive appearance with limited cytoplasmic organelles, accumulations of glycogen, and prominent nucleoli (Fig. 8.22B); a few embryonal carcinoma ABC specimens have been examined ultrastructurally.[24,164] Endodermal-sinus (yolk sac) tumors have two additional diagnostic features: conspicuous amounts of basal lamina material focally between tumor cells (Fig. 8.22B), and relatively large globules of densely staining secretory material within the cytoplasm; the latter appears to represent AFP.[171] The ultrastructural features of four ABC specimens of endodermal-sinus tumor have been reported[8] and are illustrated in the following example of a combined embryonal carcinoma and endodermal-sinus tumor.

A 36-year-old man presented with a 6-week history of increasing left-sided shoulder and anterior chest-wall pain, and a chest radiograph revealed a large anterior mediastinal mass. ABC was consistent with a germ-cell tumor (Fig. 8.22A). Electron microscopy of this aspirate revealed glandular lumens, poorly developed tight junctions, and desmosomes, confirming an embryonal carcinoma (Fig. 8.22B,C). In addition, however, the presence of prominent amounts of basal lamina–like intercellular materials (Fig. 8.22B) indicated endodermal-sinus tumor differentiation.

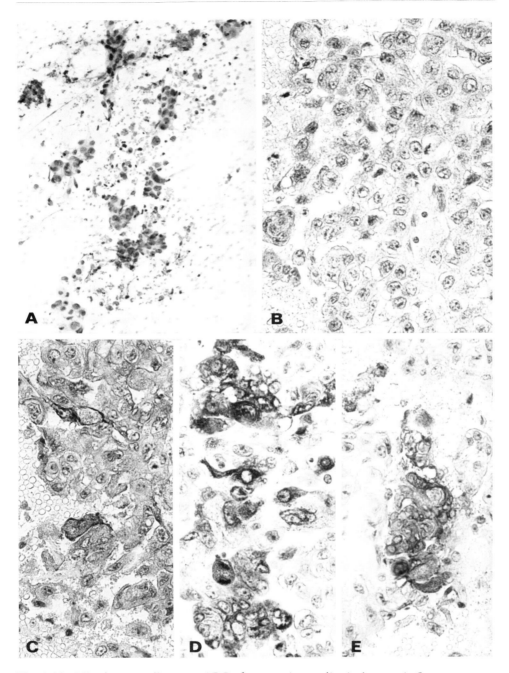

Fig. 8.20. Mixed germ-cell tumor, ABC of an anterior mediastinal mass. **A.** Smear preparation. Note acinus and papillary-like structures. × 104. **B.** Cell block section. Note large nuclei, some with irregular nuclear profiles and prominent nucleoli. × 260. **C.–E.** The neoplastic cells strongly expressed placental alkaline phosphatase (PLAP), alpha-fetoprotein (AFP), and carcinoembryonic antigen (CEA). (C) PLAP. Note cytoplasmic and membranous pattern of staining. × 260. (D) AFP. × 260. (E) CEA. × 260.

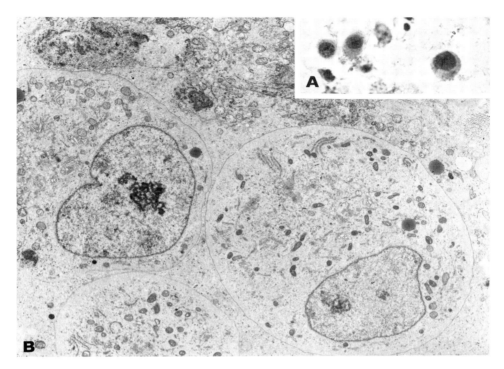

**Fig. 8.21.** Seminoma, ABC of an anterior mediastinal mass. **A.** Smear preparation. Isolated tumor cells with moderate amounts of cytoplasm. × 340. **B.** Tightly apposed tumor cells with considerable pale staining cytoplasm containing limited organelles but no accumulation of glycogen. Nuclei have little condensed chromatin and fenestrated-appearing nucleoli. × 4,100.

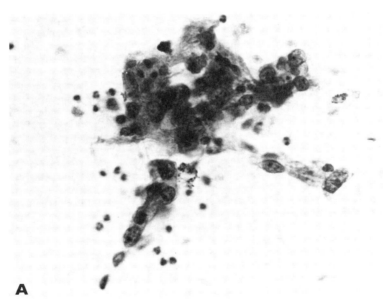

**Fig. 8.22.** Mixed embryonal carcinoma or endodermal sinus tumor, ABC of an anterior mediastinal mass. **A.** Smear preparation. Clusters of tumor cells with moderate amounts of cytoplasm and variably sized nuclei. × 320.

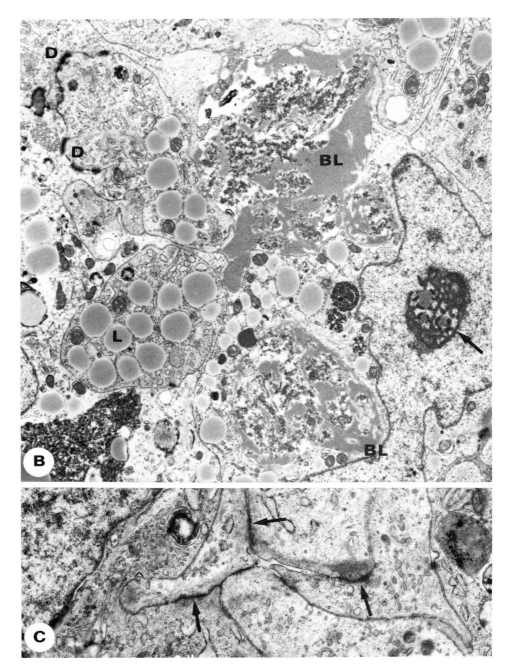

Fig. 8.22. **B.** Variously sized tumor cells have irregularly shaped nuclei with fenestrated nucleoli (arrow). Lipid droplets (L) are plentiful and desmosomes (D) readily identified. Accumulations of basal lamina (BL) following the contours of and filling irregular extracellular spaces between tumor cells and their processes is a distinctive feature indicating an endodermal sinus component to this germ-cell tumor. × 8,300. **C.** Multiple tight junctions (arrows) forming a microlumen-like space indicates the adenocarcinoma-type differentiation of an embryonal carcinoma. × 17,000.

On the basis of this finding, clinical testing included a serum AFP determination that was abnormally elevated and a beta-hCG level that was normal. Chemotherapy produced a considerable reduction in tumor volume, and an attempt was subsequently made to resect the residual tumor.

# REFERENCES

1. Abenoza P, Manivel JC, Swanson PE, et al: Synovial sarcomas: Ultrastructural study and immunohistochemical analysis by a combined peroxidase-antiperoxidase/avidin-biotin-peroxidase complex procedure. *Hum Pathol* 17:1107–1115, 1986.
2. Akhtar M, Ali MA, Huq M, Bakry M: Fine-needle aspiration biopsy of seminoma and dysgerminoma: Cytologic, histologic, and electron microscopic correlations. *Diagn Cytopathol* 6:99–105, 1990.
3. Akhtar M, Ali MA, Owen EW: Application of electron microscopy in the interpretation of fine-needle aspiration biopsies. *Cancer* 48:2458–2463, 1981.
4. Akhtar M, Ali MA, Sabbah R, Bakry M, Nash JE: Fine-needle aspiration biopsy of round cell malignant tumors of childhood: A combined light and electron microscopic approach. *Cancer* 55:1805–1817, 1985.
5. Akhtar M, Ali MA, Sabbah R: Aspiration cytology of Ewing's sarcoma: Light and electron microscopic correlations. *Cancer* 56:2051–2060, 1985.
6. Akhtar M, Ali MA, Sabbah RS, et al: Aspiration cytology of neuroblastoma: Light and electron microscopic correlations. *Cancer* 57:797–803, 1986.
7. Akhtar M, Ali MA, Sabbah R, Sackey K, Bakry M: Aspiration cytology of retinoblastoma: light and electron microscopic correlations. *Diagn Cytopathol* 4:306–311, 1988.
8. Akhtar M, Ali MA, Sackey K, Jackson D, Bakry M: Fine-needle aspiration biopsy diagnosis of endodermal sinus tumor: Histologic and ultrastructural correlations. *Diagn Cytopathol* 6:184–192, 1990.
9. Albores-Saavedra J, Angeles-Angeles A, Nadji M, Henson DE, Alvarez L: Mucinous cystadenocarcinoma of the pancreas, morphologic and immunocytochemical observations. *Am J Surg Pathol* 11:11–20, 1987.
10. Allen BC, Herrera GA: Phenotypic cytoskeletal heterogeneity in melanoma, a challenge to the surgical pathologist evaluating a poorly differentiated neoplasm. An illustrative case. *Ultrastruct Pathol* 15:87–97, 1991.
11. Azumi N, Battifora H: The distribution of vimentin and keratin in epithelial and nonepithelial neoplasms; a comprehensive immunohistochemical study on formalin-and alcohol-fixed tumors. *J Clin Pathol* 88:286–296, 1987.
12. Azumi N, Ben-Ezra J, Battifora H: Immunophenotypic diagnosis of leiomyosarcoma and rhabdomyosarcoma with monoclonal antibodies to muscle-specific actin and desmin in formalin-fixed tissue. *Mod Pathol* 1:469–474, 1988.
13. Battifora H, Sheibani K, Tubbs RR, Kopinski MI, Sun T-T: Antikeratin antibodies in tumor diagnosis; distinction between seminoma and embryonal carcinoma. *Cancer* 54:843–848, 1984.
14. Battifora H, Silva EG: The use of antikeratin antibodies in the immunohistochemical distinction between neuroendocrine (Merkel cell) carcinoma of the skin, lymphoma, and oat cell carcinoma. *Cancer* 58:1040–1046, 1986.
15. Bergh J, Esscher T, Steinholtz L, Nilsson K, Påhlman S: Immunocytochemical demon-

stration of neuron-specific enolase (NSE) in human lung cancers. *Am J Clin Pathol* 84:1–7, 1985.

16. Blaauwgeers JL, Troost D, Dingemans KP, Taat CW, van den Tweel JG: Multifocal rhabdomyoma of the neck: Report of a case studied by fine needle aspiration, light and electron microscopy, histochemistry and immunohistochemistry. *Am J Surg Pathol* 13:791–799, 1989.

17. Blobel GA, Gould VE, Moll R, et al: Coexpression of neuroendocrine markers and epithelial cytoskeletal proteins in bronchopulmonary neuroendocrine neoplasms. *Lab Invest* 52:39–51, 1985.

18. Bolen JW, Thorning D: Benign lipoblastoma and myxoid liposarcomas: A comparative light and electron microscopic study. *Am J Surg Pathol* 4:163–169, 1980.

19. Brecher ME, Franklin WA: Absence of mononuclear phagocyte antigens in malignant fibrous histiocytoma. *Am J Clin Pathol* 86:344–348, 1986.

20. Brooke PK, Wakely Jr PE, Frable WL: Utility of electron microscopy as an adjunct in aspiration cytology diagnosis (abstract). *Lab Invest* 64:22A, 1991.

21. Brown DC, Theaker JM, Banks PM, Gatter KC, Mason DY: Cytokeratin expression in smooth muscle and smooth muscle tumours. *Histopathology* 11:477–486, 1987.

22. Bulinski JC, Kumar S, Titani K, Hauschka SD: Peptide antibody specific for the amino terminus of skeletal muscle—actin. *Proc Natl Acad Sci USA* 80:1506–1510, 1983.

23. Bunn PA Jr, Linnoila I, Minna JD, Carney D, Gazdar AF: Small cell lung cancer, endocrine cells of the fetal bronchus, and other neuroendocrine cells express the Leu-7 antigenic determinant present on natural killer cells. *Blood* 65:764–768, 1985.

24. Burns BF, Dardick I: Mixed germ cell tumour of the mediastinum (seminoma, embryonal carcinoma, choriocarcinoma and teratoma): Light and electron microscopic cytology and histological investigation—Letter to the case. *Pract Res Pathol* 185:511–513, 1989.

25. Bussolati G, Gugliotta P, Sapina A, Eusebi V, Lloyd RV: Chromogranin-reactive endocrine cells in argyrophilic carcinomas ("carcinoids") and normal tissue of the breast. *Am J Pathol* 120:186–192, 1985.

26. Capo V, Ozzello L, Fenoglio CM, Lombardi L, Rilke F: Angiosarcomas arising in edematous extremities: Immunostaining for factor VIII-related antigen and ultrastructural features. *Hum Pathol* 16:144–150, 1985.

27. Carstens PHB: The Weibel-Palade body in the diagnosis of endothelial tumors. *Ultrastruct Pathol* 2:315–325, 1981.

28. Carter RL, McCarthy KP, Machin LG, et al: Expression of desmin and myoglobin in rhabdomyosarcomas and in developing skeletal muscle. *Histopathology* 15:585–595, 1989.

29. Chang T-K, Li C-Y, Smithson WA: Immunocytochemical study of small round cell tumors in routinely processed specimens. *Arch Pathol Lab Med* 113:1343–1348, 1989.

30. Chejfec G, Falkmer S, Grimelius L, et al: Synaptophysin; a new marker for pancreatic neuroendocrine tumors. *Am J Surg Pathol* 11:241–247, 1987.

31. Chess Q, Hajdu SI: The role of immunoperoxidase staining in diagnostic cytology. *Acta Cytol* 130;1–7, 1986.

32. Coindre J-M, De Mascarel A, Trojani M, De Mascarel I, Pages A: Immunohistochemical study of rhabdomyosarcoma. Unexpected staining with S100 protein and cytokeratin. *J Pathol* 155:127–132, 1988.

33. Dardick I, Hammar SP, Scheithauer B: Ultrastructural spectrum of hemangiopericy-

toma: A comparative study of fetal, adult, and neoplastic pericytes. *Ultrastruct Pathol* 13:111–154, 1989.

34. Dardick I, Ramjohn S, Thomas MJ, Jeans D, Hammar SP: Synovial sarcoma: interrelationship of the biphasic and monophasic subtypes. *Pathol Res Pract* 1991 (in press).

35. de Jong ASH, van Kessel-van Vark M, van Heerde P: Fine needle aspiration biopsy diagnosis of rhabdomyosarcoma: An immunocytochemical study. *Acta Cytol* 31:573–577, 1987.

36. de Jong ASH, Van Vark M, Albus-Lutter Ch E, Van Raamsdonk W, Voûte PA: Myosin and myoglobin as tumor markers in the diagnosis of rhabdomyosarcoma, a comparative study. *Am J Surg Pathol* 8:521–528, 1984.

37. Desai J, Katz RL, Ordóñez N: Rhabdomyosarcoma: A cytologic and immunocytochemical study. *Acta Cytol* 32:783, 1988.

38. Dickersin GR: The contributions of electron microscopy in the diagnosis and histogenesis of controversial neoplasms. *Clin Lab Med* 4:123–164, 1984.

39. Dickersin GR: The electron microscopic spectrum of nerve sheath tumors. *Ultrastruct Pathol* 11:103–146, 1987.

40. Dickersin GR, Rosenberg AE: The ultrastructure of small-cell osteosarcoma, with a review of the light microscopy and differential diagnosis. *Hum Pathol* 22:267–275, 1991.

41. Domagala W, Lubinski J, Lasota J, et al: Neuroendocrine (Merkel-cell) carcinoma of the skin: Cytology, intermediate filament typing and ultrastructure of tumor cells in fine needle aspirates. *Acta Cytol* 31:267–275, 1987.

42. du Boulay CEH: Anti-alpha-1-antichymotrypsin specificity. *Am J Surg Pathol* 11:820–825, 1987.

43. du Boulay CEH: Demonstration of alpha-1-antitrypsin and alpha-1-antichymotrypsin in fibrous histiocytomas using the immunoperoxidase technique. *Am J Surg Pathol* 6:559–564, 1982.

44. Erlandson RA: *Diagnostic Transmission Electron Microscopy of Human Tumors*. New York, Masson, 1981.

45. Erlandson RA: Ultrastructural diagnosis of amelanotic malignant melanoma: Aberrant melanosomes, myelin figures or lysosomes? *Ultrastruct Pathol* 11:191–208, 1987.

46. Erlandson RA: The ultrastructural distinction between rhabdomyosarcoma and other undifferentiated "sarcomas." *Ultrastruct Pathol* 11:83–101, 1987.

47. Eusebi V, Bondi A, Rosai J: Immunohistochemical localization of myoglobin in nonmuscular cells. *Am J Surg Pathol* 8:51–55, 1984.

48. Eusebi V, Rilke F, Ceccarelli C, et al: Fetal heavy chain skeletal myosin: An oncofetal antigen expressed by rhabdomyosarcoma. *Am J Surg Pathol* 10:680–686, 1986.

49. Fine RE, Blitz AL: A chemical comparison of tropomyosins from muscle and non-muscle tissues. *J Mol Biol* 95:447–454, 1975.

50. Finley JL, Silverman JF, Dabbs DJ, et al: Chordoma: Diagnosis by fine-needle aspiration biopsy with histologic, immunocytochemical, and ultrastructural confirmation. *Diagn Cytopathol* 2:330–337, 1986.

51. Fisher C: The value of electron microscopy and immunohistochemistry in the diagnosis of soft tissue sarcomas: A study of 200 cases. *Histopathology* 16:441–454, 1990.

52. Fisher C: Epithelioid sarcoma: The spectrum of ultrastructural differentiation in seven immunohistochemically defined cases. *Hum Pathol* 19:265–275, 1988.

53. Foa C, Aubert C: Ultrastructural comparison between cultured and tumor cells of human malignant melanomas. *Cancer Res* 37:3957–3963, 1977.
54. Fukuda T, Tsuneyoshi M, Enjoji M: Malignant fibrous histiocytoma of soft parts: An ultrastructural quantitative study. *Ultrastruct Pathol* 12:117–129, 1988.
55. Garrels JI, Gibson W: Identification and characterization of multiple forms of actin. *Cell* 9:793–805, 1976.
56. Ghadially FN: The role of electron microscopy in diagnostic pathology. *Histopathology* 4:245–262, 1981.
57. Ghadially FN: Alterations in melanosomes in melanomas and pigmentary disorders. *Ultrastructural Pathology of the Cell and Matrix,* 3rd ed, vol 2. London, Butterworths, 1988. Pp. 794–803.
58. González-Cámpora R, Otal-Salaverri C, Hevia-Vázquez A, et al: Fine needle aspiration in myxoid tumors of the soft tissues. *Acta Cytol* 34:179–191, 1990.
59. Gould VE: Synaptophysin: A new and promising panneuroendocrine marker. *Arch Pathol Lab Med* 111:791–794, 1987.
60. Gould VE, DeLellis RA: The neuroendocrine system, in Silverberg SG (ed): *Principles and Practice of Surgical Pathology.* New York, Churchill Livingstone, 1990. Pp 1981–1995.
61. Gould VE, Lee I, Wiedenmann B, et al: Synaptophysin: A novel marker for neurons, certain neuroendocrine cells, and their neoplasms. *Hum Pathol* 17:979–983, 1986.
62. Gould VE, Wiedenmann B, Lee I, et al: Synaptophysin expression in neuroendocrine neoplasms as determined by immunocytochemistry. *Am J Pathol* 126:243–257, 1987.
63. Gray MH, Rosenberg AE, Dickersin GR, Bhan AK: Glial fibrillary acidic protein and keratin expression by benign and malignant nerve sheath tumors. *Hum Pathol* 20:1089–1096, 1989.
64. Gray MH, Rosenberg AE, Dickersin GR, Bhan AK: Cytokeratin expression in epithelioid vascular neoplasms. *Hum Pathol* 21:212–217, 1990.
65. Hajdu SI, Hajdu EO: *Cytopathology of Sarcomas and Other Nonepithelial Tumors.* Philadelphia, W. B. Saunders, 1976.
66. Hammar S, Bockus D, Remington F: Metastatic tumors of unknown origin: An ultrastructural analysis of 265 cases. *Ultrastruct Pathol* 11:209–250, 1987.
67. Höfler H, Kerl H, Lackinger E, Helleis G, Denk H: The intermediate filament cytoskeleton of cutaneous neuroendocrine carcinoma (Merkel cell tumour): Immunohistochemical and biochemical analysis. *Virchows Arch {A}* 406:339–350, 1985.
68. Inoue M, Ueda G, Yamasaki M, Tanaka Y, Hirmatsu K: Immunohistochemical demonstration of peptide hormones in endometrial carcinomas. *Cancer* 54:2127–2131, 1984.
69. Irie T, Watanabe H, Kawaoi A, Takeuchi J: Alpha-fetoprotein (AFP), human chorionic gonadotropin (HCG), and carcinoembryonic antigen (CEA) demonstrated in the immature glands of mediastinal teratocarcinoma: A case report. *Cancer* 50:1160–1165, 1982.
70. Jabi M, Jeans D, Dardick I: Ultrastructural heterogeneity in malignant fibrous histiocytoma of soft issue. *Ultrastruct Pathol* 11:583–592, 1987.
71. Jansson D, Gould VE, Gooch GT, et al: Immunohistochemical analysis of colon carcinomas applying exocrine and neuroendocrine markers. *Acta Pathol Microbiol Immunol Scand* 96:1129–1139, 1988.
72. Jockers-Wretou E, Pfleider G: Quantitation of creatinine kinase isozymes in human tissues and sera by an immunological method. *Clin Chim Acta* 58:223–232, 1975.

73. Johnson TL, Zarbo RJ, Lloyd RV, Crissman JD: Paragangliomas of the head and neck: Immunohistochemical neuroendocrine and intermediate filament typing. *Mod Pathol* 1:216–223, 1988.

74. Jürgens H, Bier V, Harms D, et al: Malignant peripheral neuroectodermal tumors: A retrospective analysis of 42 patients. *Cancer* 61:349–357, 1988.

75. Kawahara E, Oda Y, Ooi A, et al: Expression of glial fibrillary acidic protein (GFAP) in peripheral nerve sheath tumors; a comparative study of immunoreactivity of GFAP, vimentin, S-100 protein, and neurofilament in 38 schwannomas and 18 neurofibromas. *Am J Surg Pathol* 12:115–120, 1988.

76. Keller JM, Listrom MB, Hart B, Olson NJ, Jordan SW: Cytologic detection of penile malignant melanoma of soft parts in pleural effusion using monoclonal antibody HMB-45. *Acta Cytol* 34:393–396, 1990.

77. Khorsand J, Katz RL, Savaraj N: Malignant carcinoid of the pancreas: A cytologic, ultrastructural, and immunocytochemical study of a case diagnosed by fine-needle aspiration of a supraclavicular node metastatsis. *Diagn Cytopathol* 3:222–227, 1987.

78. Kimura N, Sasano N, Yamada R, Satoh J: Immunohistochemical study of chromogranin in 100 cases of pheochromocytoma, carotid body tumour, medullary thyroid carcinoma and carcinoid tumour. *Virchows Arch {A}* 413:33–38, 1988.

79. Kindblom L-G: Light and electron microscopic examination of embedded fine-needle aspiration biopsy specimens in the preoperative diagnosis of soft tissue and bone tumors. *Cancer* 51:2264–2277, 1983.

80. Kindblom L-G, Walaas L, Widéhn S: Ultrastructural studies in the preoperative diagnosis of soft tissue tumors. *Semin Diagn Pathol* 3:317–344, 1986.

81. Koivuniemi A, Nickels J: Synovial sarcoma diagnosed by fine needle aspiration biopsy. *Acta Cytol* 22:515–518, 1978.

82. Koss LG, Woyke S, Olszewski W: *Aspiration Biopsy: Cytologic Interpretation and Histologic Bases.* New York, Igaku-Shoin, 1984.

83. Kurman RJ, Ganjei P, Nadji M: Contribution of immunocytochemistry to the diagnosis and study of ovarian neoplasms. *Int J Gynecol Pathol* 3:3–26, 1984.

84. Lagacé R, Leroy J-P: Comparative electron microscopic study of cutaneous and soft tissue angiosarcomas, post-mastectomy angiosarcoma (Stewart-Treves syndrome) and Kaposi's sarcoma. *Ultrastruct Pathol* 11:161–173, 1987.

85. Lawson CW, Fisher C, Gatter KC: An immunohistochemical study of differentiation in malignant fibrous histiocytoma. *Histopathology* 11:375–383, 1987.

86. Lazarides E, Balzer DR Jr: Specificity of desmin to avian and mammalian muscle cells. *Cell* 14:429–438, 1978.

87. Leader M, Collins PM, Henry K: Anti-α1-antichymotrypsin staining of 194 sarcomas, 38 carcinomas, and 17 malignant melanomas: Its lack of specificity as a tumour marker. *Am J Surg Pathol* 11:133–139, 1987.

88. Lehto V-P, Miettinen M, Dahl D, Virtanen I: Bronchial carcinoid cells contain neural-type intermediate filaments. *Cancer* 54:624–628, 1984.

89. Leong AS-Y, Kan AE, Milios J: Small round cell tumors in childhood: Immunohistochemical studies in rhabdomyosarcoma, neuroblastoma, Ewing's sarcoma, and lymphoblastic lymphoma. *Surg Pathol* 2:5–17, 1989.

90. Leong AS-Y, Milios J: An assessment of a melanoma-specific antibody (HMB-45) and other immunohistochemical markers of malignant melanoma in paraffin-embedded tissues. *Surg Pathol* 2:137–145, 1989.

91. Liem KH, Yen S-H, Salomon GD, Shelanski ML: Intermediate filaments in nervous tissues. *J Cell Biol* 79:637–645, 1978.
92. Linnoila RI, Gazdar AF: Non-small cell lung carcinoma with neuroendocrine features. *American Society of Clinical Pathologists* Check Sample AP 90-3 (AP-195):1–5, 1990.
93. Linnoila RI, Mulshine JL, Steinberg SM, et al: Neuroendocrine differentiation in endocrine and nonendocrine lung carcinomas. *Am J Clin Pathol* 90:641–652, 1988.
94. Lloyd RV: Immunohistochemical localization of chromogranin in normal and neoplastic endocrine tissues. *Pathol Annu* 22(Part 2):69–90, 1987.
95. Lloyd RV, Mervak T, Schmidt K, Warner TFCS, Wilson BS: Immunohistochemical detection of chromogranin and neuron-specific enolase in pancreatic endocrine neoplasms. *Am J Surg Pathol* 8:607–614, 1984.
96. Lloyd RV, Sisson JC, Shapiro B, Verhofstad AAJ: Immunohistochemical localization of epinephrine, norepinephrine, catecholamine-synthesizing enzymes, and chromogranin in neuroendocrine cells and tumors. *Am J Pathol* 125:45–54, 1986.
97. Lloyd RV, Wilson BS, Kovacs K, Ryan N: Immunohistochemical localization of chromogranin in human hypophyses and pituitary adenomas. *Arch Pathol Lab Med* 109:515–517, 1985.
98. Longtine JA, Pinkus GS, Fujiwara K, Carson JM: Immunocytochemical localization of smooth muscle myosin in normal human tissues. *J Histochem Cytochem* 33:179–184, 1985.
99. Mackay B, Fanning T, Bruner JM, Steglich MC: Diagnostic electron microscopy using fine needle aspiration biopsies. *Ultrastruct Pathol* 11:659–672, 1987.
100. Mackay B, Ordóñez NG, Huang W-L: Ultrastructural and immunocytochemical observations on angiosarcomas. *Ultrastruct Pathol* 13:97–110, 1989.
101. Mackay B, Ro J, Floyd C, Ordóñez NG: Ultrastructural observations on smooth muscle tumors. *Ultrastruct Pathol* 11:593–607, 1987.
102. Mazur MT, Katzenstein A-LA: Metastatic melanoma: The spectrum of ultrastructural morphology. *Ultrastruct Pathol* 1:337–356, 1980.
103. Mierau G: Extraskeletal Ewing's sarcoma (peripheral neuroepithelioma). *Ultrastruct Pathol* 9:91–98, 1985.
104. Miettinen M: Synaptophysin and neurofilament proteins as markers for neuroendocrine tumors. *Arch Pathol Lab Med* 111:813–818, 1987.
105. Miettinen M: Immunoreactivity for cytokeratin and epithelial membrane antigen in leiomyosarcoma. *Arch Pathol Lab Med* 112:637–640, 1988.
106. Miettinen M: Keratin subsets in spindle cell sarcomas; keratins are widespread but synovial sarcoma contains a distinctive keratin polypeptide pattern and desmoplakins. *Am J Pathol* 138:505–513, 1991.
107. Miettinen M, Franssila K: Immunochemical spectrum of malignant melanoma: The common presence of keratins. *Lab Invest* 61:623–628, 1989.
108. Miettinen M, Lehto V-P, Dahl D, Virtanen I: Varying expression of cytokeratin and neurofilaments in neuroendocrine tumors of the human gastrointestinal tract. *Lab Invest* 52:429–436, 1985.
109. Miettinen M, Rapola J: Immunohistochemical spectrum of rhabdomyosarcoma and rhabdomyosarcoma-like tumors; expression of cytokeratin and the 68-kD neurofilament protein. *Am J Surg Pathol* 13:120–132, 1989.
110. Miettinen M, Virtanen I, Talerman A: Intermediate filament proteins in human testis and testicular germ-cell tumors. *Am J Pathol* 120:402–410, 1985.

111. Miller TR, Bottles K, Abele JS, Beckstead JH: Neuroblastoma diagnosed by fine needle aspiration biopsy. *Acta Cytol* 29:461–468, 1985.

112. Miralles TG, Gosalbez F, Menéndez P, et al: Fine needle aspiration cytology of soft-tissue lesions. *Acta Cytol* 30:671–678, 1986.

113. Moll R, Franke WW: Cytoskeletal differences between human neuroendocrine tumors: A cytoskeletal protein of molecular weight 46,000 distinguishes cutaneous from pulmonary neuroendocrine neoplasms. *Differentiation* 30:165–175, 1985.

114. Moll R, Lee I, Gould VE, et al: Immunocytochemical analysis of Ewing's tumors: Patterns of expression of intermediate filaments and desmosomal proteins indicate cell type heterogeneity and pluripotential differentiation. *Am J Pathol* 127:288–304, 1987.

115. Moll R, Osborn M, Hartschuh W, et al: Variability of expression and arrangement of cytokeratin and neurofilaments in cutaneous neuroendocrine carcinomas (Merkel cell tumors): Immunocytochemical and biochemical analysis of twelve cases. *Ultrastruct Pathol* 10:473–495, 1986.

116. Monaghan P, Raghavan D, Neville M: Ultrastructural studies of xenografted human germ cell tumors. *Cancer* 49:683–697, 1982.

117. Mooi WJ, Dewar A, Springall D, Polak JM, Addis BJ: Non-small cell lung carcinomas with neuroendocrine features. A light microscopic, immunohistochemical and ultrastructural study of 11 cases. *Histopathology* 13:329–337, 1988.

118. Mukai M, Torikata C, Iri H, et al: Expression of neurofilament triplet proteins in human neural tumors, an immunohistochemical study of paraganglioma, ganglioneuroma, ganglioneuroblastoma, and neuroblastoma. *Am J Pathol* 122:28–35, 1986.

119. Nakhleh RE, Wick MR, Rocamora A, Swanson PE, Dehner LP: Morphologic diversity in malignant melanomas. *Am J Clin Pathol* 93:731–740, 1990.

120. Nemes Z, Thomázy V: Factor XIIIa and the classic histiocytic markers in malignant fibrous histiocytoma: A comparative immunohistochemical study. *Hum Pathol* 19:822–829, 1988.

121. Nesland JM, Holm R, Johannessen JV, Gould VE: Neuron specific enolase immunostaining in the diagnosis of breast carcinomas with neuroendocrine differentiation: Its usefulness and limitations. *J Pathol* 148:35–43, 1986.

122. Neuhold N, Artlieb U, Wimmer M, Krisch I, Schratter M: Aspiration cytology, immunocytochemistry and electron microscopy of a malignant peripheral neuroectodermal tumor: A case report. *Acta Cytol* 33:74–79, 1989.

123. Nguyen G-K: What is the value of fine-needle aspiration biopsy in the cytodiagnosis of soft-tissue tumors? *Diagn Cytopathol* 4:352–355, 1988.

124. Nordgren H, Åkerman M: Electron microscopy of fine needle aspiration biopsy from soft tissue tumors. *Acta Cytol* 26:179–188, 1982.

125. Norton AJ, Thomas JA, Isaacson PG: Cytokeratin-specific monoclonal antibodies are reactive with tumours of smooth muscle derivation. An immunocytochemical and biochemical study using antibodies to intermediate filament cytoskeletal proteins. *Histopathology* 11:487–499, 1987.

126. O'Connor DT, Deftos LJ: Secretion of chromogranin A by peptide-producing endocrine neoplasms. *N Engl J Med* 314:1145–1151, 1986.

127. Ohashi K, Maruyama K: A new structural protein located in the Z lines of chicken skeletal muscle. *J Biochem* 85:1103–1105, 1979.

128. Om A, Ghose T: Use of anti-skeletal muscle antibody from myasthenic patients in the diagnosis of childhood rhabdomyosarcomas. *Am J Surg Pathol* 11:272–276, 1987.

129. Ordóñez NG, Mahfouz SM, Mackay B: Synovial sarcoma: An immunohistochemical and ultrastructural study. *Hum Pathol* 21:733–749, 1990.
130. Osborn M, Hill C, Altmannsberger M, Weber K: Monoclonal antibodies to titin in conjunction with antibodies to desmin separate rhabdomyosarcomas from other tumor types. *Lab Invest* 55:101–108, 1986.
131. Periasamy M, Wieczorek DF, Nadal-Ginard B: Characterization of a developmentally regulated perinatal myosin heavy-chain gene expressed in skeletal muscle. *J Biol Chem* 259:13573–13578, 1984.
132. Persson S, Willems JS, Kindblom LG, Angervall L: Alveolar soft part sarcoma: An immunohistochemical, cytologic and electron-microscopic study and quantitative DNA analysis. *Virchows Arch {A}* 412:499–513, 1988.
133. Pettinato G, De Chiara A, Insabato L, et al: Neuroendocrine (Merkel cell) tumor of the skin: Fine-needle aspiration cytology, histology, electron microscopy and immunohistochemistry of 12 cases. *Appl Pathol* 6:17–27, 1988.
134. Pettinato G, De Chiara A, Insabato L: Diagnostic significance of intermediate filament buttons in fine needle aspirates of neuroendocrine (Merkel-cell) carcinoma of the skin. *Acta Cytol* 33:420–421, 1989.
135. Pettinato G, Manivel JC, Petrella G, De Chiara A, Cali A: Primary osteogenic sarcoma and osteogenic metaplastic carcinoma of the breast, immunocytochemical identification in fine needle aspirates. *Acta Cytol* 33:620–626, 1989.
136. Plaza JA, Ballestín C, Pérez-Barrios A, Martínez MA, de Agustín P: Cytologic, cytochemical, immunocytochemical and ultrastructural diagnosis of a sacrococcygeal chordoma in a fine needle aspiration biopsy specimen. *Acta Cytol* 33:89–92, 1989.
137. Rosai J: *Ackerman's Surgical Pathology,* ed 7. St. Louis, CV Mosby, 1989.
138. Rupp M, Hafiz MA, Khalluf E, Sutula M: Fine needle aspiration in stromal sarcoma of the breast: Light and electron microscopic findings with histologic correlation. *Acta Cytol* 32:72–74, 1988.
139. Said JW, Vimadalal S, Nash G, et al: Immunoreactive neuron-specific enolase, bombesin, and chromogranin as markers for neuroendocrine lung tumors. *Hum Pathol* 16:236–240, 1985.
140. Sápi Z, Bodó M, Megyesi J, Rahóty P: Fine needle aspiration cytology of biphasic synovial sarcoma of soft tissue: Report of a case with ultrastructural, immunologic and cytophotometric studies. *Acta Cytol* 34:69–73, 1990.
141. Schmid KW, Fischer-Colbrie R, Hagn C, et al: Chromogranin A and B and secretogranin II in medullary carcinomas of the thyroid. *Am J Surg Pathol* 11:551–556, 1987.
142. Schmidt D, Steen A, Voss C: Immunohistochemical study of rhabdomyosarcoma: Unexpected staining with S-100 protein and cytokeratin. *J Pathol* 157:83, 1989.
143. Schulz A, Jundt G, Berghäuser K-H, Gehron-Robey P, Termine JD: Immunohistochemical study of osteonectin in various types of osteosarcoma. *Am J Pathol* 132:233–238, 1988.
144. Schwartz JC, Zollars PR: Fine needle aspiration cytology of malignant melanoma of soft parts, report of two cases. *Acta Cytol* 34:397–400, 1990.
145. Seidal T, Walaas L, Kindblom L-G, Angervall L: Cytology of embryonal rhabdomyosarcoma: A cytologic, light microscopic, electron microscopic, and immunohistochemical study of seven cases. *Diagn Cytopathol* 4:292–299, 1988.
146. Shaw JA, Vance RP, Geisinger KR, Marshall RB: Islet cell neoplasms, a fine-needle aspiration cytology study with immunocytochemical correlations. *Am J Clin Pathol* 94:142–149, 1990.

147. Shoup SA, Johnston WW, Siegler HF, et al: A panel of antibodies useful in the cytologic diagnosis of metastatic melanoma. *Acta Cytol* 34:385–391, 1990.
148. Sibley RK, Dahl D: Primary neuroendocrine (Merkel cell?) carcinoma of the skin: II. An immunohistochemical study of 21 cases. *Am J Surg Pathol* 9:109–116, 1985.
149. Sikri KL, Varndel IM, Hamid QA, et al: Medullary carcinoma of the thyroid: An immunocytochemical and histochemical study of 25 cases using eight separate markers. *Cancer* 56:2481–2491, 1985.
150. Silverman JF, Lannin DL, Larkin EW, Feldman P, Frable WJ: Fine-needle aspiration cytology of postirradation sarcomas, including angiosarcoma, with immunocytochemical confirmation. *Diagn Cytopathol* 5:275–281, 1989.
151. Silverman JF, Dabbs DJ, Ganick DJ, Holbrook CT, Geisinger KR: Fine needle aspiration cytology of neuroblastoma, including peripheral neuroectodermal tumor, with immunocytochemical and ultrastructural confirmation. *Acta Cytol* 32:367–376, 1988.
152. Skalli O, Gabbiani G, Babaï F, et al: Intermediate filament proteins and actin isoforms as marker of soft tissue tumor differentiation and origin. II. Rhabdomyosarcomas. *Am J Pathol* 130:515–531, 1988.
153. Skoog L, Schmitt FC, Tani E: Neuroendocrine (Merkel-cell) carcinoma of the skin: Immunocytochemical and cytomorphologic analysis on fine-needle aspirates. *Diagn Cytopathol* 6:53–57, 1990.
154. Sneige N, Ordóñez NG, Veanattukalathil S, Samaan NA: Fine-needle aspiration cytology in pancreatic endocrine tumors. *Diagn Cytopathol* 3:35–40, 1987.
155. Srigley JR, Mackay B, Toth P, Ayala A: The ultrastructure and histogenesis of male germ neoplasia with emphasis on seminoma with early carcinomatous features. *Ultrastruct Pathol* 12:67–86, 1988.
156. Stefansson K, Wollmann R, Jerkovic M: S-100 protein in soft-tissue tumors derived from schwann cells and melanocytes. *Am J Pathol* 106:261–268, 1982.
157. Steiner GC: Neuroectodermal tumor versus Ewing's sarcoma—immunohistochemical and electron microscopic observations, in Roessner A (ed): *Current Topics in Pathology. Biological Characterization of Bone Tumors.* Berlin, Springer-Verlag, 1989. Pp 1–29.
158. Swanson PE: Heffalumps, Jagulars, and Cheshire cats; a commentary on cytokeratins and soft tissue sarcomas. *Am J Clin Pathol* 95:S2–S7, 1991.
159. Swanson PE, Manivel JC, Wick MR: Immunoreactivity for Leu-7 in neurofibrosarcoma and other spindle cell sarcomas of soft tissue. *Am J Pathol* 126:546–560, 1987.
160. Swanson PE, Scheithauer BW, Manivel JC, Wick MR: Epithelial membrane antigen reactivity in mesenchymal neoplasms: An immunohistochemical study of 306 soft tissue sarcomas. *Surg Pathol* 2:313–322, 1989.
161. Szpak CA, Shelburne J, Linder J, Klintworth GK: The presence of stage II melanosomes (premelanosomes) in neoplasms other than melanomas. *Mod Pathol* 1:35–43, 1988.
162. Szyfelbein WM, Ross JS: Carcinoids, atypical carcinoids, and small-cell carcinomas of the lung: differential diagnosis of fine-needle aspiration specimens. *Diagn Cytopathol* 4:1–8, 1988.
163. Tabatowski K, Vollmer RT, Tello JW, et al: The use of a panel of monoclonal antibodies in ultrastructurally characterized small cell carcinoma of the lung. *Acta Cytol* 32:667–674, 1988.
164. Taccagni GL, Parafioriti A, Dell'Antonio G, Crespi G: Mixed germ cell tumour of the mediastinum (seminoma, embryonal carcinoma, choriocarcinoma and teratoma): Light

and electron microscopic cytology and histological investigation. *Pract Res Pathol* 185:506–510, 1989.

165. Têtu B, Ordóñez NG, Ayala AG, Mackay B: Chondrosarcoma with additional mesenchymal component (dedifferentiated chondrosarcoma) II. An immunohistochemical and electron microscopic study. *Cancer* 58:287–298, 1986.

166. Triche TJ, Askin FB, Kissane JM: Neuroblastoma, Ewing's sarcoma, and the differential diagnosis of small-, round-, blue-cell tumors, in Finegold M, (ed): *Pathology of Neoplasia in Children and Adolescents*. Philadelphia, W. B. Saunders, 1986: Pp 145–195.

167. Triche TJ, Askin FB: Neuroblastoma and differential diagnosis of small-, round-, blue-cell tumors. *Hum Pathol* 14:569–595, 1983.

168. Truong LD, Rangdaeng S, Cagle P, et al: The diagnostic utility of desmin; a study of 584 cases and review of the literature. *Am J Clin Pathol* 93:305–314, 1990.

169. Tsokos M, Linnoila RI, Chandra RS, Triche TJ: Neuron-specific enolase in the diagnosis of neuroblastoma and other small, round-cell tumors in children. *Hum Pathol* 15:575–584, 1984.

170. Tsokos M, Howard R, Costa J: Immunohistochemical study of alveolar and embryonal rhabdomyosarcoma. *Lab Invest* 48:148–155, 1983.

171. Ulbright TM, Roth LM, Brodhecker CA: Yolk sac differentiation in germ cell tumors. A morphologic study of 50 cases with emphasis on hepatic, enteric and parietal yolk sac features. *Am J Surg Pathol* 10:151–164, 1986.

172. Van Muijen GNP, Ruiter DJ, Van Leeuwen C, et al: Cytokeratin and neurofilament in lung carcinomas. *Am J Pathol* 116:363–369, 1984.

173. Vandekerckhove J, Weber K: At least six different actins are expressed in a higher mammal: An analysis based on the amino acid sequence of the amino-terminal tryptic peptide. *J Mol Biol* 126:783–802, 1978.

174. Vinores SA, Bonnin JM, Rubinstein LJ, Marangos PJ: Immunohistochemical demonstration of neuron-specific enolase in neoplasms of the CNS and other tissues. *Arch Pathol Lab Med* 108:536–540, 1984.

175. Visscher DW, Zarbo RJ, Trojanowski JQ, Sakr W, Crissman JD: Neuroendocrine differentiation in poorly differentiated lung carcinomas: A light microscopic and immunohistologic study. *Mod Pathol* 3:508–512, 1990.

176. Vuzevski VD, van der Heul RO: Comparative ultrastructure of soft-tissue myxoid tumors. *Ultrastruct Pathol* 12:87–105, 1988.

177. Walaas L, Kindblom L-G: Light and electron microscopic examination of fine-needle aspirates in the preoperative diagnosis of osteogenic tumors: A study of 21 osteosarcomas and two osteoblastomas. *Diagn Cytopathol* 6:27–38, 1990.

178. Walaas L, Kindblom L-G: Fine-needle aspiration biopsy in the preoperative diagnosis of chordoma: A study of 17 cases with application of electron microscopic, histochemical, and immunocytochemical examination. *Hum Pathol* 22:22–28, 1991.

179. Walaas L, Angervall L, Hagmar et al: A correlative cytologic and histologic study of malignant fibrous histiocytoma, an analysis of 40 cases examined by fine-needle aspiration cytology. *Diagn Cytopathol* 2:46–54, 1986.

180. Walaas L, Kindblom L-G, Gunterberg B, Bergh P: Light and electron microscopic examination of fine-needle aspirates in the preoperative diagnosis of cartilaginous tumors. *Diagn Cytopathol* 6:396–408, 1990.

181. Walts AE, Said JW, Shintaku IP, Lloyd RV: Chromogranin as a marker of neuroendocrine cells in cytologic material—An immunocytochemical study. *Am J Clin Pathol* 84:273–277, 1985.

182. Weiss SW, Bratthauer GL, Morris PA: Postirradiation malignant fibrous histiocytoma expressing cytokeratin; indications for the immunodiagnosis of sarcomas. *Am J Surg Pathol* 12:554–558, 1988.

183. Weiss SW, Langloss JM, Enzinger FM: Value of S-100 protein in the diagnosis of soft tissue tumors with particular reference to benign and malignant schwann cell tumors. *Lab Invest* 49:299–308, 1983.

184. Whalen RG, Schwartz K, Bouveret P, Sell SM, Gros F: Contractile protein isozymes in muscle development: Identification of the embryonic form of myosin heavy chain. *Proc Natl Acad Sci USA* 76:5197–5201, 1979.

185. White VA, Fanning CV, Ayala AG, et al: Osteosarcoma and the role of fine-needle aspiration: A study of 51 cases. *Cancer* 62:1238–1246, 1988.

186. Wick MR, Swanson PE, Scheithauer BW, Manivel JC: Malignant peripheral nerve sheath tumor, an immunohistochemical study of 62 cases. *Am J Clin Pathol* 87:425–433, 1987.

187. Wilander E, Påhlman S, Sällström J, Lindgren A: Neuron-specific enolase expression and neuroendocrine differentiation in carcinomas of the breast. *Arch Pathol Lab Med* 111:830–832, 1987.

188. Wilson BS, Lloyd RV: Detection of chromogranin in neuroendocrine cells with a monoclonal antibody. *Am J Pathol* 115:458–468, 1984.

189. Wood GS, Beckstead JH, Turner RR, et al: Malignant fibrous histiocytoma tumor cells resemble fibroblasts. Am J Surg Pathol 10:323–335, 1986.

190. Young JL Jr, Miller RW: Incidence of malignant tumors in US children. J Pediatr 86:254–258, 1975.

191. Zarbo RJ, Gown AM, Nagle RB, Visscher DW, Crissman JD: Anomalous cytokeratin expression in malignant melanoma: One- and two-dimensional Western blot analysis and immunohistochemical survey of 100 melanomas. Mod Pathol 3:494–501, 1990.

192. Zbieranowski I, Bedard YC: Fine needle aspiration of schwannomas: Value of electron microscopy and immunocytochemistry in the preoperative diagnosis. Acta Cytol 33:381–384, 1989.

# 9
# Malignant Lymphoma

Needle aspiration biopsy (NAB) has become an integral part of the initial diagnostic procedure for patients with lymphadenopathy. The diagnosis of metastatic carcinoma or other neoplasms with known primary sites is very reliable in lymph node aspirates; there is a high degree of sensitivity and specificity.[29-31] Aspiration biopsy cytology (ABC) has also been used for diagnosis of malignant lymphoma with a generally good correlation between histological and cytological (on ABC specimen) classification.[9] The major limitation of ABC is evaluation of the architectural patterns. Follicular or diffuse patterns of malignant lymphoma cannot be distinguished on cytological specimens. ABC is particularly useful for staging of lymphoma, documentation of recurrences with possible evolution to a more aggressive type, and diagnosis of new malignancies or infectious disease in patients with a history of malignant lymphoma.[9,35,43,55,60] NAB is an ideal method of obtaining tissue for special studies in referred patients with a diagnosis of malignant lymphoma based on light microscopy of surgically removed lymph node when no immunophenotyping was performed. It is also valuable for documentation of recurrent Hodgkin's disease. In new cases of malignant lymphoma, where the diagnosis is suggested or even confidently made on ABC specimens, histological confirmation by open biopsy is mandatory. The exceptions are patients whose medical conditions preclude surgery, elderly patients for whom major surgical intervention is not desirable, or when the site of the tumor is difficult to access by surgery. We have had cases where the treatment initiated after ABC diagnosis of intraabdominal and intrathoracic malignant lymphomas without histological confirmation.

The diagnosis of malignant lymphoma in ABC specimens can be difficult and at times is impossible.[5,9,30,31,46,48] This difficulty is particularly true for low- and intermediate-grade lymphomas.[35,49,57,60] When there is a heterogeneous population of lymphoid cells in an aspirate the diagnosis is difficult. Small-cell or mixed small-cleaved and large-cell lymphomas may be indistinguishable from benign lymphoproliferative disorders. In certain small lymphocytic lymphomas the cytological atypia may be minimal or totally absent. Large-cell lymphomas may be difficult to differentiate from nonlymphomas, such as poorly differentiated or undifferentiated carcinomas and malignant melanomas. Similarly, it may be difficult to distinguish

TABLE 9.1. ABC of Lymphoproliferative Disorders—Cytologic Categories

1. Positive for malignant cells, non-Hodgkin's lymphoma
    Large cell
    Small cleaved cell
    Mixed small cleaved and large cell
    Small noncleaved cell
    Small lymphocytic cell
    Immunoblastic cell
    Lymphoblastic cell
2. Positive for malignant cells, consistent with non-Hodgkin's lymphoma
3. Positive for malignant cells, consistent with Hodgkin's lymphoma
4. Lymphoma versus benign hyperplasia
5. Negative for malignant cells
6. Nondiagnostic (e.g., inadequate sample, necrosis)

small-cell lymphomas from other small-cell neoplasms, particularly in childhood tumors. Immunocytochemical study, therefore, in conjunction with cytomorphology, is essential for an accurate diagnosis. We try to group ABC findings of lymphoproliferative disorders into one of six cytologic categories (Table 9.1). The degree of certainty depends on the experience of the observer and proper immunophenotyping. Other studies such as hybridization techniques or gene rearrangement studies may also be required.[9,23]

# LYMPHOID MARKERS

## Leukocyte Common Antigen

Leukocyte common antigen (LCA) is a predominantly membranous antigen with a molecular weight of 200 kd, which is expressed by lymphohemopoietic cells including B- and T-lymphocytes, macrophages, histiocytes, and polymorphs.[19,32] Some cytoplasmic staining is also seen. Monocytes, mast cells, and, occasionally, Reed-Sternberg cells show positive staining.[32] Plasma cells are generally nonreactive. LCA is a glycoprotein composed of a group of structurally related molecules.[32] Monoclonal antibodies to LCA are very reliable and give consistent results; it is therefore a very useful antibody to be used in the panel to characterize undifferentiated neoplasms. Antibody to LCA is a very good screening marker for lymphoid neoplasms (see Plate 5.4B) and can be followed by more specific lymphoid markers.[63] Nearly all non-Hodgkins lymphomas and chronic lymphocytic leukemias are immunoreactive with antibody to LCA,[13,32] and are therefore of great value for supporting the diagnosis of lymphoma and differentiating it from poorly differentiated nonlymphoid neoplasms.[19,32,34,47,64] LCA is especially useful in null-cell malignant lymphomas. Not all lymphomas, however, demonstrate positivity.[19,32,63] Occasionally, false cytoplasmic positivity is seen in nonlymphoid neoplasms. Warnke and Rouse[64] emphasized the requirement of strong membrane staining. Antibody to LCA, therefore, should be used in a panel with other antibodies to characterize a poorly differentiated neoplasm.

## Immunoglobulins

Both monoclonal and polyclonal antibodies to immunoglobulins (light and heavy chain) are available to demonstrate the B-cell nature of a lymphoid population, as well as to determining the clonality of the B cells. Demonstration of monoclonality supports the diagnosis of a B-cell neoplasm (see Plate 5.4C). Monoclonal antibodies work nicely on properly fixed direct smears, cytospin preparations, and frozen tissue sections. They are not generally reactive, however, with paraffin-embedded cells or tissues, mainly due to the denaturation or masking of the antigen during processing. Our experience with polyclonal antibodies has been poor.

## B- and T-cell Markers

Several specific monoclonal antibodies are available that are capable of detecting subpopulation of lymphoid cells (Table 9.2). The majority of the available antibodies only work on appropriately fixed cytospin or direct smear preparations or frozen tissue sections and do not react with paraffin-embedded cells or tissues. There are, however, a few monoclonal antibodies, such as L26 (a pan-B-cell marker) (see Plate 5.4D) and UCHL1 (a pan-T-cell marker) (see Plate 5.4E), that react with epitopes resistant to fixation/processing; therefore, they work on paraffin-embedded cell or tissue block sections. In contrast to L26, which is a sensitive marker for B lymphocytes, UCHL1 is not very sensitive. Introduction of more of these lineage-specific antibodies reactive on routinely prepared paraffin-embedded cell or tissue block sections is needed,[13,14,44] which could facilitate immunophenotyping of malignant lymphoma in more laboratories.

## Ki-1 Antigen

Ki-1 antigen is a transmembrane glycoprotein with a molecular weight of 90 to 120 kd that is expressed by Reed-Sternberg cells, a minor population of "activated" T and B cells, and Ki-1–positive large-cell lymphomas (see Plate 5.4F).[10,21,41,52] Immunoreactivity is reliable only when it is dot-like or on the surface of neoplastic cells, or both. A weak, diffuse cytoplasmic staining is of no diagnostic relevance.[53] One of the available monoclonal antibodies (Ber-H$_2$) is a high-avidity antibody reactive with a fixative-resistant epitope of Ki-1 antigen; it therefore works on alcohol- or formalin-fixed paraffin-embedded cell or tissue block sections.[53] Proteolytic digestion is required for formalin-fixed cells or tissues.[19] Ki-1 antigen expression is not restricted to lymphoid cells, but is also demonstrated in other cells and neoplasms using monoclonal antibodies to Ki-1 antigen (e.g., benign and malignant mesenchymal tumors,[37] and late-stage macrophage primary cultures[3]) and monoclonal antibody Ber-H$_2$ (e.g., embryonal carcinoma and embryonal elements of mixed germ tumor,[41] some carcinomas, and malignant melanoma[53]). Antibodies to Ki-1 antigen are therefore useful in conjunction with antibodies that are reactive with other cell-specific antigens.

## Leu-M1

Leu-M1 is a myelomonocytic-related marker that is expressed in Reed-Sternberg cells of Hodgkin's disease and some other benign and malignant lymphoid prolifera-

TABLE 9.2. Monoclonal Antibodies Used in Our Laboratory for Lymphoproliferative Disorders

| Cluster Designation | Antibody | Reactivity | Source |
|---|---|---|---|
| $CD1_a$ | Leu-6 | Cortical thymocytes, Langerhans' cells | Becton-Dickinson (San Jose, CA) |
| CD2 | Leu-5b | T cells | Becton-Dickinson |
| CD3 | Leu-4 | T cells | Becton-Dickinson |
| CD4 | Leu-3a | T subset (helper/inducer) | Becton-Dickinson |
| CD5 | Leu-1 | T cells, some low grade B-cell lymphomas | Becton-Dickinson |
| CD8 | Leu-2a | T subset (cytotoxic/suppressor) | Becton-Dickinson |
| CD10 | CALLA | cALL, Pre-B, granulocytes | Becton-Dickinson |
| CD14 | Leu-M3* | Monocytes | Becton-Dickinson |
| CD15 | Leu-M1*† | Reed-Sternberg cells, granulocytes | Becton-Dickinson |
| CD22 | To15 | B cells | Dako Corporation (Carpinteria, CA) |
| CD30 | Ki-1 | Activated T and B cells, Reed-Sternberg cells | Dako Corporation |
| CD30 | Ki-1 (Ber-$H_2$)*† | Activated T and B cells, Reed-Sternberg cells | Dako Corporation |
| CD45RB | LCA*† | B cells, T subset, monocytes, macrophages | Dako Corporation |
| CD45RO | UCHL1*† | T cells, B subset, monocytes, macrophages | Dako Corporation |
| CD57 | Leu-7*† | Natural killer cells, T subset | Becton-Dickinson |
| CD71 | Transferrin receptor | Activated T and B cells, macrophages, proliferating cells | Becton-Dickinson |
| | L-26† | B cells | Dako Corporation |
| | HLA-DR | B cells, monocytes, activated T cells | Becton-Dickinson |
| | DRC | Dendritic reticulum cells | Dako Corporation |
| | Kappa | B cells | Becton-Dickinson |
| | Lambda | B cells | Becton-Dickinson |
| | IgM | B cells | Becton-Dickinson |

*Protease digestion if formalin-fixed.
†Reactive with paraffin-embedded cells or tissues.
CD = cluster designation nomenclature.

tion, such as peripheral T-cell lymphoma.[54,59] The antigen is also expressed in nonlymphoid tumor cells (a variety of carcinomas). In contrast, mesotheliomas, germ-cell tumors, and malignant melanomas generally stain negative with monoclonal antibody to Leu-M1.[54] Positive staining of neoplastic cells with monoclonal antibody to Leu-M1, in conjunction with other antibodies, supports the diagnosis of adenocarcinoma rather than malignant mesothelioma. Most cytomegalovirus-infected cells express Leu-M1 antigen,[50] and recognition of this cross-reactivity is important to avoid misdiagnosis of cytomegalovirus lymphadenitis as Hodgkin's disease.[50]

## Leu-7

Monoclonal antibody to Leu-7 was originally characterized as a lymphoid marker recognizing natural killer cells, a subset of normal lymphocytes.[58] However, it has since been reported in a number of other normal and neoplastic nonlymphoid cells.[6,33,36,38,51,58] Among these are myelin-associated cells in the central and peripheral nervous system and their corresponding neoplasms, various neuroectodermal neoplasms, and prostatic carcinomas (see Plate 5.4G).

Various neuroendocrine cells and neoplams, such as carcinoid tumor, pheochromocytoma, paraganglioma, neuroblastoma, primitive neuroectodermal neoplasms, and small-cell carcinoma of the lung may express Leu-7.[6,38] In a study of 283 small-cell neoplasms, Michels and colleagues[38] demonstrated positive staining in 44% of neuroendocrine carcinomas. Nonendocrine carcinomas were uniformly negative.

Leu-7 expression is also demonstrated in normal, hyperplastic, and malignant prostatic epithelium and has been suggested as a marker for detection of prostatic carcinoma,[36,38,51] especially when used in conjunction with antibody to prostate-specific antigen (see Fig. 10.2). Focal Leu-7 positivity, largely confined to cell membrane, is also observed in some ovarian, endometrial, renal, lung, and breast adenocarcinomas.[36]

# IMMUNOCYTOCHEMISTRY

## Processing

Most cell surface lymphoid markers are destroyed by paraffin embedding; however, these markers can be easily detected in cytospin preparations. In our laboratory, whenever there is a clinical suspicion of malignant lymphoma (deep or superficial site), the aspirate is processed in such a way that proper material for immunophenotyping is available, if there is a need (Table 9.3; Fig. 9.2). Two to four direct smears are prepared by a cytotechnologist, half of the smears are immediately fixed in 95% ethanol, and the other half are air dried. The needle is then rinsed in 10 ml of balanced electrolyte solution (BES; Abbott laboratories, Chicago, IL). Usually two to four passes are needed to obtain adequate material. The alcohol-fixed smears are stained by a rapid Papanicolaou method (see Chapter 2), and the air-dried smears are stained by Leishman (a Romanovsky-type stain) method (Table 9.4). After cytological evaluation of the direct smears, the material in BES is processed (see Table 9.3). Preparation of a monolayer cytospin preparation is essential for immunophenotyping of lymphoid proliferations. In an optimal preparation, the cells are separate from each other but still adequately cellular for immunocytochemical study. Although it is possible to quantitate the number of cells on each slide,[55] we use a semiquantitative technique to adjust the number of drops per cytospin chamber (see Table 9.3). One of the cytospins is stained with Leishman stain to evaluate the cytomorphology and cellularity of the smear before proceeding to immunophenotyping. In contrast to flow cytometry, immunophenotyping of malignant lymphomas

**TABLE 9.3. Processing of Needle Rinse for Immunophenotyping of Lymphoid Proliferations**

1. Place 4 ml of ficoll-hypaque (Sigma's Histopaque-1077) in a 15-ml centrifuge tube (bring to room temperature).
2. Gently overlay the needle rinse specimen onto the ficoll solution.
3. Centrifuge at 1,350 rpm (400 g) for 20 minutes.
4. Using a pipette, carefully remove the supernatant to 1 ml above the interface. Then carefully collect the interface solution and transfer to a 15-ml tube.
5. Add fresh balanced electrolyte solution (BES) and centrifuge at 1,350 rpm for 10 minutes.
6. Remove the supernatant with a pipette without disrupting the cell button.
7. Repeat steps 5 and 6.
8. Pour off the supernatant, leaving 1–2 ml of BES in the tube. Then thoroughly resuspend the cells by vortexing.
9. Put 1 drop of the cell suspension (unstained) on a glass slide, cover with a 24 × 50 mm cover glass and view under a 40× objective lens to determine the number of drops of cell suspension per cytocentrifuge chamber.*
10. Cytocentrifuge at 800 rpm for 5 minutes.
11. Air dry the smears for at least 1 hour.
12. Fix the smears in acetone for 2 minutes (room temperature).
13. Drain and rinse with phosphate buffer saline (PBS).
14. Fix in freshly prepared PLP (periodate lysine paraformaldehyde) for 8 minutes.
15. Rinse in PBS.
16. Start immunostaining (see Table 2.9).

*Refer to the guidelines in Cytospin 2 manual (Shandon Inc.).

in cytospin preparation has the advantage of simultaneous evaluation of the cytomorphology.[55]

Proper fixation is one of the critical aspects of processing of lymphoid infiltrate aspirates for immunophenotyping. Most laboratories use acetone or a combination of acetone and formaldehyde.[4,8,35,39,49,55,60] In our experience, a combination of acetone and periodate lysine paraformaldehyde (PLP) fixation produces optimal results with lymphoid markers. Acetone fixation provides excellent preservation of antigens, whereas the addition of a low concentration of formaldehyde results in good nuclear morphology, which is essential for cytodiagnosis.

**TABLE 9.4. Leishman Staining Method for Air-Dried Preparations**

1. Prepare air-dried smears or cytospins.
2. Apply Leishman stain* to the entire slide and let sit for 8 minutes.
3. Apply buffer mixture (BDH chemicals) directly over the stain. Allow the mixture to remain on the slide 15 minutes.
4. Rinse in tap water.
5. Allow to dry completely.
6. Clear in Xylol for a few seconds.
7. Mount.

*Leishman stain: Gurr (BDH Chemicals), 0.75 gm; Aldrich (Aldrich Chemical Co.), 0.75 gm; and absolute methanol, 1,000 ml. Let stand for 4–5 days. Filter before use.

For most aspirates it is necessary to block the endogenous peroxidase with 0.3% solution of hydrogen peroxide methanol for 10 minutes after the secondary antibody incubation. This step is not required when the red blood cells are removed by ficoll-hypaque (see Table 9.3) or when using other methods such as alkaline phosphatase antialkaline phosphatase or immunogold-silver/Romanovsky techniques.[24]

## Malignant Lymphoma Versus Nonlymphoma Neoplasms

High-grade non-Hodgkin's lymphomas are relatively easy to diagnose in ABC specimens. They are characterized by a dispersed population of monotonous single cells, lack of three-dimensional arrangement, and the presence of lymphoglandular bodies. The distinction, however, between large-cell lymphomas and nonlymphoid neoplasms such as poorly differentiated carcinomas, sarcomas, and malignant melanomas may be difficult. For example, Ki-1–positive large-cell lymphomas can be misdiagnosed as malignant melanoma or carcinoma due to the bizarre cytological features of the neoplastic cells.[10,16,19,62] Similarly, small-cell lymphomas may be difficult to distinguish from other small-cell neoplasms, especially in childhood tumors. In contrast, undifferentiated or poorly differentiated carcinomas may be predominantly composed of isolated cells simulating malignant lymphomas (Fig. 9.1). Also, nonlymphoid neoplasms may be extensively infiltrated by reactive lymphoid cells, which should not be mistaken as tumor cells.

Immunocytochemical study allows confirmation and subclassification of malignant lymphomas and exclusion of nonlymphoma neoplasms.[34,35,47,49,55,57,60,61] A simple panel of diagnostic antibodies to LCA, cytokeratins, vimentin, and S-100 protein/HMB-45 can be useful for a definite classification or it may shorten the list of differential diagnoses. Nearly all non-Hodgkin's lymphomas and chronic lymphocytic leukemias are immunoreactive with antibody to LCA,[13,32] which is of great value in differentiating lymphomas from poorly differentiated nonlymphoid neoplasms.[19,32,34,64] However, not all lymphomas express LCA. A large number of lymphoblastic lymphomas/leukemias and Ki-1 positive large cell lymphomas, as well as occasional immunoblastic lymphomas have been reported to be non-reactive with antibody to LCA.[19,32,63] Similarly, only occasionally myeloma cells may show LCA positivity.[32] When staining for LCA is negative, the nonlymphoid nature of the tumor should be confirmed by other markers, such as cytokeratins, in the panel. In contrast, Warnke and Rouse[64] demonstrated occasional false cytoplasmic positivity in nonlymphoid neoplasms with antibody to LCA and emphasized the requirement of strong membrane staining for an accurate interpretation.

## Immunophenotyping

Cytological diagnosis of low- and intermediate-grade malignant lymphomas can be difficult or impossible, and immunocytochemical study is required for a definite diagnosis on specimens obtained by NAB. Most low-grade lymphomas are of monoclonal B-cell origin;[35] therefore, the demonstration of monoclonality of a B-cell population, expressing predominantly lambda or kappa light chains, is a reliable marker supporting a diagnosis of malignancy over a reactive process, particularly

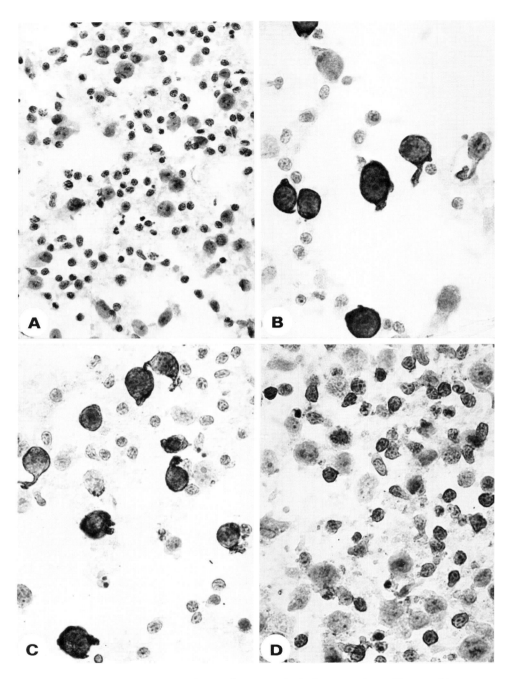

Fig. 9.1. Metastatic carcinoma, ABC of a supraclavicular mass in a 67-year-old woman. A. Smear preparation. Cellular aspirate composed of isolated mixed small and large cells simulating lymphoma. × 260. B.–D. Immunocytochemical stains showed strong staining of large cells with antibodies to cytokeratins (AE1 and AE3). Small cells expressed only leukocyte common antigen (LCA). This immunophenotype confirmed the diagnosis of metastatic carcinoma. (B) Anti-cytokeratin AE1, × 416; (C) Anti-cytokeratin AE3, × 416; (D) LCA, × 416. Further diagnostic workup revealed a poorly differentiated adenocarcinoma of the stomach.

in nonimmunologically compromised patients.[8,15,25,34,35,55,57] The value of immunophenotyping of malignant lymphoma on cytospin preparations is illustrated in Figure 9.2. Light-chain ratios of kappa to lambda or lambda to kappa that are greater than 6 : 1, as defined by Sneige and colleagues,[57] are considered light-chain restriction in ABC specimens and are supportive of a B-cell neoplasm. Ratios between 3 : 1 are 6 : 1 are considered atypical or suggestive of monoclonality. In a study of 220 lymph node aspirates, Sneige and colleagues[55,57] demonstrated monoclonality in all 24 cases of small lymphocytic lymphomas, and the majority (78.6%) of 103 cases of small cleaved-cell, mixed-cell, large-cell, and small noncleaved-cell lymphomas. A variable number (5–40%) of reactive T cells are usually present.[55] Anomalous expression of a pan T-cell antigen (Leu-1) is another characteristic of low-grade lymphomas.[8,35] Diagnosis of plasma-cell tumors may be difficult on ABC specimens. The main differential diagnosis includes reactive infiltrate, immunoblastic lymphoma, and amelanotic melanoma.[45] Using a panel of antibodies to light- and heavy-chain immunoglobulins, LCA, and S-100 protein/HMB-45 can resolve the problem. The demonstration of monoclonality, expressing predominantly kappa or lambda light chain in the cytoplasm, in conjunction with cytomorphology confirm the diagnosis of plasma-cell tumor. Unlike plasma-cell tumor, malignant lymphomas generally express LCA.[45]

Immunophenotyping of malignant lymphoma, of B- or T-cell lineage, is important and dictates the therapeutic approach in some institutions, including ours.[8,40] B- or T-cell lineage is also supportive of a lymphoid rather than a nonlymphoid pathway of differentiation.[25] Immunophenotyping of lymphoma is less expensive and easier to perform on cytospin preparations than by flow cytometry.[39] It is also easier to interpret the results in ABC specimens with mixed cell types, because they can be directly correlated with cytomorphology of the neoplastic cells.[39] Our initial panel consists of monoclonal antibodies to kappa and lambda light chains, a pan T-cell antigen (Leu-4), and a pan B-cell antigen (To15). In some cases, additional markers may be required (see Table 9.2). In contrast to cell or tissue block sections, however, the number of immunostains that can be done on ABC specimens is limited to the number of prepared cytospins, and reaspiration of the lesion may be necessary. It may be particularly difficult to perform a complete T-cell marker analysis on ABC specimens.[28] In contrast to B-cell lymphomas, no clonal marker is available for T-cell lymphomas. The suppressor and helper T cells are not clonal markers. The expression of an aberrant phenotype, however, such as a dominant T-cell subset (helper/inducer or cytotoxic/suppressor) or failure to express some pan T-cell antigens or both in morphologically atypical cells is supportive of a T-cell neoplasm.[15,25,26,35,55,57] The value of immunophenotyping of malignant lymphoma on alcohol-fixed, paraffin-embedded ABC specimens is illustrated in Figure 9.3. Extensive necrosis and inflammatory cell infiltrate cause difficulty in cytomorphological and immunocytochemical interpretation, and perhaps immunocytochemical study should not even be attempted in such cases because of the potential for misinterpretation. Immunophenotyping may be difficult or it may not be representative when the lymph node is focally involved by malignant lymphoma or when numerous nonneoplastic T cells are present in a B-cell lymphoma.[55]

Ki-1–positive large-cell lymphomas have a characteristic cytological picture (Fig. 9.4) consisting of large isolated cells with abundant dense or vacuolated cytoplasm and large nuclei with irregular profiles.[65] Occasionally, binucleated and multilobed

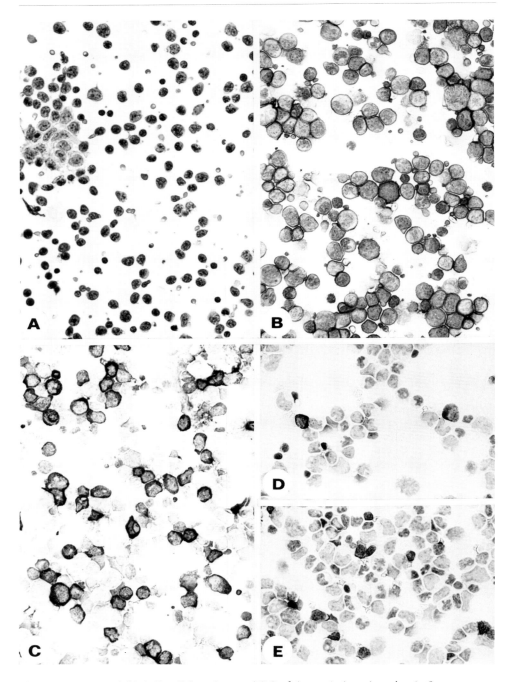

Fig. 9.2. Non-Hodgkin's B-cell lymphoma, ABC of the groin lymph node. **A.** Smear preparation. Cellular aspirate composed of dispersed single cells with occasional tight cell clusters. × 260. **B.–E.** The majority of the cells strongly expressed leukocyte common antigen (LCA) and lambda light chain diagnostic of a B-cell malignant lymphoma. Only occasional cells expressed kappa light chain. A small percentage of reactive T cells, expressing Leu-1, were also present. (B) LCA, × 260; (C) Lambda, × 260; (D) Kappa, × 260; (E) Leu-1, × 260.

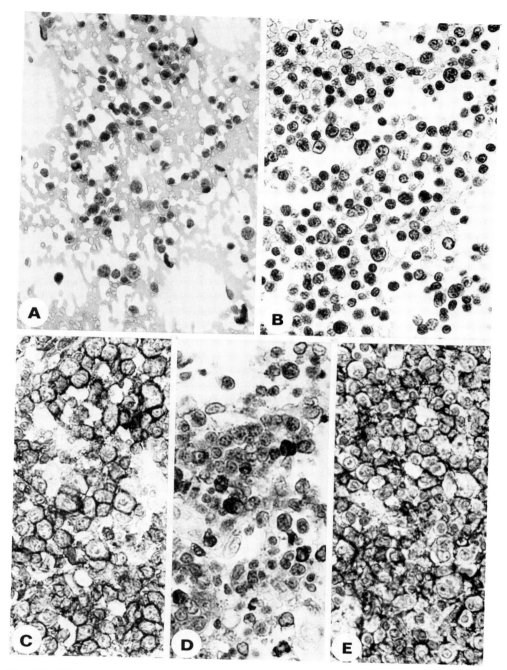

**Fig. 9.3.** Non-Hodgkin's B-cell lymphoma, ABC of the paraaortic lymph node in a 76-year-old woman with a past history of colonic adenocarcinoma and malignant lymphoma. **A.** Smear preparation. Dispersed population of single lymphoid cells. × 260. **B.** Cell block section. Note narrow rim of cytoplasm and irregular nuclear profile. × 416. **C.–E.** The majority of the neoplastic cells stained positively with antibodies to leukocyte common antigen (LCA), kappa light chain, and a pan B-cell marker (L26), confirming the diagnosis of malignant lymphoma. Occasional cells expressed lambda light chain. (C) LCA, × 416; (D) Kappa, × 416; (E) Monoclonal antibody L26, × 416.

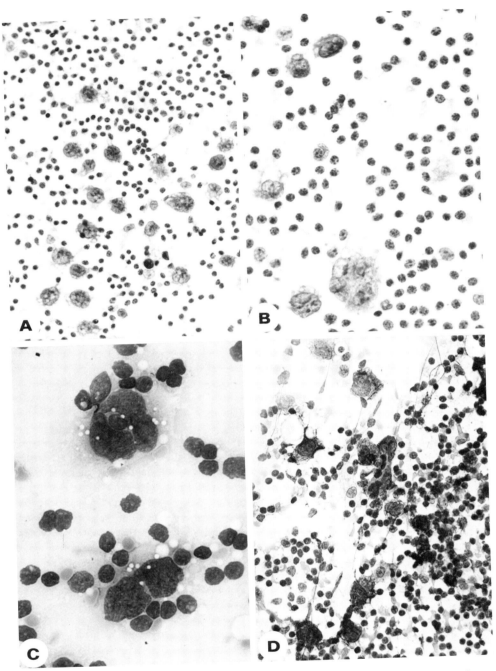

Fig. 9.4. Ki-1–positive large-cell lymphoma, ABC of a lymph node, in a 25-year-old man with inguinal lymphadenopathy of two months' duration. **A.–C.** Smear preparations. (A) Many large single cells in a background of numerous small lymphocytes. × 260. (B) Vacuolated cytoplasm and large nuclei with irregular nuclear profiles. × 416. (C) Note the presence of small cytoplasmic vacuoles in two neoplastic cells. Leishman stain, × 416. **D.** Large cells strongly expressed Ki-1 antigen supporting the diagnosis of Ki-1–positive lymphoma.

or multinucleated cells may also be present. Hodgkin's disease is the main differential diagnosis in ABC smears due to the heterogeneous nature of the cells and presence of atypical large cells resembling Hodgkin's cells or Reed-Sternberg cells.[16,62,65] This misinterpretation is more likely to occur when the lymph node is partially involved and the ratio of atypical large cells to mature lymphoid cells is low.[62] The pleomorphic asppearance of the large cells may suggest malignant melanoma,[16] particularly if the cells express vimentin and S-100 protein and are negative for cytokeratins. The large cells may also be mistaken for metastatic carcinoma.[10,16,19,62] This confusion is often compounded by their tendency to express epithelial membrane antigen and negative staining with antibody to LCA in 38% of cases.[2,10,19,22] Even histologically they may be confused with epithelial tumors.[1] Cytologically, Ki-1–positive large-cell lymphoma should be considered in the differential diagnosis when (1) there is a heterogeneous population of lymphoid cells, (2) the diagnosis of Hodgkin's disease is suspected, and (3) when large-cell pleomorphic neoplasms fail to express LCA and cytokeratins.[16] The cytological suspicion should be confirmed by immunocytochemistry. Because of their diverse immunophenotype,[7,11,65] a panel of antibodies reactive to Ki-1 antigen, B- and T-cell antigens, LCA, cytokeratins, and S-100 protein/HMB-45 is essential for correct diagnosis. The most important marker for the diagnosis is the expression of Ki-1 antigen.[11,19] Diffuse, weak cytoplasmic staining is reported occasionally in carcinomas (e.g., pancreatic and salivary gland) and malignant melanoma.[53] Ki-1 antigen expression is also demonstrated in other neoplasms, including mesenchymal tumors and some germ-cell neoplasms.[37,41] Therefore, the diagnosis must also be supported by demonstration of absence of cytokeratins, melanocyte antigen, and other appropriate markers.[19] More than one-third of cases do not express LCA.[19] The majority of Ki-1–positive large-cell lymphomas are of T-cell phenotype; some are of B-cell origin and some express neither B- nor T-cell markers.[1,11]

ABC is helpful in the primary diagnosis and staging of Hodgkin's disease as well as in documenting the presence of the recurrent or residual disease after therapy.[9,20,27] Cytologically, Hodgkin's disease is characterized by the presence of multilobed or multinucleated and binucleated Reed-Sternberg cells in a polymorphous background, consisting predominantly of activated lymphocytes. A combination of other cells such as mononuclear Hodgkin's cells, plasma cells, eosinophils, and neutrophils; granulomatous elements; metachromatic material; and necrosis are also seen. Reed-Sternberg–like cells may be seen in aspirates of non-Hodgkin's lymphomas, carcinomas, and malignant melanomas. Having the characteristic background is important for accurate interpretation. Reed-Sternberg cells and their mononuclear variants express Leu-M1 and Ki-1 antigens, which supports the diagnosis.[10,12,55] The immunophenotype of the cells in the background, many helper T cells and small populations of polyclonal B cells,[55] is not conclusive, and the immunocytochemical findings should be interpreted in conjunction with cytological findings.

A typical aspirate of reactive lymphoid hyperplasia is composed of a polymorphous cell population, consisting predominantly of small lymphocytes admixed with lymphocytes in different stages of maturation (small and large cells), plasma cells, tangible-body macrophages, eosinophils, and neutrophils. Such aspirates are relatively easy to diagnose. In some cases, however, distinction from malignant lymphoma is difficult and immunocytochemical study may be helpful. Reactive lymph nodes display a polyclonal B-cell phenotype. In Sneige's[55] study of 23 cases, the ratios of kappa to lambda or lambda to kappa were less than 3:1 and the number of T cells ranged from 30 to 60% of the cell population. The T cells consist of a mixture of suppressor and helper cells in variable proportions. Polyclonality does not completely exclude the possibility of malignancy, especially in suspicious lesions, and open biopsy is required in such cases. Genotyping is also required if available.[25] T-cell lymphomas, Hodgkin's disease, and focal involvement of lymph made by a B-cell lymphoma may also show a polyclonal immunophenotype in ABC specimens.[55] In some cases, specific causes of lymphadenopathy, such as infectious mononucleosis and sinus histiocytosis with massive lymphadenopathy, may be suggested by combination of cytomorphology and immunophenotyping.[26,42]

# ELECTRON MICROSCOPY

Generally, whether in surgical pathology or cytopathology, electron microscopy has little application in the routine diagnosis of Hodgkin's and non-Hodgkin's lympho-

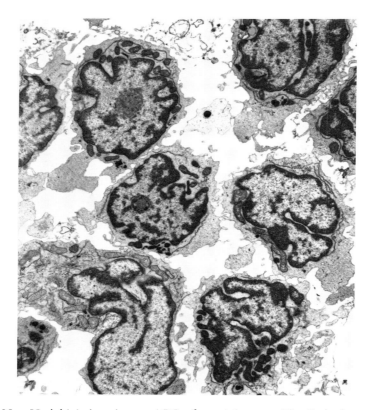

Fig. 9.5. Non-Hodgkin's lymphoma, ABC of a pelvic mass. The limited cytoplasm and organelle content, nuclear features with multiple profile indentations, and the tumor-cell separation provided a diagnosis of non-Hodgkin's lymphoma (likely small cleaved-cell–type). × 4,300.

mas. Because non-Hodgkin's lymphoma is a frequent inclusion in the differential diagnosis of small-, large-, look-alike, and undifferentiated tumors, ultrastructural studies frequently disclose this entity. Within the large-cell category, electron microscopy can also effectively diagnose certain infrequently encountered and often unsuspected lymphomas.[7,18] Lymphocytes, both normal and neoplastic, have certain readily recognized ultrastructural features that are central to the diagnosis of lymphomas.[17] These features include a noncohesive growth pattern without intercellular junctions or basal lamina formation. Characteristic nuclear configurations and chromatin organization, a high nuclear-to-cytoplasmic ratio (primarily due to limited cytoplasm), and scanty cytoplasmic organelles (except for many polyribosomes) also help in the recognition of non-Hodgkin's lymphomas. This was the case where on ABC of a pelvic mass in a 39-year-old woman, diagnosed as a small-cell tumor, proved to be a non-Hodgkin's lymphoma when the glutaraldehyde-fixed tissue fragments from the needle rinse were examined by electron microscopy (Fig. 9.5).

# REFERENCES

1. Agnarsson BA, Kadin ME: Ki-1 positive large cell lymphoma; a morphologic and immunologic study of 19 cases. *Am J Surg Pathol* 12:264–274, 1988.
2. Al Saati T, Caveriviere P, Gorguet B, et al: Epithelial membrane antigen in hematopoietic neoplasms. *Hum Pathol* 17:533–534, 1986.
3. Andreesen R, Brugger W, Löhr GW, Bross KJ: Human macrophages can express the Hodgkin's cell-associated antigen Ki-1 (CD 30). *Am J Pathol* 134:187–192, 1989.
4. Aratake Y, Tamura K, Kotani T, Ohtaki S: Application of the avidin-biotin-complex method for the light microscopic analysis of lymphocyte subsets with monoclonal antibodies on air-dried smears. *Acta Cytol* 32:117–122, 1988.
5. Betsill WL Jr, Hajdu SI: Percutaneous aspiration biopsy of lymph nodes. *Am J Clin Pathol* 73:471–479, 1980.
6. Bunn PA Jr, Linnoila I, Minna JD, Carney D, Gazdar AF: Small cell lung cancer, endocrine cells of the fetal bronchus, and other neuroendocrine cells express the Leu-7 antigenic determinant present on natural killer cells. *Blood* 65:764–768, 1985.
7. Burns BF, Dardick I: Ki-1-positive non-Hodgkin's lymphomas; an immunophenotypic, ultrastructural, and morphometric study. *Am J Clin Pathol* 93:327–332, 1990.
8. Cafferty LL, Katz RL, Ordonez NG, Carrasco CH, Cabanillas FR: Fine needle aspiration diagnosis of intraabdominal and retroperitoneal lymphomas by a morphologic and immunocytochemical approach. *Cancer* 65:72–77, 1990.
9. Carter TR, Feldman PS, Innes DJ Jr, Frierson HF Jr, Frigy AF: The role of fine needle aspiration cytology in the diagnosis of lymphoma. *Acta Cytol* 32:848–853, 1988.
10. Chittal SM, Caveriviere P, Schwarting R, et al: Monoclonal antibodies in the diagnosis of Hodgkin's disease; the search for a rational panel. *Am J Surg Pathol* 12:9–21, 1988.
11. Chott A, Kaserer K, Augustin I, et al: Ki-1-positive large cell lymphoma; a clinicopathologic study of 41 cases. *Am J Surg Pathol* 14:439–448, 1990.
12. Corrigan C, Sewell C, Martin A: Recurrent Hodgkin's disease in the breast; diagnosis of a case by fine needle aspiration and immunocytochemistry. *Acta Cytol* 34:669–672, 1990.

13. Davey FR, Elighetany MT, Kurec AS: Immunophenotyping of hematologic neoplasms in paraffin-embedded tissue sections. *Am J Clin Pathol* 93:517–526, 1990.
14. Davey FR, Gatter KC, Ralfkiaer E, et al: Immunophenotyping of non-Hodgkin's lymphomas using a panel of antibodies on paraffin-embedded tissues. *Am J Pathol* 129:54–63, 1987.
15. Deegan MJ: Membrane antigen analysis in the diagnosis of lymphoid leukemias and lymphomas; differential diagnosis, prognosis as related to immunophenotype, and recommendations for testing. *Arch Pathol Lab Med* 113:606–618, 1989.
16. Dekmezian RH, Sneige N, Siddiqui S, Katz RL: Ki-1-positive anaplastic lymphomas: Fine needle aspiration cytologic features and cytogenetic, immunochemical, molecular and ultrastructural findings in seven cases. *Acta Cytol* 33:731, 1989.
17. Dickersin GR: Electron microscopy of leukemias and lymphomas. *Clin Lab Med* 7:199–247, 1987.
18. Erlandson RA, Filippa DA: Unusual non-Hodgkin's lymphomas and true histiocytic lymphomas. *Ultrastruct Pathol* 13:249–273, 1989.
19. Falini B, Pileri S, Stein H, et al: Variable expression of leukocyte-common (CD45) antigen in CD30 (Ki-1)-positive anaplastic large-cell lymphoma: Implications for the differential diagnosis between lymphoid and non-lymphoid malignancies. *Hum Pathol* 21:624–629, 1990.
20. Friedman M, Kim U, Shimaoka K, et al: Appraisal of aspiration cytology in management of Hodgkin's disease. *Cancer* 45:1653–1663, 1980.
21. Froese P, Lemke H, Gerdes J, et al: Biochemical characterization and biosynthesis of the Ki-1 antigen in Hodgkin's disease and virus-transformed human B and T lymphoid cell lines. *J Immunol* 139:2081–2087, 1987.
22. Fujimoto J, Hata J, Ishii E, et al: Ki-1 lymphoma in childhood: Immunohistochemical analysis and the significance of epithelial membrane antigen (EMA) as a new marker. *Virchows Archiv {A}* 412:307–314, 1988.
23. Hu E, Horning S, Flynn S, et al: Diagnosis of B cell lymphoma by analysis for immunoglobulin gene rearrangements in biopsy specimens obtained by fine needle aspirations. *J Clin Oncol* 4:278–283, 1986.
24. Hughes DA, Kempson MG, Carter NP, Morris PJ: Immunogold-silver/Romanowsky staining: Simultaneous immunocytochemical and morphologic analysis of fine-needle aspirate biopsies. *Transplant Proc* 20:575–576, 1988.
25. Kamat D, Laszewski MJ, Kemp JD, et al: The diagnostic ultility of immunophenotyping and immunogenotyping in the pathologic evaluation of lymphoid proliferations. *Mod Pathol* 3:105–112, 1990.
26. Kardos TF, Kornstein MJ, Frable WJ: Cytology and immunocytology of infectious mononucleosis in fine needle aspirates of lymph nodes. *Acta Cytol* 32:722–726, 1988.
27. Kardos TF, Vinson JH, Behm FG, Frable WJ, O'Dowd GJ: Hodgkin's disease: Diagnosis by fine-needle aspiration biopsy; analysis of cytologic criteria from a selected series. *Am J Clin Pathol* 86:286–291, 1986.
28. Katz RL, Gritsman A, Cabanillas F, et al: Fine-needle aspiration cytology of peripheral T-cell lymphoma; a cytologic, immunologic and cytometric study. *Am J Clin Pathol* 91:120–131, 1989.
29. Kline TS: Aspiration biopsy of the lymph node—yea or nay. *Diagn Cytopathol* 1:3, 1985.
30. Kline TS, Kannan V, Kline IK: Lymphadenopathy and aspiration biopsy cytology; review of 376 superficial nodes. *Cancer* 54:1076–1081, 1984.

31. Koss LG, Woyke S, Olszewski W: *Aspiration Biopsy: Cytologic Interpretation and Histologic Bases.* New York, Igaku-Shoin, 1984. Pp 105–153.
32. Kurtin PJ, Pinkus GS: Leukocyte common antigen—a diagnostic discriminant between hematopoietic and nonhematopoietic neoplasms in paraffin sections using monoclonal antibodies: Correlation with immunologic studies and ultrastructural localization. *Hum Pathol* 16:353–365, 1985.
33. Leong A S-Y, Millios J: An assessment of a melanoma-specific antibody (HMB-45) and other immunohistochemical markers of malignant melanomas in paraffin-embedded tissues. *Surg Pathol* 2:137–145, 1989.
34. Levitt S, Cheng L, DuPuis MH, Layfield LJ: Fine needle aspiration diagnosis of malignant lymphoma with confirmation by immunoperoxidase staining. *Acta Cytol* 29:895–902, 1985.
35. Martin SE, Zhang H-Z, Magyarosy E, et al: Immunologic methods in cytology: Definitive diagnosis of non-Hodgkin's lymphomas using immunologic markers for T-and B-cells. *Am J Clin Pathol* 82:666–673, 1984.
36. May EE, Perentes E: Anti-Leu 7 immunoreactivity with human tumours: Its value in the diagnosis of prostatic adenocarcinoma. *Histopathology* 11:295–304, 1987.
37. Mechtersheimer G, Möller P: Expression of Ki-1 antigen (CD 30) in mesenchymal tumors. *Cancer* 66:1732–1737, 1990.
38. Michels S, Swanson PE, Robb JA, Wick MR: Leu-7 in small cell neoplasms; an immunohistochemical study with ultrastructural correlations. *Cancer* 60:2958–2964, 1987.
39. Miller RT, Groothuis CL: Improved avidin-biotin immunoperoxidase method for terminal deoxyribonucleotidyl transferase and immunophenotypic characterization of blood cells. *Am J Clin Pathol* 93:670–674, 1990.
40. Osborne BM, Butler JJ, Pugh WC: The value of immunophenotyping on paraffin sections in the identification of T-cell rich B-cell large-cell lymphomas. *Am J Surg Pathol* 14:933–938, 1990.
41. Pallesen G, Hamilton-Dutoit SJ: Ki-1 (CD 30) antigen is regularly expressed by tumor cells of embryonal carcinoma. *Am J Pathol* 133:446–450, 1988.
42. Pettinato G, Manivel JC, d'Amore ESG, Petrella G: Fine needle aspiration cytology and immunocytochemical characterization of the histiocytes in sinus histiocytosis with massive lymphadenopathy (Rosai-Dorfman Syndrome). *Acta Cytol* 34:771–777, 1990.
43. Pontifex AH, Klimo P: Application of aspiration biopsy cytology to lymphomas. *Cancer* 53:553–556, 1984.
44. Poppema S, Hollema H, Visser L, Vos H: Monoclonal antibodies (MT1, MT2, MB1, MB2, MB3) reactive with leukocyte subsets in paraffin-embedded tissue sections. *Am J Pathol* 127:418–429, 1987.
45. Powers CN, Wakely PE, Silverman JF, Kornstein MJ, Frable WJ: Fine needle aspiration biopsy of extramedullary plasma cell tumors. *Mod Pathol* 3:648–653, 1990.
46. Qizilbash AH, Elavathil LA, Chen V, Young JEM, Stuart DA: Aspiration biopsy cytology of lymph nodes in malignant lymphoma. *Diagn Cytopathol* 1:18–22, 1985.
47. Rajasekhar A, Kapila K, Verma K: Value of demonstration of cytokeratin and leukocyte common antigen in poorly differentiated malignant tumors in fine needle aspirates. *Acta Cytol* 33:385–389, 1989.
48. Ramzy I, Rone R, Schultenover SJ, Buhaug J: Lymph node aspiration biopsy; diagnostic reliability and limitations—an analysis of 350 cases. *Diagn Cytopathol* 1:39–45, 1985.
49. Robey SS, Cafferty LL, Beschorner WE, Gupta PK: Value of lymphocyte marker studies in diagnostic cytopathology. *Acta Cytol* 31:453–459, 1987.

50. Rushin JM, Riordan GP, Heaton RB, et al: Cytomegalovirus-infected cells expressed Leu-M1 antigen; a potential source of diagnostic error. Am J Pathol 136:989–995, 1990.

51. Rusthoven JJ, Robinson JB, Kolin A, Pinkerton PH: The natural-killer-cell-associated HNK-1 (Leu-7) antibody reacts with hypertrophic and malignant prostatic epithelium. Cancer 56:289–293, 1985.

52. Schwab H, Stein H, Gerdes J, et al: Production of a monoclonal antibody specific for Hodgkin and Sternberg-Reed cells of Hodgkin's disease and a subset of normal lymphoid cells. Nature 299:65–67, 1982.

53. Schwarting R, Gerdes J, Dürkop H, et al: Ber-H2: A new anti-Ki-1 (CD 30) monoclonal antibody directed at a formol-resistant epitope. Blood 74:1678–1689, 1989.

54. Sheibani K, Battifora H, Burke JS, Rappaport H: Leu-M1 antigen in human neoplasms; an immunohistologic study of 400 cases. Am J Surg Pathol 10:227–236, 1986.

55. Sneige N: Diagnosis of lymphoma and reactive lymphoid hyperplasia by immunocytochemical analysis of fine-needle aspiration biopsy. Diagn Cytopathol 6:39–43, 1990.

56. Sneige N, Dekmezian R, El-Naggar A, Manning J: Cytomorphologic, immunocytochemical, and nucleic acid flow cytometric study of 50 lymph nodes by fine-needle aspiration; comparison with results obtained by subsequent excisional biopsy. Cancer 67:1003–1007, 1991.

57. Sneige N, Dekmezian RH, Katz RL, et al: Morphologic and immunocytochemical evaluation of 220 fine needle aspirates of malignant lymphoma and lymphoid hyperplasia. Acta Cytol 34:311–322, 1990.

58. Swanson PE, Manivel JC, Wick MR: Immunoreactivity for Leu-7 in neurofibrosarcoma and other spindle cell sarcomas of soft tissue. Am J Pathol 126:546–560, 1987.

59. Swerdlow SH, Wright SA: The spectrum of Leu-M1 staining in lymphoid and hematopoietic proliferations. Am J Clin Pathol 85:283–288, 1986.

60. Tani EM, Christensson B, Porwit A, Skoog L: Immunocytochemical analysis and cytomorphologic diagnosis on fine needle aspirates of lymphoproliferative disease. Acta Cytol 32:209–215, 1988.

61. Tani E, Liliemark J, Svedmyr E, et al: Cytomorphology and immunocytochemistry of fine needle aspirates from blastic non-Hodgkin's lymphoma. Acta Cytol 33:363–371, 1989.

62. Tani E, Lowhagen T, Nasiell K, Öst Å, Skoog L: Fine needle aspiration cytology and immunocytochemistry of large cell lymphomas expressing the Ki-1 antigen. Acta Cytol 33:359–362, 1989.

63. Van Eyken P, De Wolf-Peeters C, Van Den Oord, Tricot G, Desmet V: Expression of leukocyte common antigen in lymphoblastic lymphoma and small noncleaved undifferentiated non-Burkitt's lymphoma: An immunohistochemical study. J Pathol 151:257–261, 1987.

64. Warnke RA, Rouse RV: Limitations encountered in the application of tissue section immunodiagnosis to the study of lymphomas and related disorders. Hum Pathol 16:326–331, 1985.

65. Yazdi HM, Burns BF: Fine needle aspiration biopsy of Ki-1-positive large-cell "anaplastic" lymphoma. Acta Cytol 35:306–310, 1991.

# 10

# Determination of the Primary Site

Determining the primary site of a metastatic neoplasm by aspiration biopsy cytology (ABC) is one of the most difficult and challenging aspects of cytology. In most cases, cytomorphological interpretation in conjunction with immunocytochemical and electron microscopic studies can broadly define the neoplasm as lymphoid, squamous, glandular, or neuroendocrine. These pathways of differentiation, however, are shared by a variety of neoplasms, and these kinds of tumor can occur at different sites and organs.[11] On the basis of the development of chemotherapeutic regimens that are fairly specific for certain primary sites,[18] providing a cytological diagnosis of "adenocarcinoma" may not be sufficient clinically. For example, the treatment plan for a patient with metastatic prostate carcinoma is quite different from that for metastatic carcinomas of other primary sites.[25] Thus, clinicians increasingly require more precise diagnoses.

For diagnostic purposes in cytopathology, metastatic neoplasms can be grouped into three categories. In the first, the ABC specimen is from a metastatic neoplasm in which historically there is a known primary site. If possible, the neoplastic cells in the current aspirate should be compared with the previous cytologic or histologic material or both. If the previous material is not available for review, interpretation of the cytological features should be governed by the possibility that the information and history given by the radiologist or the clinician may be unintentionally incorrect. The second category includes patients with an undetermined primary site, but in whom a particular site is suspected based on clinical, radiological, or other findings. The last category consists of metastatic neoplasms where, despite clinical investigations, the primary site remains occult; this category constitutes 1.5 to 15% of malignant neoplasms.[78,113,140] Regardless of the circumstance, cytopathologists are expected to determine if the metastatic neoplasm is consistent with the known or suspected primary tumor or, in the latter case, to suggest a probable primary site.

The first step is to formulate a differential diagnosis based on cytological criteria, clinical data (e.g., known or suspected primary site, presence of tumor elsewhere, site of aspiration, age, sex), radiological findings, and statistical probabilities. Familiarity with the light microscopic appearance of different neoplasms in ABC specimens provides an essential guideline. For example, metastatic colonic adenocarcinomas usually show characteristic features in aspirates that are suggestive of the disease. At this stage experience in cytopathology is a great asset. Radiological findings may be of assistance. Some metastatic neoplasms, such as those from prostate and kidney, may have distinctive radiological features.[18] Routine diagnostic radiological procedures such as upper gastrointestinal series, barium enema, chest radiographs, and intravenous pyelograms are infrequently helpful and, because of the high number of false-positive and false-negative results, may be misleading.[113] Although diagnostic imaging studies may be essential for the initial recognition or confirmation of metastatic neoplasms, they generally fail to produce information that alters a patient's clinical course.[67,144] In one study, computed tomographic (CT) scanning only located the primary site of the tumor in approximately one-third of adenocarcinomas with an unknown primary source.[101]

Familiarity with statistical probabilities is certainly helpful. Common primary sites such as breast, lung, and prostate are always initial considerations. Metastatic melanoma, which are cytologically and histologically diverse, should be considered in all metastatic neoplasm with no known primary site. The location of a lymph node involved with a metastatic tumor may be of critical value in determining a primary site. For example, when dealing with a metastatic tumor of an inguinal lymph node, the most common primary sites include cervix, endometrium, and ovary in women, and prostate, anus, and rectum in men. Further information of this type can be gained from the excellent review by Bosman and colleagues.[18]

After formulating a list of differential diagnoses, a panel of diagnostic antibodies or an ultrastructural study should be used to address the probabilities. In many cases, immunocytochemistry or electron microscopy can determine the cell type or pathway of differentiation. Unfortunately, there are only a few current markers, such as prostate-specific antigen (PSA), muscle-specific actin, thyroglobulin, and melanoma-specific antigen, as well as certain ultrastructural features, that are capable of specifying a primary site. Using a panel of diagnostic antibodies or a combination of ultrastructural features, however, may help point toward the primary site.

# SQUAMOUS-CELL CARCINOMA VERSUS ADENOCARCINOMA

Differentiation of squamous-cell carcinoma from adenocarcinoma narrows the differential diagnosis and may point to a particular primary site in a particular clinical setting. For example, diagnosing squamous-cell carcinoma in a liver aspirate usually indicates a lung primary site. In contrast, colon, breast, or lung are the initial considerations for a metastatic adenocarcinoma in the liver. Diagnosis of a well-differentiated squamous-cell carcinoma, especially when keratinization is present, from a well-differentiated adenocarcinoma is very easy in ABC specimens, whereas poorly differentiated neoplasms are difficult to classify on cytomorphology alone.

Different epithelia express variable combinations of cytokeratins (See "Cytokeratins" section in Chapter 5). It has been suggested that using monoclonal antibodies, such as 34βE12 and 35βH11, to high- and low-molecular-weight cytokeratins, respectively, is helpful to subclassify carcinomas.[54,135] Whether this is practical is questionable. In some reports, squamous-cell carcinomas were generally negative with 35βH11 and positive with 34βE12,[54,135] whereas in other studies squamous-cell carcinomas usually expressed both high- and low-molecular-weight cytokeratins.[161] Immunocytochemical staining with monoclonal antibodies to different cytokeratins appears to be of value in differential diagnosis of pulmonary squamous carcinoma and adenocarcinoma in ABC specimens;[20] however, this is currently not widely used.

From personal experience based on the ultrastructural assessment of ABC specimens from lung tumors, it is usually not difficult to distinguish adenocarcinomas from squamous-cell carcinomas in these minibiopsies. In adenocarcinomas, features such as tight junctions, true and intracytoplasmic lumens, microvilli, mucus-type secretory granules, prominent Golgi complexes, and conspicuous amounts of rough endoplasmic reticulum are all diagnostically useful. Squamous-cell carcinomas generally have scantier amounts of rough endoplasmic reticulum, less prominent Golgi complexes, more tonofilaments (often preferentially perinuclear) that are compactly arranged in a curvilinear fashion (Fig. 6.27B,C), desmosomes associated with blunt cytoplasmic extensions forming the "intercellular bridges" (Fig. 6.3), and comma-shaped tonofilament bundles emanating from the cytoplasmic aspect of the dense plaques of desmosomes (Figs. 6.3B, 6.4A). An important characteristic such as the intercellular bridging may not be apparent cytologically, but the arrangement and form of the tumor cells ultrastructurally even in poorly differentiated lesions may suggest this feature (Fig. 6.3C; Fig. 10.1E). Tumor cells in adenocarcinomas generally are closely apposed with some intricate interdigitations of the cell membranes, whereas squamous cells are more loosely arranged, often with multiple intercellular gaps produced by the cytoplasmic projections from adjacent cells. This latter feature, seen best in low-power micrographs, can even occur in the absence of or with poorly formed desmosomes (Fig. 6.3C). These distinguishing features of squamous epithelium are also evident in primary and metastatic tumors of cervix, esophagus, ureter, and urinary bladder.

The general rules for diagnosing squamous-cell carcinomas by electron microscopy are somewhat altered in nasopharyngeal carcinomas. These carcinomas often initially present as a mass in the neck and may be subject to needle aspiration biopsy (NAB). Since nasopharyngeal carcinomas ("lymphoepitheliomas") are poorly differentiated squamous-cell carcinomas, the tumor cells infrequently form intercellular bridges, and the polygonal cells are in close apposition with little cell membrane interlocking.[34] Moderately to well-developed desmosomes, however, are evident and they often have readily apparent tufts of tonofilaments projecting from them. Tonofilaments, on the other hand, are rarely seen. The moderate amount of cytoplasm of the tumor cells in nasopharyngeal carcinomas contain few organelles except for many free ribosomes.

The following case (Fig. 10.1) shows how electron microscopy of an ABC specimen can detect a second primary site, in this case a squamous-cell carcinoma of lung, by excluding the diagnosis of metastatic adenocarcinoma. Two years after a mastectomy for infiltrating duct carcinoma, this 56-year-old woman was found to have a 2-cm nodule in the left lower lung. ABC was interpreted cytologically as "large-cell, poorly differentiated carcinoma" (Fig. 10.1A). Ultrastructurally, how-

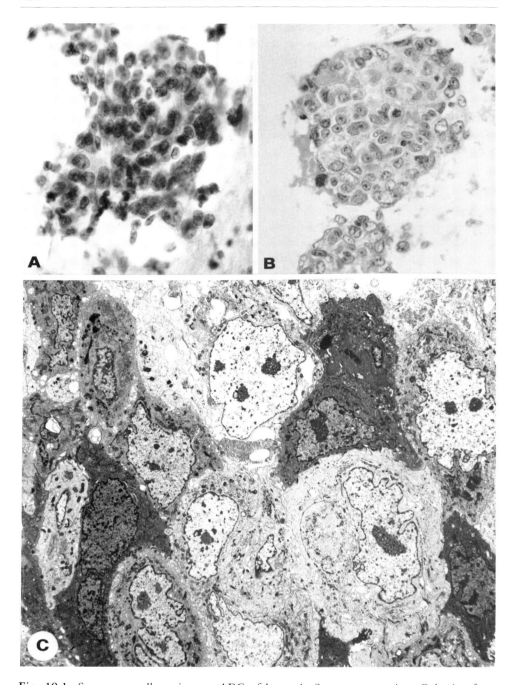

Fig. 10.1. Squamous-cell carcinoma, ABC of lung. **A.** Smear preparation. Cohesive fragments of tumor cells with the cytology and arrangement of a large-cell carcinoma. × 260. **B.** Cell block section. No distinctive growth pattern is seen. × 260. **C.** Closely associated tumor cells have limited cytoplasmic organelles and do not form intercellular lumens. × 1,700.

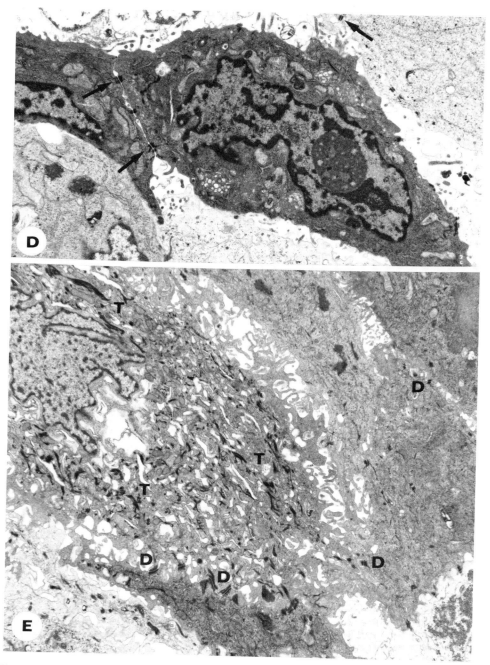

**Fig. 10.1. D.** Typical desmosomes (arrows) join adjacent cells. × 5,000. **E.** One tumor cell has innumerable tonofilaments (T) and many classic desmosomes (D). The combination of ultrastructural features established this "large-cell carcinoma" as a squamous-cell carcinoma, a second primary neoplasm in this patient. × 4,000.

ever, the frequency and form of the desmosomes, some with comma-shaped tonofilaments (Fig. 10.1B,C), and the occasional tumor cell with many desmosomes at right angles to the cell surface and numerous tonofilaments (see Fig. 10.1D), along with individual tumor-cell keratinization, indicated squamous-cell differentiation. A diagnosis of squamous-cell carcinoma primary in the lung was rendered and was confirmed by subsequent resection of the tumor.

Using electron microscopic features detailed in Chapter 6 and illustrated in Figure 10.3, it may be possible to localize a metastatic adenocarcinoma to its tissue or organ of origin. The experience of Hammar and associates[58] with a large series of metastatic tumors of unknown primary origin provides instructive ultrastructural guidelines that can be applied profitably to ABC specimens.

If a previous ABC or surgical pathology specimen is available for comparison, a distinct difference in the cytomorphology or pathway of differentiation between the initial neoplasm and the second neoplasm strongly suggests two separate primary sites. In contrast, cytological similarity between the two aspirates suggests metastatic disease, but in this circumstance the possibility of a second primary site cannot be completely ruled out. Only an adjunctive technique such as electron microscopy has the potential to determine whether such lesions are primary or metastatic.

# PROSTATIC ADENOCARCINOMA

Prostatic adenocarcinoma has become the most common noncutaneous malignant disease diagnosed in American men, and the age-adjusted incidence and mortality of prostate cancer appear to be increasing in most countries.[105,169] It is not infrequent in ABC specimens from men to be confronted with an adenocarcinoma or poorly differentiated carcinoma in patients with a history of prostatic carcinoma. In some of these cases, the diagnosis of metastatic prostatic carcinoma may be considered clinically unlikely due to the unusual sites of the metastasis or the clinical and radiological findings that favor a new primary site. Less frequently there is no history of a known primary neoplasm. In a male patient the possibility of an occult or unsuspected prostatic cancer should be considered and ruled out. Prostatic cancer may be undetectable by rectal examination or serum prostatic acid phosphatase (PrAP) levels.[145] In all these situations the distinction between prostatic carcinoma and other cancers is critical because of the choices of therapy.[145,157] Metastatic prostatic carcinoma can be better treated with hormonal manipulation with good results and minimal toxic effect as compared with chemotherapy.[145,157]

Antibodies to PSA and PrAP, as a small panel, are extremely reliable to support prostatic origin of a neoplasm in ABC specimens (Fig. 10.2). This small panel is not only useful in metastatic carcinomas but also in aspirates obtained from the prostatic region. Sometimes it is necessary to rule out the possibility of neoplasms of the adjacent organs, such as transitional-cell carcinoma and colonic adenocarcinoma, with or without involvement of the prostate gland. These tumors do not react with antibodies to PSA and PrAP.

Most prostatic carcinomas are PrAP-positive.[3,45,72,108,110,136,156] However, a num-

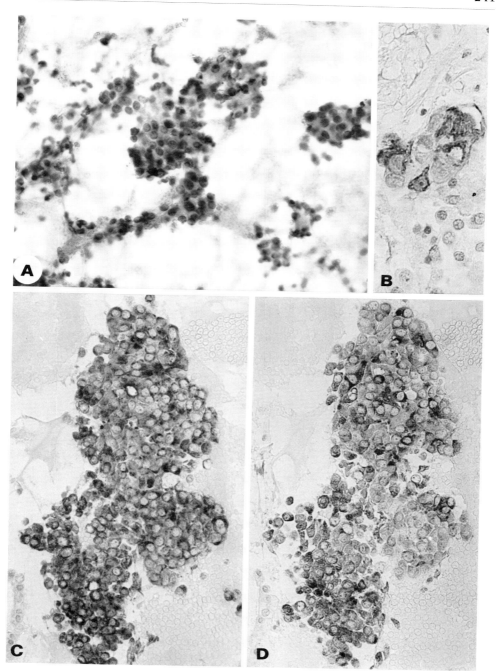

Fig. 10.2. Metastatic prostatic adenocarcinoma, ABC of lung. A 67-year-old man with a history of prostatic adenocarcinoma presented with a 3.5-cm lung nodule. **A.** Smear preparation. Three-dimensional fragments of large tumor cells and glandular arrangements. × 260. **B.–D.** The neoplastic cells expressed prostate-specific antigen (PSA), prostatic acid phosphatase (PrAP), and Leu-7, supporting the prostatic origin of the tumor. (B) Leu-7, × 416; (C) PrAP, × 260; (D) PSA, × 260.

ber of nonprostatic neoplasms, including renal-cell carcinoma, pancreatic islet-cell tumor, rectal carcinoid tumor, breast carcinoma, and adenocarcinoma of the urinary bladder, may also express PrAP.[26,30,45,46,49,72,168] The majority of prostate carcinomas show positive staining with antibody to PSA.[3,45,72,109,110,119] So far we have not encountered any positivity in nonprostatic neoplasms in our institution. May and Perentes,[98] however, demonstrated weak to moderately positive cytoplasmic staining with antibody to PSA in some adenocarcinomas of paranasal sinus, gallbladder, liver, gastrointestinal tract, kidney, endometrium, ovary, breast, and lung. Weak cytoplasmic staining was also seen in one of several malignant melanomas examined.[98] Although these results are not supported by others, they suggest the possible nonspecific nature of the antibody and the need to switch to monoclonal antibodies. Positive staining with antibody to PSA is more intense and extensive than PrAP.[45,130] In summary, PrAP, although a sensitive marker, is not specific for prostate cancer. In contrast, PSA is more specific but it is not as sensitive as PrAP. One can confirm the prostatic origin of a neoplasm with these antibodies, particularly when both show immunoreactivity. Overall there is a good correlation between ABC smears, cell block sections and histological specimens for the presence of PSA and PrAP.[72] The results should be interpreted in the context of clinical and radiological findings, especially when tumor cells are PrAP-positive but PSA-negative. PrAP is more sensitive in undifferentiated prostatic carcinomas.[48] Absence of PSA in a poorly differentiated neoplasm of undetermined histogenesis does not rule out the possibility of the prostate as the primary site.[45,48,145]

Some variants of prostatic carcinoma such as signet-ring cell carcinoma and mucinous adenocarcinoma also express PSA and PrAP.[47,127,128] This may be an important finding for differentiation of prostatic carcinoma from other carcinomas such as mucin-producing colonic adenocarcinoma. Immunoperoxidase stains for PSA and PrAP should be done in signet-ring cell neoplasms, particularly when staining for mucin is negative. Other variants of prostatic carcinoma such as small-cell carcinoma show variable results. Ro and colleagues[129] studied 18 small-cell carcinomas of the prostate with antibodies to PSA and PrAP. Only three neoplasms stained positively; this finding again emphasizes that negative staining does not rule out the possibility of the prostate as the primary site.

Normal, hyperplastic, and neoplastic prostatic cells express cytokeratin proteins; however, the immunoreactivity varies with antibodies to different types of cytokeratins. For example, normal, hyperplastic, and neoplastic secretory epithelial cells react with monoclonal antibody EAB902, a panepithelial marker, which recognizes 54-kd cytokeratin.[19] Immunoreactivity with EAB902 is seen in all prostatic carcinomas regardless of degree of differentiation.[19] In contrast, monoclonal antibody EAB903, which recognizes 49-, 51-, 57-, and 66-kd cytokeratins, only reacts with the basal cells of normal and hyperplastic prostatic epithelium and shows no reactivity in prostatic carcinoma cells.[19] The latter antibody has potential value in ABC of the prostate. It is the experience of most cytopathologists that the majority of well-differentiated adenocarcinomas of the prostate cannot be definitely diagnosed on ABC specimens. The main differential diagnosis is atypical hyperplasia.[169] Basal cells are generally present and intermingle with secretory epithelial cells in the same tissue fragments in atypical hyperplasia and should be absent in prostatic carcinoma. Ostrzega and colleagues[116] suggested that negative staining of cells with antibody to EAB903, originally interpreted as suspicious for malignancy, may represent underdiagnosed adenocarcinoma. However, further study with a large number of histologically verified cases is necessary to support this observation.

Prostatic carcinoma as well as normal and hyperplastic prostatic cells may express Leu-7, a lymphoid marker recognizing natural killer cells. May and Perentes[98] demonstrated cytoplasmic Leu-7 positivity in all prostatic carcinomas tested, regardless of degree of differentiation. Monoclonal antibody to Leu-7 may be used in conjunction with PSA and PrAP for detection of prostatic carcinomas (Fig. 10.2B); however, Leu-7 is not a specific marker and may be expressed in neuroendocrine carcinomas and neural neoplasms. The majority of prostatic carcinomas express epithelial membrane antigen (EMA).[45] Carcinoembryonic antigen (CEA) is expressed in a small percentage of higher grade prostatic adenocarcinomas, but well-differentiated adenocarcinomas are negative.[45]

Although, ultrastructurally, the normal and hyperplastic prostatic epithelium often has many cytoplasmic secretory granules and multivesicular bodies,[142] they are neither sufficiently distinctive nor seen frequently enough in prostatic adenocarcinomas to be of diagnostic value in distinguishing metastatic neoplasms from this organ. Lysosomes are also plentiful (the location for prostatic-specific acid phosphatase) but are not specific morphologically.

# GASTROINTESTINAL ADENOCARCINOMA

Gastrointestinal carcinomas are among the most common causes of death in adults.[120] In the United States, colonic carcinoma is both the most common and the most curable carcinoma of the gastrointestinal tract, and its incidence appears to be increasing.[130] The liver is the most common site of distant metastasis, followed by lung and ovary. Rarely, gastrointestinal tumors metastasize to other organs or sites such as the central nervous system and bone.[130] The majority of cases that are seen in ABC specimens obtained from distant metastasis or local abdominal recurrence or spread in patients with a known history of colorectal carcinoma.

Except for poorly differentiated tumors, colorectal adenocarcinomas have a characteristic appearance in ABC specimens. The aspirate is usually cellular, consisting of cohesive fragments of carcinoma cells with acinus formation and palisading of columnar cells. Necrosis is usually present. The majority of colorectal adenocarcinomas express CEA. In fact, the lack of staining suggests a primary tumor other than large bowel. Many adenocarcinomas, however, from a variety of organs and sites can express CEA. Therefore, there is a need for a more specific marker to distinguish colorectal adenocarcinoma from other adencarcinomas.

Jothy and colleagues[66] produced several monoclonal antibodies reacting with CEA; antibodies D-14 and D-18 were found to have a high degree of specificity for colonic carcinoma. Using monoclonal antibody D-14, specific for an epitope of CEA, Pavelic and colleagues[120] demonstrated positivity in all 61 primary and 29 metastatic colorectal carcinomas they studied. Fourteen of 22 gastric adenocarcinomas and 2 of 5 pancreatic carcinomas were also reactive. Only 6 of 100 nongastrointestinal neoplasms expressed weak-to-moderate immunoreactivity. There was also no cross-reactivity with nonspecific cross-reacting antigen and biliary glycoprotein. Still, there is need for further studies with a large number of gastrointestinal and nongastrointestinal neoplasms to confirm the specificity of D-14 and other antibodies to CEA.

On the basis of ultrastructural findings described and illustrated in Chapter 6, it is possible to use specific features (e.g., the glycocalyx and microvilli) to determine the gastrointestinal tract as the primary site for a metastatic neoplasm. The filamentous "rootlets"projecting down from some microvilli into the apical cytoplasm of tumor cells is a particularly valuable determinant. This feature can be quite striking even in relatively poorly differentiated adenocarcinomas of the gastrointestinal tract, where the cells may bear few if any microvilli (Fig. 6.12D). Because of the relative frequency of colorectal and gastric carcinomas (tumor types that not infrequently present clinically initially as a metastasis), the ability to accurately specify these organs as primary sites is of practical diagnostic and therapeutic importance. The following example illustrates such an application by electron microscopy to the ABC of a bone lesion.

Due to increasing pain in the lower left leg, this 47-year-old woman with a history of mastectomy for breast carcinoma had a radiograph that revealed an osteolytic lesion in the tibia (Fig. 10.3A). ABC of this lesion showed compact epithelial cells

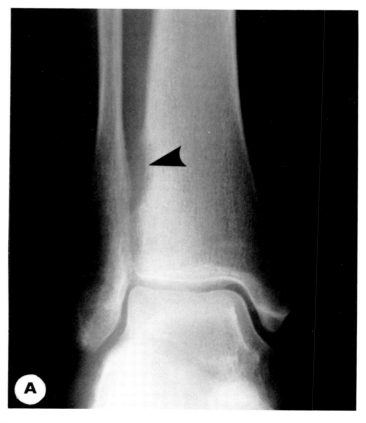

**Fig. 10.3.** Metastatic gastrointestinal adenocarcinoma, ABC of bone. **A.** Radiolucent zone (arrowhead) in the lower end of the tibia.

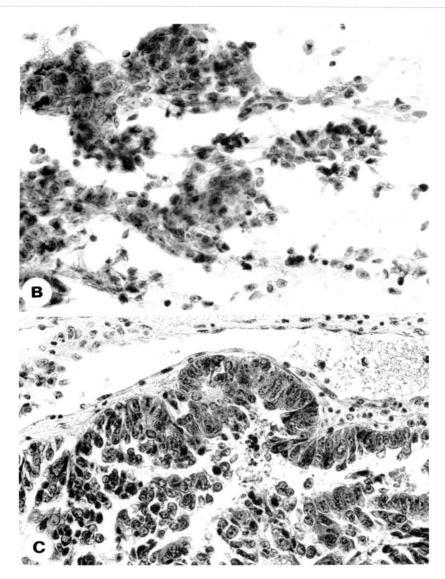

Fig. 10.3. B. Smear preparation. Cellular aspirate of tumor fragments, some of which have an acinar arrangement. × 260. C. Cell block section. Glandular architecture of the tumor is confirmed. × 260.

with the cytology and architectural pattern of an adenocarcinoma (Fig. 10.3B,C). Electron microscopy (Fig. 10.3D) confirmed the adenocarcinomatous nature of the bone tumor. The presence of readily identifiable microfilamentous core rootlets projecting down from the microvilli into the apical cytoplasm of the tumor cells (Fig. 10.3E) indicated that this neoplasm was of gastrointestinal origin. This patient initially had presented with a pelvic mass eight months prior to this NAB, and a surgical biopsy at that time showed a metastatic adenocarcinoma suspected to originate in the colon.

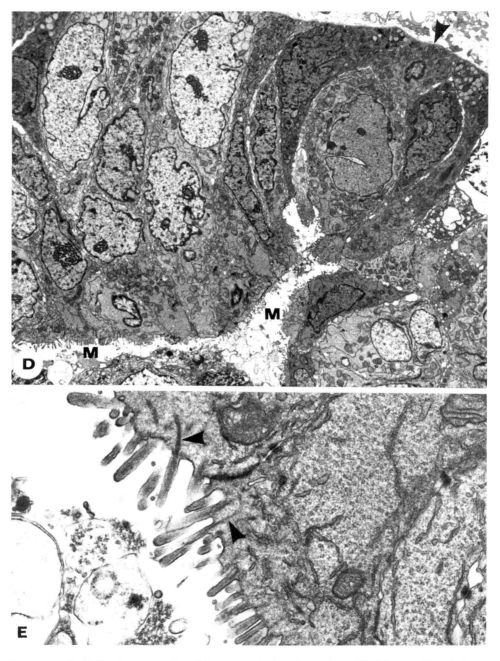

Fig. 10.3. D. Tall columnar cells, with many apical surface microvilli (M), are aligned along a basal lamina (arrowhead). × 3,000. E. Microvilli have a central dense-staining region that extends down from some (arrowheads) into the apical cytoplasm. × 23,000. (From Dardick et al: A quantitative comparison of light and electron microscopic diagnosis in specimens obtained by fine-needle aspiration biopsy. Ultrastruct Pathol 15:105–129, 1991. Used with permission.)

# HEPATOCELLULAR CARCINOMA VERSUS ADENOCARCINOMA

ABC is a simple and accurate method for the diagnosis of malignant neoplasms of the liver. The majority of the neoplasms are metastatic in nature. In fact, the liver is one of the most frequent sites of metastases from cancers of various organs and histological types, particularly adenocarcinoma. Hepatocellular carcinoma is the most common primary malignant tumor of the liver; it accounts for approximately 90% of all primary liver cancers.[167] Although hepatocellular carcinoma is one of the most common tumors worldwide with markedly different geographic incidences, it constitutes only 2 to 3% of all cancers in the United States.[13,25,32,147] Hepatocellular carcinoma, with the exception of the fibrolamellar variant, carries a poor prognosis and is a rapidly fatal disease.[13] Therefore, accurate distinction between hepatocellular carcinoma and metastatic neoplasms is essential to establish appropriate therapy and to assess prognosis.

The cytomorphological features of hepatocellular carcinoma in ABC specimens depend on the degree of tumor differentiation, and diagnostic difficulties occur especially at either end of the spectrum of tumor differentiation. Well-differentiated hepatocellular carcinomas exhibit some but not all nuclear criteria of malignancy. Architectural abnormalities, such as thickened cell cords and general organizational disarray, when present, are useful indicators of neoplasia. In these tumors, determination of the histogenesis of the cells in ABC specimens is easy, but the diagnosis of malignancy can be extremely difficult if not impossible. Benign lesions such as adenoma, cirrhosis, or hepatitis may show similar changes and should be considered in the differential diagnosis.[50,55,147,155] Immunocytochemical and electron microscopic studies are not helpful in these cases and a surgical biopsy is needed.

Moderately and poorly differentiated hepatocellular carcinoma may share cytological features with a variety of carcinomas and even sarcomas.[27] Sarcomatous appearance occurs in a small percentage of hepatocellular carcinomas.[69] An acinar-type of hepatocellular carcinoma may be mistaken as metastatic adenocarcinoma.[147] Ancillary techniques may supply additional information required for classification and can improve diagnostic accuracy. Currently, the most specific antigen is alpha-fetoprotein (AFP). The detection of this antigen strongly supports the diagnosis of hepatocellular carcinoma and excludes the possibility of cholangiocarcinoma and metastatic neoplasms, with the exception of germ-cell tumors. Positive staining also helps to differentiate sarcomatous hepatocellular carcinoma from sarcomas.[69] In our experience, AFP is expressed in approximately 45% of hepatocellular carcinomas in ABC specimens. This is similar to what is reported by others in ABC and histological specimens.[12,25] The staining may be focal and the number of positive cells may be as low as 5% of the neoplastic cells.[156] Although AFP is rather specific for hepatocellular carcinoma, it is not a very sensitive marker. Negative staining does not rule out the possibility of hepatocellular carcinoma, and diagnosis requires use of the panel of antibodies outlined in Table 10.1.

CEA expression is not a specific marker, because hepatocellular carcinoma, cholangiocarcinoma, and other carcinomas metastatic to the liver may show immunoreactivity.[80,156] In hepatocellular carcinoma, however, immunoreactivity for CEA is

TABLE 10.1. Antigenic Profiles Supporting the Diagnosis of Hepatocellular Carcinoma and Adenocarcinoma

| | Hepatocellular Carcinoma | Adenocarcinoma |
|---|---|---|
| AFP | + | − |
| CEA | +* | +** |
| Cytokeratins | | |
|   AE1 | − | + |
|   AE3 | + | ± |
|   CAM 5.2 | + | + |
| HBsAg | + | − |

*Canalicular pattern of staining.
**Diffuse cytoplasmic staining.
AFP = alpha-fetoprotein. CEA = carcinoembryonic antigen. HBsAg = hepatitis B−surface antigen.

predominantly concentrated on the surface of the cells, outlining bile canalicular structures (Fig. 10.4C). In our experience, this characteristic bile canalicular staining is seen in approximately 90% of hepatocellular carcinomas in ABC specimens.[167] This phenomenon is also reported by other authors.[5,27,53,80,103,156,158] In contrast, adenocarcinomas and large-cell carcinomas show diffuse cytoplasmic staining when CEA is expressed (Fig. 10.5C).

Typing with monoclonal antibodies to lower and higher molecular weight cytokeratins may be of great value. In our experience, the majority of hepatocellular carcinomas and metastatic carcinomas express the cytokeratins detected by monoclonal antibodies AE3 and CAM 5.2. Hepatocellular carcinomas are consistently negative when stained with the anticytokeratin antibody AE1. In contrast, almost all carcinomas stain positive with antibody AE1. Johnson and associates[63] reported similar findings of cytokeratin expression using the monoclonal antibodies AE1 and CAM 5.2 on formalin-fixed tissue.

Expression of hepatitis B−surface antigen (HBsAg) has also been suggested as a reliable marker for hepatocellular carcinoma.[12] We have stained six hepatocellular carcinomas with an antibody to HBsAg. The neoplastic cells in three aspirates stained positive, one was negative, and in two cases the results were equivocal. Alpha1-antitrypsin (A1-AT) was expressed in 81% of 16 hepatocellular carcinomas in ABC specimens studied in our laboratory, which is similar to what is reported in the literature.[53,80,156,158] Expression of A1-AT or alpha1-antichymotrypsin, however, is not specific for hepatocellular carcinoma and may be expressed in sarcoma, carcinoma, and malignant melanoma.[83,86,106,152,156] Both sinusoidal cells and most hepatocellular carcinoma cells express vimentin and *Ulex europaeous* agglutinin-1. Because the staining for both of these markers is more intense in sinusoidal cells, the sinusoidal cells traversing groups of malignant hepatocytes or investing clusters of hepatocytes are accentuated, which may assist in the diagnosis of liver tumors (see Fig.10.4E,F).

In summary, in ABC of the liver, when the cytological features are consistent with hepatocellular carcinoma or this tumor is considered in the differential diagnosis, we use a panel of antibodies to AFP, CEA, and, for cytokeratin analysis, antibodies

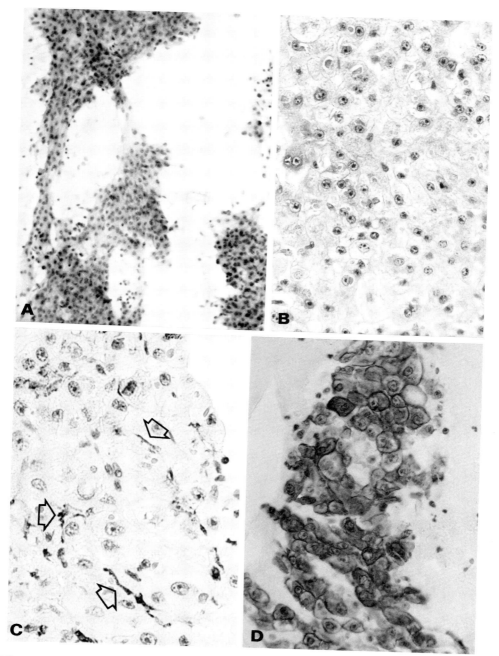

Fig. 10.4. Hepatocellular carcinoma, ABC of the liver in a 76-year-old man with cirrhosis and a 5-cm mass. **A.** Smear preparation. Large cohesive fragments of tumor cells focally surrounded by endothelial cells. × 104. **B.** Cell block section. Note the granularity of the cytoplasm and large nucleoli. × 260. **C.** Carcinoembryonic antigen (CEA). Note canalicular pattern of staining (arrows). × 416. **D.** Strong staining of tumor cells with anticytokeratin AE3. × 260. Positive staining with antibody AE3, expression of CEA (canalicular pattern), and negative staining with anticytokeratin AE1 supports the diagnosis of hepatocellular carcinoma. Staining for alpha-fetoprotein was negative.

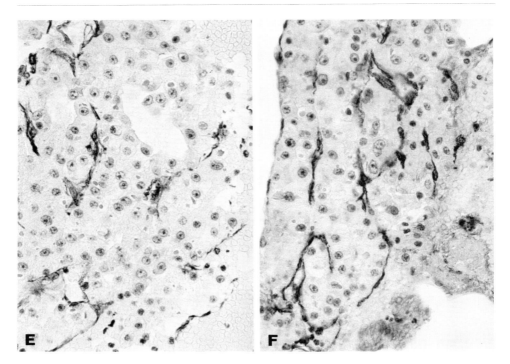

Fig. 10.4. E.,F. Vimentin and *Ulex europaeous* agglutinin-1 (UEA-1) stains help to accentuate the sinusoidal cells traversing hepatocellular carcinoma cells. (E) Vimentin, × 260; (F) UEA-1, × 260.

AE1 and AE3. In most cases, the results are helpful in arriving at an accurate diagnosis (Table 10.1). The value of immunocytochemical study in ABC of the liver is illustrated in Figures 10.4 and 10.5. Cholangiocarcinoma, which is a less-often encountered primary hepatic neoplasm, shows an immunocytochemical profile similar to metastatic carcinomas, making their distinction difficult by ABC specimens alone.

A combination of electron microscopic features assist in distinguishing primary hepatic tumors from adenocarcinomas in ABC specimens. Neoplastic hepatocytes generally mimic their normal counterpart with respect to the numbers and structure of the mitochondria; the formation of bile canaliculi; and the amounts of smooth and rough endoplasmic reticulum, glycogen, lipid, and lysosomes. Of these organelles, the smooth endoplasmic reticulum is a particularly useful marker because it is an unusual finding in the metastatic adenocarcinomas commonly encountered in the liver. Bile canaliculi are a further distinctive marker of hepatic differentiation (Fig. 10.6). They present as microlumens at the lateral margins or the corners of two to three adjacent hepatocytes. Although such intercellular lumens show a greater variation in form and size in primary liver tumors compared with the normal liver, they still reveal tight junctions, a few to moderate numbers of microvilli, and, on occasion, even evidence of bile secretion; the latter in the form of irregularly shaped, concentric rings of membranous materials (Fig. 10.6B). The bile canaliculi formed by neoplastic hepatocytes are generally smaller than the lumens seen in

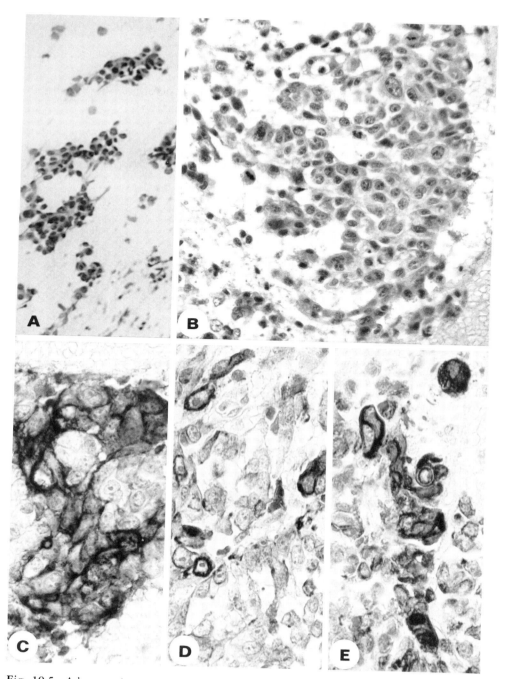

Fig. 10.5. Adenocarcinoma, ABC of the liver in a 76-year-old woman. **A.** Smear preparation. Tumor fragments, some with well-defined outlines. × 104. **B.** Cell block section. Large malignant cells with dense, somewhat granular, cytoplasm. × 260. The diagnosis of hepatocellular carcinoma was considered in the differential diagnosis. The neoplastic cells expressed carcinoembryonic antigen (CEA) (diffuse cytoplasmic staining), stained positive with anticytokeratin antibodies AE1 and AE3, and were negative for alpha-fetoprotein. This immunophenotype supported the diagnosis of adenocarcinoma. **C.** CEA, × 416; **D.** Antibody AE1, × 416; **E.** Antibody AE3, × 416.

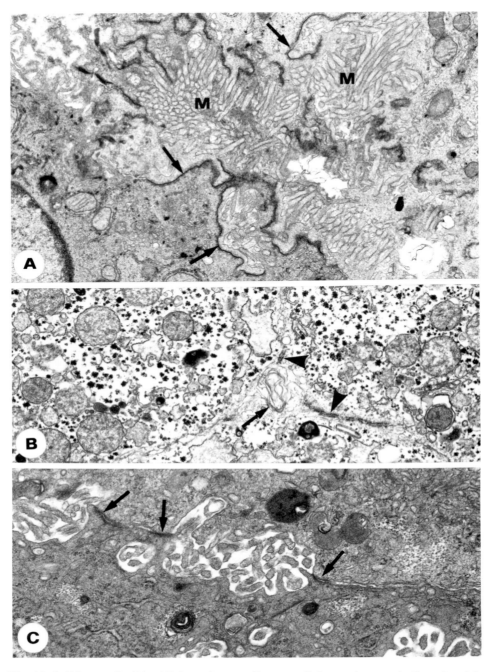

**Fig. 10.6.** Bile canaliculi in ABC specimens of hepatocellular carcinoma. **A.** Complex tight junctions (arrows) are associated with a lumen that is packed with well-aligned microvilli (M). × 9,500. **B.** An intercellular space at the junction of three neoplastic hepatocytes has few microvilli, poorly formed junctions (arrowheads), and the lamellar formation indicative of bile secretion (arrow). × 5,500. **C.** A series of widened intercellular spaces contain microvilli and have the cell membranes at opposite poles joined by tight junctions (arrows). × 12,000.

metastatic carcinomas, which, coupled with the cytoplasmic characteristics noted above, assists in separating the two diagnostic possibilities. Immature hepatocytes of four examples of hepatoblastoma of infancy sampled by NAB have shown this spectrum of diagnostic features.[38,162] In well-differentiated primary liver tumors, the architectural association of hepatocytes and the sinusoid-lining endothelial cells (Fig. 10.7) are important diagnostic features that can be retained in ABC specimens.

A 72-year-old man presented with abdominal distention and mild jaundice. A liver mass was detected by ultrasound and an NAB was done. The smears showed small fragments and single cells with prominent nucleoli, general features, and an organization suggestive of hepatocellular carcinoma (Fig. 10.7A). Ultrastructurally, fragments from the ABC specimen were composed of polygonal cells with many darkly staining mitochondria, complements of smooth and rough endoplasmic reticulum, some glycogen, and lipid droplets (Fig. 10.7B). Bordering these cells were endothelial-type cells forming sinusoidal spaces. In addition, bile canaliculus–like lumens, complete with tight junctions and microvilli, developed between adjacent tumor cells (Fig. 10.7C). These features of hepatocellular carcinoma correlate with some of the immunocytochemical findings described. The endothelial-type cells lining the sinusoids are responsible for the positive *Ulex europaeus* agglutinin-1 staining, whereas the CEA localizes immunocytochemically as a network to the luminal surfaces of neoplastic bile canaliculi (Figs. 10.4, 10.6, 10.7).

At the ultrastructural level, distinguishing hepatocellular carcinoma, particularly when they are well differentiated, from normal liver can pose a problem in assessing ABC specimens of the liver. To ensure that the sample for electron microscopy is representative of the tumor, it is essential to review the toluidine blue–stained plastic sections and the smear preparations with a cytopathologist.

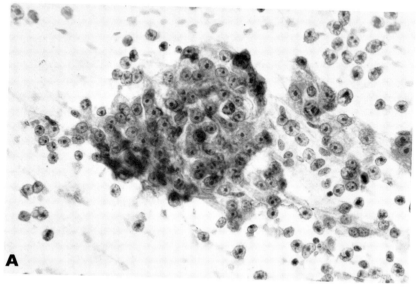

**Fig. 10.7.** Hepatocellular carcinoma, ABC of liver. **A.** Smear preparation. A cellular fragment of compact tumor cells with prominent nucleoli was interpreted as representing hepatocellular carcinoma. × 260.

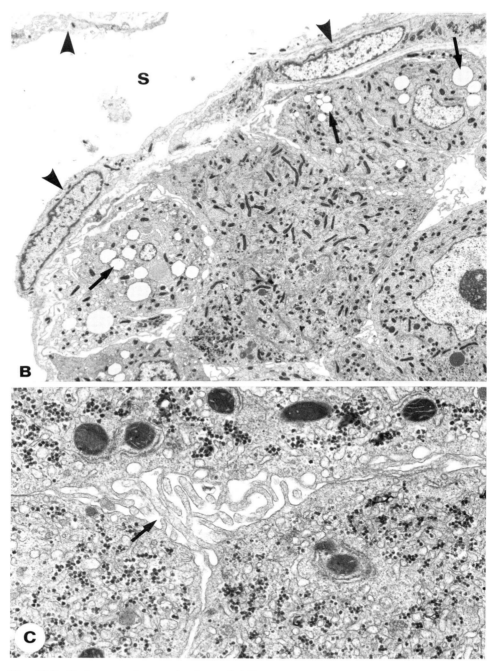

Fig. 10.7. **B.** Survey micrograph of a well-differentiated hepatocellular carcinoma displays a sinusoid (S) lined by endothelial-type cells (arrowheads). Hepatocyte cytoplasm contains dark-staining mitochondria, many profiles of rough endoplasmic reticulum and a few lipid droplets (arrows). × 1,800. **C.** A microvillus-ladened bile canaliculus–like region (arrow) is formed by the interaction of three tumor cells, but tight junction differentiation is limited. × 11,200.

In the ultrastructural assessment of an ABC of a liver mass, there are some general and a few specific findings that assist in distinguishing the less-differentiated hepatocellular carcinoma from a metastatic adenocarcinoma. The somewhat unique combination of cytoplasmic organelles described for hepatocellular tumors (Fig. 10.7B) is not duplicated in the more common sources for metastatic adenocarcinoma, namely lung, gastrointestinal tract, and pancreas. The latter all tend to grow in liver as duct- or gland-like nests of tumor cells forming lumens that, at least at the ultrastructural level, tend to be larger and more distinctive, as well as having a richer complement of microvilli, than the lumens formed in hepatocellular carcinomas. Additional features separating these entities are mucin- or zymogen-type secretory granules, central cores of actin filaments in the microvilli that at times extend down into the apical cytoplasm (microfilamentous rootlets), glycocalyceal bodies, and, occasionally, small numbers of tonofilaments.

A 58-year-old woman presented with abdominal discomfort, anorexia, and weight loss. Abdominal ultrasound revealed multiple nodules in the liver as well as a mass in the left kidney. ABC produced compactly organized tumor cells consistent with an adenocarcinoma (Fig. 10.8A,B), a diagnosis that was confirmed by electron microscopy (Fig. 10.8C,D). Small extracellular, as well as occasional intracellular (Fig. 10.8D), lumens were present. Microvilli extended into these lumens, but microfilamentous core rootlets were not identified. These features, along with numerous cytoplasmic organelles, including dilated endoplasmic reticulum, indicated the glandular differentiation of the tumor; secretory products such as mucin or zymogen granules were absent. The intracellular lumens suggested breast, lung, gastrointestinal tract, ovary, and endometrium as possible primary sites, but none of the ob-

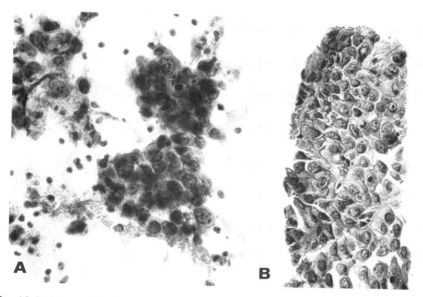

Fig. 10.8. Metastatic adenocarcinoma, ABC of liver. **A.** Smear preparation. Compactly arranged fragments of tumor suggested an adenocarcinoma. × 260. **B.** Cell block section. Details of the organization and cytology of the tumor cells in this adenocarcinoma are better appreciated. × 260.

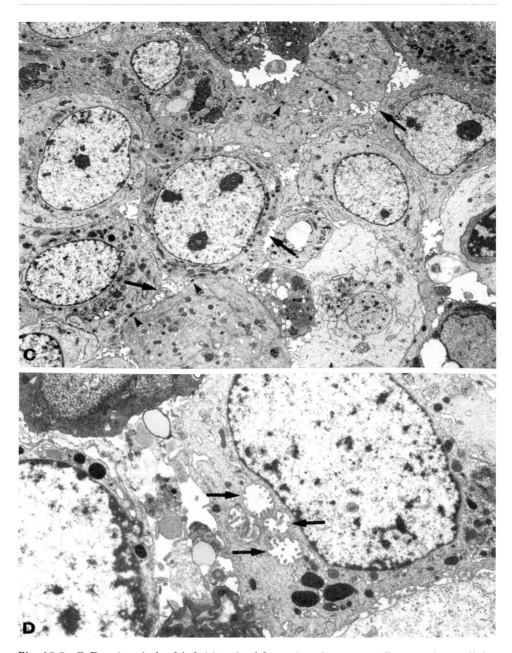

Fig. 10.8. C. Despite a lack of definitive gland formation ultrastructurally, some intercellular spaces do have tumor-cell surfaces with microvilli (arrows), and occasional tight junctions and desmosomes (arrowheads) indicate microlumen formation. × 2,200. D. Some tumor cells have one or more intracellular lumens (arrows). × 5,200.

served ultrastructural features allowed a specific diagnosis. The advanced degree of metastatic disease in the liver precluded benefit from chemotherapy, and palliative care was provided to the patient.

# MESOTHELIOMA VERSUS ADENOCARCINOMA

Mesothelioma is a tumor derived from the mesothelium of serous cavities, predominantly the pleura. It usually occurs in adults, has a poor prognosis, and survival beyond two years is uncommon.[130] Although malignant mesothelioma is uncommon it constitutes an important differential diagnosis in aspiration biopsy and effusion cytology. In ABC, especially of peripheral lung lesions and pleural-based nodules, the main differential diagnoses include malignant mesothelioma and primary or metastatic adenocarcinomas. This distinction is critical not only because of the prognosis and different treatment methods, but also due to the potential compensation and litigation regarding asbestos exposure.[115] The presence of monolayered sheets of cells with slits between the neoplastic cells, abundant dense cytoplasm, dispersed spindle cells, and multinucleated cells in an ABC are in favor of a diagnosis of malignant mesothelioma.[154] However, diagnosis may be extremely difficult. Some malignant mesotheliomas are sarcomatoid or biphasic in nature and may be difficult to differentiate from other spindle-cell tumors.[23] These tumors may not be associated with effusions. Differentiation between malignant mesothelioma and adenocarcinoma and distinction between reactive mesothelial hyperplasia and malignant mesothelioma are common diagnostic problems in effusion cytology. Therefore, there is a need for more objective ancillary techniques for accurate diagnosis.

In most cases, it is possible to differentiate malignant mesothelioma from adenocarcinoma in ABC specimens based on cytomorphology, cytochemistry, immunocytochemistry, and electron microscopic studies in conjunction with clinical and radiological findings.[154] Cytochemical stains have been useful. Most of these stains are based on the presence of hyaluronic acid in mesotheliomas and neutral mucin in adenocarcinomas. A positive mucicarmine stain or diastase-resistant periodic acid-Schiff stain in a poorly differentiated tumor is strongly in favor of adenocarcinoma. On the other hand, mesotheliomas are characterized by large amounts of hyaluronic acid, which can be demonstrated by alcian blue or colloidal iron stains. There are, however, exceptions. For example, occasionally mucin can be demonstrated in mesothelioma, and hyaluronic acid can be demonstrated in small amounts in adenocarcinoma.[115] Hyaluronic acid is easier to demonstrate in alcohol-fixed ABC specimens, probably due to its water-soluble nature and at least partial extraction during formalin fixation of surgical specimens.[146]

Immunocytochemical studies can provide additional information helpful in distinguishing malignant mesotheliomas from adenocarcinomas in most cases. Although there is currently no specific antibody for this purpose, CEA appears to be the most useful and most accepted marker. There is some controversy in the literature regarding CEA expression in malignant mesothelioma. Although some authors report focal or weak CEA staining in a small percentage of mesotheliomas,[117,159] others report no staining.[17,28,76,115,165,166] This discrepancy may be partly due to the

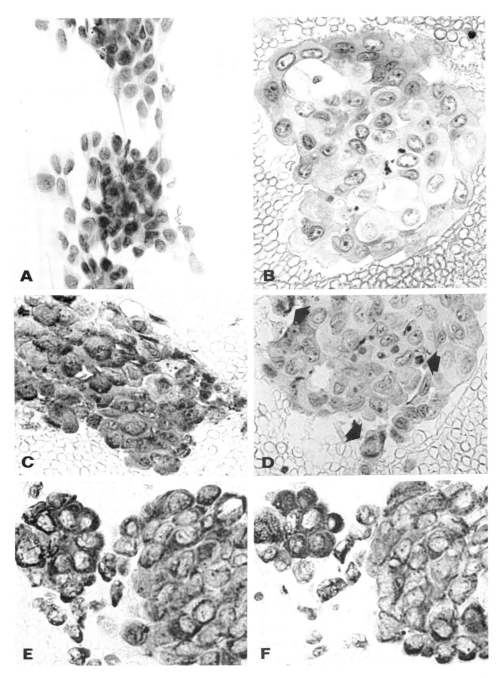

Fig. 10.9. Adenocarcinoma, ABC of a pleural thickening. **A.** Smear preparation. Loosely cohesive fragments of tumor cells with large nuclei and small nucleoli. × 260. **B.** Cell block section. Note large amount of relatively dense cytoplasm and scalloped cell borders. × 416. Because of the cytomorphology of the tumor cells, the diagnosis of malignant mesothelioma was considered in the differential diagnosis. However, strong expression of carcinoembryonic antigen (CEA) and B72.3 supported the diagnosis of adenocarcinoma. **C.** CEA. × 416. **D.** B72.3. Note focal but strong staining (arrows). × 416. **E.** Anticytokeratin AE1. × 416. **F.** Anticytokeratin AE3. × 416. Strong immunostaining with antibodies to cytokeratins did not contribute to the differential diagnosis.

TABLE 10.2. Antigenic Profiles Supporting the Diagnosis of Mesothelioma and Adenocarcinoma

|  | Mesothelioma | Adenocarcinoma |
|---|---|---|
| Carcinoembryonic antigen | − | + |
| B72.3 | − | + |
| Leu-M1 | − | + |
| Cytokeratins |  |  |
| AE1 | + | + |
| AE3 | + | − |
| Secretory component | − | + |
| Pregnancy–specific beta-1 glycoprotein | − | + |

use of different monoclonal or polyclonal antibodies in these studies. Judging from our experience, strong CEA staining with a monoclonal antibody is quite supportive of adenocarcinoma, because it excludes a malignant mesothelioma (Fig. 10.9C); staining is usually diffuse in adenocarcinomas. On the other hand, if the staining is weak or focal, the result should be interpreted in conjunction with other markers, such as B72.3 and Leu-M1 (Table 10.2).

The majority of adenocarcinomas and epithelial tumors stain positively with monoclonal antibody B72.3, whereas malignant mesotheliomas stain negatively in a high proportion of cases.[17,76,115,117,150,165,166] When staining is positive in mesotheliomas, it is very focal in nature.[76] B72.3 in combination with CEA is the best limited panel of antibodies to distinguish malignant mesotheliomas from adenocarcinomas in ABC specimens (Fig. 10.9C,D) and effusion cytology. Another useful marker is Leu-M1. This antigen is expressed in many adenocarcinomas and is generally negative in malignant mesotheliomas,[76,115,117,165] but focal staining is reported in a minority of the latter.[17,166] For practical purposes, however, strong Leu-M1 positivity in conjunction with supportive results from other antibodies excludes the possibility of malignant mesothelioma.

Unfortunately, the diagnosis of malignant mesothelioma with these antibodies, because it is based on a negative result that may occur in adenocarcinomas, is not ideal. What is required is an antibody that detects an antigen specifically present in mesotheliomas and absent in adenocarcinomas. Donna and associates,[42] using an antimesothelial-cell polyclonal antibody, demonstrated positive staining in 16 unequivocal mesotheliomas and no immunoreactivity in 31 lung carcinomas. Similarly, O'Hara and colleagues[114] demonstrated positive staining in 40 malignant mesotheliomas using monoclonal antibody $ME_1$; however, six of 88 nonmesothelioma malignancies also stained positivity. Confirmation of these results in a large series of cases is required before accepting them as specific markers for malignant mesothelioma.

Leong and associates,[87] using an anti-EMA antibody with cytological smears, cell blocks, and surgical biopsy sections, demonstrated staining predominantly of the cell membrane in malignant mesotheliomas. They suggested that the demonstration of thick cell membrane may be an additional diagnostic clue in favor of malignant mesothelioma. However, we have seen this phenomenon in some adenocarcinomas. Monoclonal antibody to human milk fat globule-related antigen (HMFG-2), an antigen that is related to EMA, has also been used in an attempt to differentiate mesothelioma from adenocarcinoma with controversial results. Although it is claimed that most adenocarcinomas express HMFG-2 and normal, reactive, and neoplastic mesothelial cells stain negatively,[10] others have demonstrated HMFG-2 positivity in malignant mesotheliomas.[14,28,96,166] Other markers that may be expressed in adenocarcinomas but are usually absent in malignant mesotheliomas include secretory component and beta-1 pregnancy-specific glycoprotein.[115] Antigens such as vimentin and S-100 protein have no significant discriminative value.

Both malignant mesothelioma and adenocarcinoma strongly express cytokeratins. The majority of malignant mesotheliomas express both low- and high-molecular-weight cytokeratins.[117] Most adenocarcinomas, however, only express low-molecular-weight cytokeratins. A lack of staining with an antibody to high-molecular-weight cytokeratins favors adenocarcinoma, but this finding is neither practical nor reliable (Fig. 10.9E,F). Some mesotheliomas are entirely or predominantly composed of spindle cells, which may be difficult to distinguish from other spindle-cell neoplasms, particularly sarcomas, in ABC specimens. The distinction can have therapeutic, prognostic, and legal implications. Positive cytokeratin staining strongly supports the diagnosis of mesothelioma.[23,130] Cibas and associates[28] demonstrated cytokeratin expression in all the malignant mesotheliomas and metastatic adenocarcinomas they studied in effusions, and suggested that a preferential peripheral staining of the tumor cell cytoplasm was a characteristic feature of adenocarcinomas. In contrast, most mesotheliomas had an accentuated perinuclear cytokeratin distribution. A similar observation is reported in tissue sections.[68] The case illustrated in Figure 10.9 demonstrates the value of immunocytochemistry in differentiating mesothelioma from adenocarcinoma.

The principal electron microscopic feature used in the differential diagnosis of mesothelioma and adenocarcinoma is the microvilli on the cell surfaces of tumor cells; details of the structural and placement characteristics in these two types of tumors have been provided in the "Microvilli" section in Chapter 6. Briefly, microvilli tend to be more sinuous, longer, branching, and have a greater density in mesotheliomas compared with adenocarcinomas. In the series reported by Wick and associates,[165] this combination of microvillous features was apparent in 30 of 39 diffuse epithelial mesotheliomas and none of 52 peripheral pulmonary adenocarcinomas with pleural invasion. For diagnostic purposes, it is important to note that the tumor cells in certain forms of malignant mesothelioma can have few if any microvilli.[36] Other features of neoplastic mesothelial cells, such as the relative paucity of the usual cytoplasmic organelles (especially when compared to the rough endoplasmic reticulum–rich complement of adenocarcinomas), perinuclear tonofilaments, absence of secretory granules, and well-developed, often particularly long, desmosomes,[21] assist in separating these two diagnostic possibilities.[14,15,35,36,165]

Some of the ultrastructural characteristics of mesotheliomas were invaluable in establishing this diagnosis in the following case, which demonstrates its effectiveness in quality assurance for radiologists. Respiratory symptoms prompted a chest radiograph (Fig. 10.10A) and, subsequently, a CT scan (Fig. 10.10B) of a 56-year-old woman with a history of a previous breast "lesion." On the basis of this history, the CT scan was interpreted as showing an "intrapulmonary mass" indicative of metastatic adenocarcinoma (radiologically there was also a suggestion of tumor deposits in the anterior mediastinum and in the left adrenal gland). On the basis of

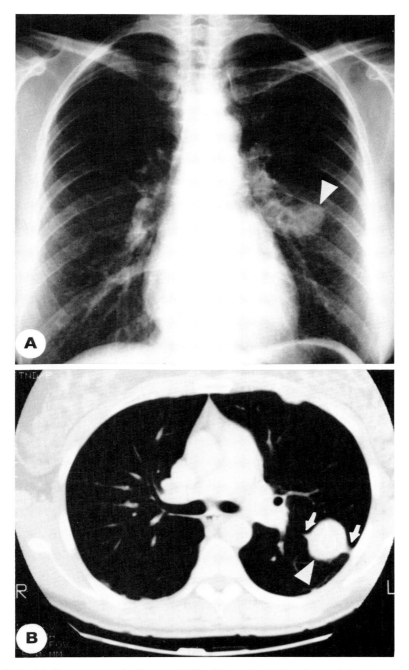

Fig. 10.10. Malignant mesothelioma, ABC of lung. Initially, the nodule (arrowheads) in both the chest radiograph A. and the computed tomographic (CT) scan B. were interpreted as indicating a primary pulmonary tumor. On the basis of the ultrastructural findings from the ABC specimen, review of the CT scan revealed that the tumor lies within the interlobar fissure (arrows).

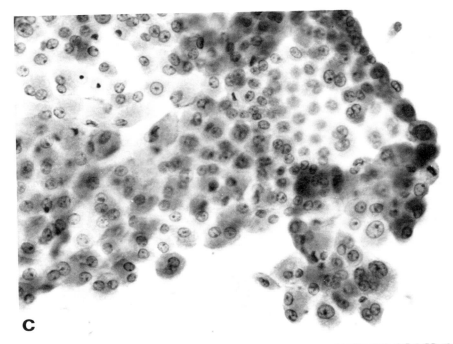

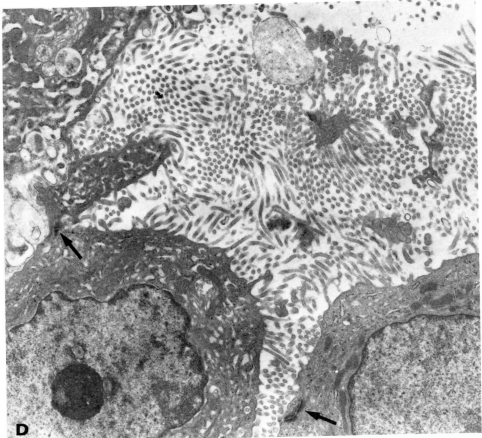

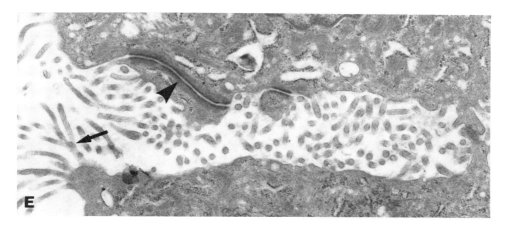

Fig. 10.10. C. Smear preparation. Flat sheets of uniform-appearing tumor cells, with a suggestion of being slightly separated, indicated a diagnosis of epithelial mesothelioma. × 240. D. Characteristic of mesothelioma, spaces between the tumor cells are occupied by numerous, relatively long microvilli, and desmosomes (arrows) join adjacent cells. × 5,600. E. Some microvilli branch (arrow) and certain desmosomes (arrowheads) are particularly lengthy. × 8,200.

the cytological features of the ABC specimen (Fig. 10.10C), mesothelioma was included in the differential diagnosis. Because the radiologist interpreted the neoplasm as not being pleural-based and pleural thickening was absent, the ABC specimen was reported as "consistent with adenocarcinoma." No material was available for immunocytochemical confirmation, but glutaraldehyde-fixed tissue fragments were provided for electron microscopy. On the basis of the overall features of the tumor cells, the profusion of long microvilli (Fig. 10.10D), their branching form, the lengthy, well-formed desmosomes (Fig. 10.10E), and the lack of secretory granules, the ultrastructural diagnosis was mesothelioma. Reappraisal of the CT scan (Fig. 10.10B) showed that the lesion was in fact present in the interlobar fissure and not within the lung parenchyma as was originally suggested by the radiological interpretation.

Electron microscopy can be of assistance in identifying certain of the subtypes of bronchioloalveolar carcinoma (Fig. 10.11) and in raising the possibility that a metastatic lesion is an adenocarcinoma (not otherwise classifiable in routine smears or by immunocytochemistry) of pulmonary origin.[16,61,151] In bronchioloalveolar carcinomas, three cells types—reflecting the normal counterparts in bronchioles and alveoli—can be identified with some precision. Of these, Clara and goblet cells have secretory granules that are distinctive (see the "Secretory Granules" section in Chapter 6 for ultrastructural details).

In a clinical setting such as differentiating pulmonary adenocarcinomas involving pleura from mesotheliomas, secretory granules can resolve this issue in problematic ABC specimens even when the production of mucin in adenocarcinomas is too limited to be detected cytologically. Secretory granules in nonneoplastic and neoplastic Clara cells generally consist of apically situated osmiophilic, nonlamellated structures. When growing along surfaces, the tumor cells of adenocarcinomas of

Clara cell-type often display another useful diagnostic feature: a bulging of the apical cytoplasm above an imaginary line drawn between the tops of the tight junctions at the lateral aspect of each tumor cell (Fig. 10.11C). Bronchioloalveolar carcinomas composed of type II pneumocytes, although an infrequent finding, also have sufficiently distinctive features for diagnostic purposes.[57,151]

In distinguishing metastatic adenocarcinomas of unknown primary origin, a certain appearance, placement, and spacing of surface microvilli, although not entirely specific, can be recognized with some degree of confidence to point to a pulmonary origin. Such microvilli are short, blunt, and moderately, but somewhat uniformly, spaced, and when projecting from adjacent cell surfaces they produce a zipper-like effect (Fig. 6.11A). Microvilli in pulmonary adenocarcinomas are not often associated with microfilamentous core rootlets or glycocalyceal bodies (see the "Micovilli" and "Glycocalyx" sections in Chapter 6).

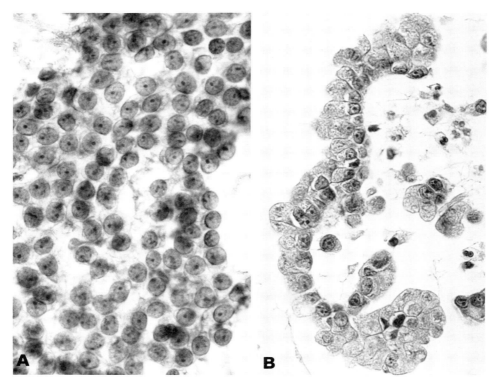

Fig. 10.11. Bronchioloalveolar carcinoma. ABC of a 3-cm opacity in the right upper lobe of lung in a 66-year-old man for whom the radiological diagnosis was organizing pneumonia versus bronchioloalveolar carcinoma. A. Smear preparation. This smear consists of uniform tumor cells with a "honeycomb" organization and a single discrete nucleolus. × 416. B. Cell block section. The regular alignment of columnar- to cuboidal-shaped tumor cells with a granular cytoplasm is evident. × 416.

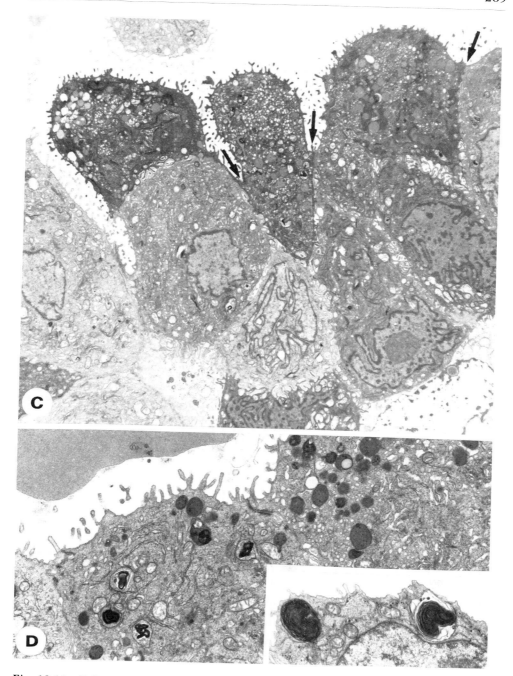

**Fig. 10.11.** C. Some tumor cells have a modest to considerable portion of the organelle-rich cytoplasm projecting above the apical regions that are joined by tight junctions (arrows). × 3,900. D. Microvillus-bearing apical regions contain a number of secretory granules that have a homogeneous-to-lamellar (inset) appearance. × 9,900 (inset, × 12,600).

# THYROID NEOPLASM

Preoperative diagnosis of medullary carcinoma and differentiation from the more common papillary and follicular neoplasms can be very difficult in ABC of the thyroid gland. Medullary carcinoma of the thyroid gland is an endocrine tumor of parafollicular C-cell origin. The aspirates are usually cellular and composed of isolated and small groups of round, oval, or polygonal cells with abundant cytoplasm and eccentric nuclei. Binucleated and multinucleated cells are commonly seen. Amyloid deposits, when present, are useful for diagnosis. Medullary carcinomas, however, are known to exhibit a wide variety of morphological features, including the presence of follicular and papillary structures, marked anaplasia, cells with granular or clear cytoplasm, cells containing mucin or melanin pigment, or composed of small, spindled, or giant cells.[62,102,130,143] Therefore, cytological classification may be difficult in ABC specimens.[123]

Demonstration of calcitonin in a thyroid neoplasm is the most sensitive and reliable method of confirming the diagnosis (Fig. 10.12C).[37,102,123,131,143] The intensity of staining varies from cell to cell and case to case. Amyloid material also stains positively with antibody to calcitonin. On rare occasions, medullary carcinomas are negative or only focally positive for calcitonin. Clinically aggressive tumors may be negative. Ruppert and colleagues[131] reported a case of calcitonin-rich medullary carcinoma that followed an indolent clinical course for seven years after surgical resection. Sudden rapid dissemination of the tumor led to death two years later. At autopsy, only slight calcitonin positivity was demonstrated in less than 5% of the neoplastic cells.[131] Staining for calcitonin is of limited use outside the thyroid gland, because some neoplasms of other organs, such as lung, colon, and pancreas, may also synthesize this hormone.[156] Although calcitonin is the most constant and prominent secretory product of medullary carcinoma, the tumor may produce a number of other substances such as somatostatin, bombesin, neuron-specific enolase, and chromogranin (Fig. 10.12D).[132,138] The latter is particularly useful as a second marker to confirm the diagnosis of medullary carcinoma, especially when calcitonin staining is negative. Other neuroendocrine neoplasms, however, such as paragangliomas, can present as a thyroid mass. This neoplasm shares morphological, immunocytochemical, and ultrastructural features with medullary carcinoma except for the lack of calcitonin. Medullary carcinomas also express CEA.

Thyroglobulin is produced by most thyroid neoplasms of follicular-cell origin. Antibody to thyroglobulin consistently produces good results, and thyroglobulin can be demonstrated in the majority of follicular neoplasms, including Hürthle-cell tumors (Fig. 10.13) and papillary carcinomas, in cytological and histological specimens.[1,37,51,60,64,156] The staining is most strong and diffuse in follicular neoplasms. The intensity and pattern of staining, however, may vary in papillary carcinoma and may only be focally positive. Most medullary carcinomas, sarcomas, or other thyroid neoplasms of nonfollicular orign stain negative.[1,37,51,156] A small panel of diagnostic antibodies to thyroglobulin, calcitonin, and chromogranin are extremely useful for differentiation of medullary carcinoma from neoplasms of follicular origin (Figs. 10.12, 10.13). Very occasionally, medullary carcinomas express thyroglobulin and calcitonin.[2,62] These tumors appear to have a better prognosis than thyroglobulin-negative medullary carcinomas.[62] Staining for thyroglobulin may also be useful to

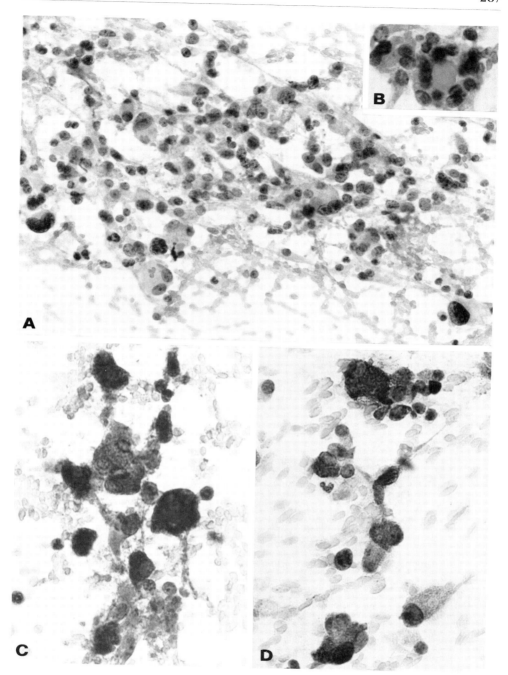

Fig. 10.12. Medullary carcinoma, ABC of thyroid. **A.** Smear preparation. Note binucleated cells and atypical nuclei. × 318. **B.** Smear preparation. Occasional follicular arrangements were seen. × 508. The neoplastic cells strongly expressed calcitonin and chromogranin, confirming the diagnosis of medullary carcinoma. Staining for thyroglobulin was negative. **C.** Calcitonin. × 508. **D.** Chromogranin. × 508.

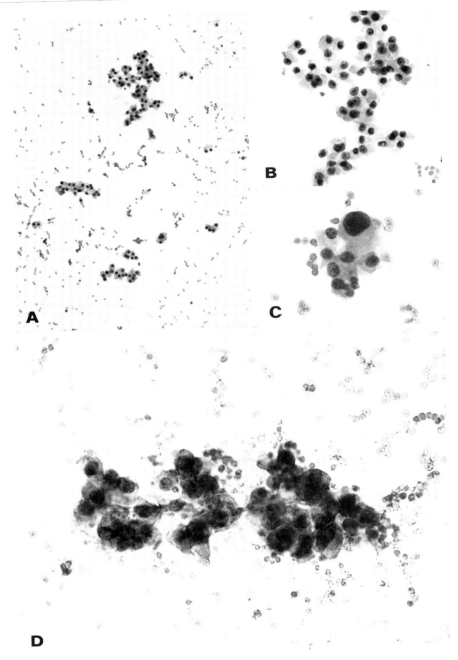

Fig. 10.13. Follicular (Hürthle cell) neoplasm, ABC of thyroid. A.–C. Smear preparation. Cellular aspirate composed of tumor cells with a large amount of rather granular dense cytoplasm. Note the eccentric location of the nuclei and occasional atypical nuclei. (A) × 104; (B) × 260; (C) × 416. The differential diagnosis included Hürthle cell neoplasm and medullary carcinoma. The neoplastic cells expressed thyroglobulin and were negative for calcitonin, supporting the diagnosis of Hürthle-cell neoplasm. A diagnosis of Hürthle-cell adenoma was established by subsequent surgery. D. Thyroglobulin. × 416.

differentiate tumors of follicular origin from metastatic carcinomas such as renal cell carcinoma, which may simulate a primary thyroid tumor.[24,143]

The expression of CEA in follicular neoplasms is a matter of debate. The differences in the reported cases, however, appear to be due to the different antibodies used to demonstrate CEA. Schröder and Klöppel[133] examined 200 primary thyroid carcinomas using three monoclonal antibodies. The monoclonal antibody reactive to an epitope unique to CEA stained only C-cell neoplasms; other antibodies common to epitopes of CEA and nonspecific cross-reacting antigen stained all types of thyroid carcinomas, including tumors of follicular origin.[133] Most thyroid carcinomas coexpress vimentin and cytokeratin,[4,41,143] a phenomenon shared by many other carcinomas and therefore of limited diagnostic use. When the thyroid neoplasm is anaplastic, especially the small cell–type, the possibility of malignant lymphoma should be considered. A panel of diagnostic antibodies to leukocyte common antigen, cytokeratins, thyroglobulin, and calcitonin may be helpful to arrive at a definite classification (see the "Malignant Lymphoma Versus Nonlymphoma Neoplasms" section in Chapter 9). In ABC specimens composed predominantly of lymphoid cells, the distinction between malignant lymphoma (particularly low and intermediate grades) and chronic lymphocytic thyroiditis is difficult and in some cases impossible.[56,153] Demonstration of light-chain restriction by immunocytochemical study supports the diagnosis of malignant lymphoma (see the "Immunophenotyping" section in Chapter 9).

Papillary carcinomas frequently metastasize to cervical lymph nodes, which may be the first manifestation of the disease. Blood-borne metastasis is less frequent. In contrast, blood-borne metastases, particularly to the lung and bone, are more common in follicular carcinomas.[130] Therefore, staining for thyroglobulin can be very useful for the confirmation of metastatic thyroid carcinomas in cytological and histological specimens,[1,37,51,89,143] particularly in metastatic tumors of unknown origin. Demonstration of thyroglobulin is virtually pathognomonic of thyroid origin of the tumor.

Electron microscopy has little or no role in the routine investigation of thyroid nodules using ABC, particularly in follicular and papillary neoplasms. In certain clinical circumstances, however, particularly when medullary carcinoma, anaplastic carcinoma, and primary non-Hodgkin's lymphoma of the thyroid are suspected, ultrastructural studies of aspiration biopsies may be diagnostically useful.[141] The ultrastructural identification of neurosecretory granules and amyloid filaments in medullary carcinoma can be a rapid method of establishing this diagnosis, even in poorly differentiated neoplasms.[70] Similarly, confirmation of epithelial features or neoplastic lymphocytes in anaplastic carcinoma and lymphoma, respectively, has a major impact on therapeutic decisions for patients with a rapidly enlarging thyroid mass.

# BREAST CARCINOMA

ABC is a safe, rapid, cost-effective, and accurate method of diagnosing breast cancer; it has a high degree of sensitivity, ranging from 89 to 98%.[79] Immunocytochemical and ultrastructural studies have no major role in the diagnosis of primary breast cancer by ABC, except occasionally for cases of poorly differentiated carcinomas, which may be difficult to distinguish from malignant lymphoma and malignant melanoma. A simple panel of diagnostic antibodies to cytokeratin, leukocyte common antigen, and S-100 protein/HMB-45 is useful to resolve the problem. Breast carcinomas generally express epithelial markers such as cytokeratin, EMA, and the tumor-associated glycoprotein referred to as TAG-72. Monoclonal antibody B72.3, recognizing the TAG-72 molecule found in neoplastic cells, has been applied to breast cancer cells in effusions and ABC specimens and has been suggested as a potentially valuable adjunct in the diagnosis of ABC of breast lesions.[65,92,112] We did a pilot study to test the value of B72.3 reactivity, to determine if positive staining with the exception of apocrine metaplastic cells supports the diagnosis of malignancy, in 29 "suspicious" ABC specimens from breast lesions.[122] B72.3 staining did not significantly contribute to the diagnosis and we do not anticipate that B72.3 staining will become a valuable adjunct for the interpretation of breast ABC specimens.

Immunocytochemistry may be of value for classifying some types of breast carcinomas. The diagnosis of carcinomas with neuroendocrine differentiation can be supported by markers such as neuron-specific enolase, chromogranin, and synaptophysin (Fig. 10.14).[160] GCDFP-15, a glycoprotein isolated from human breast gross cystic disease fluid, is a good marker for breast carcinomas with apocrine differentiation.[104] It is also expressed in metaplastic apocrine epithelium, apocrine glands, and

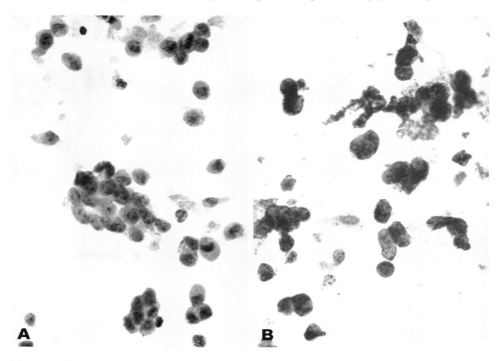

Fig. 10.14. Breast carcinoma with neuroendocrine differentiation, ABC of breast. **A.** Smear preparation. Cellular aspirate composed of single cells and tumor fragments. Note uniformity of the nuclei. × 416. **B.** The majority of the neoplastic cells expressed chromogranin, supporting neuroendocrine differentiation. × 416.

some salivary gland tumors.[149] Mazoujian and associates[99] demonstrated GCDFP-15 positivity in 75% of breast carcinomas with apocrine histological features and in 55% of 562 primary breast carcinomas.

Estrogen receptor (ER) and progesterone receptor (PR) status of breast carcinoma, in conjunction with other factors such as tumor grade and lymph node status, is known to be a good predictor of prognosis and selection of patients who are likely to respond to hormonal manipulation. Fifty-five to 60% of ER-positive breast carcinomas respond to hormonal manipulation, ablative, or additive therapy, whereas the response rate of ER-negative carcinomas is less than 10%.[73] Traditionally, quantitative analysis of these steroid hormone receptors has been done by biochemical methods that require a substantial amount of fresh tissue. This technique may not be feasible when the tumor is small and most of the tissue is used for diagnosis. Furthermore, tumor heterogeneity with respect to receptor content can not be appreciated by biochemical methods. Availability of specific monoclonal antibodies prepared against ER and PR and development of immunocytochemical assay allowed semiquantitative analysis of steroid hormone receptors on tissue sections, frozen or paraffin-embedded, of breast carcinomas with significant correlation with biochemical analyses.[33,73,121,124,126,130,134] The staining is exclusively nuclear and can be evaluated based on the intensity and percent of stained carcinoma cells, which can be done more objectively using computer-assisted image analysis.[6] ER and PR analysis performed on tissue sections using immunocytochemical assay (ICA) (ER-ICA and PR-ICA, respectively), is an acceptable substitute for biochemical analysis and may even be more predictive of prognosis.[121]

Different methods for determination of steroid hormone receptors on cytological specimens have been suggested; however, adaptation of the ER-ICA method to ABC specimens demonstrated a significant concordance between ER-ICA in ABC specimens and tissue specimens, as well as biochemical analysis.[39,74,82,91,95,97,100,125,126,163] Therefore, NAB is a simple method of obtaining representative specimens for preoperative steroid hormone receptor analysis, particularly in patients receiving preoperative adjuvant therapy. It is also useful for inoperable, recurrent, or metastatic breast carcinomas, as well as sequential determination of receptor profile during the course of treatment of advanced breast carcinomas. These uses would further expand the value of ABC. Accuracy of results depends on close attention to the processing steps (Tables 2.7, 2.11) as well as close cooperation between clinicians, cytotechnologists, immunotechnologists, and cytopathologists. Weak ER-ICA results in histological sections, low cellularity of ABC specimens, degradation of receptor, and prominent stromal component of the tumor may result in false-negative ER-ICA results in ABC specimens.[74,125,126] Semiquantitative scoring of the ER-ICA method can be done on ABC specimens using the scoring system proposed by Keshgegian (Table 10.3), with rough correlation with biochemical assay results.[73] The total score (range, 0–7) consists of the sum of the percent of positive tumor cells and the intensity of nuclear staining.

Determination of the PR content of breast carcinoma, in addition to the ER content, further enhance the specificity of prediction of hormonal response.[73,134] PR-ICA can also be performed on tissue sections as well as ABC specimens.[90,118,121,134] Breast carcinomas may also express a variety of tumor-associated antigens such as CEA, human chorionic gonadotropin, placental lactogen, alpha-lactalbumin (AL), and pregnancy-specific beta-1 glycoprotein. Some of these anti-

**TABLE 10.3. Scoring System for ER-ICA Positivity***

| Positive Cells Grade | Average Nuclear (%) | Staining Intensity |
|---|---|---|
| 0 | 0–10 | Absent |
| 1 | 10–25 | Slight |
| 2 | 25–50 | Distinct |
| 3 | 50–75 | Dark |
| 4 | >75 | ... |

*Score = grade % positive cells + grade intensity.
(0 − 7) = (0 − 4) + (0 − 3)
From Keshgegian AA: Hormone receptors and other tumor markers, in Kline TS, Kline IK (eds): Guides to Clinical Aspiration Biopsy: Breast. New York, Igaku-Shoin, 1989. Used with permission.

gens, especially CEA, have been claimed to have a significant relationship to prognosis;[94,107,137] this claim has been disputed by other studies.[9,85]

Breast carcinomas may metastasize to lymph nodes (especially axillary), bone, lung, liver, adrenal gland, and central nervous system,[130] and are frequently encountered in ABC of those organs and effusions. In some patients, metastasis may occur several years after initial diagnosis. Accurate classification of the metastatic breast carcinoma and differentiation from other primary neoplasms is therefore important for management of the patient. AL, a milk protein produced by breast epithelium, has been used as a marker for breast carcinoma. The reported frequency of AL expression in breast carcinoma is variable, ranging from 0 to 78%.[7,8,29,43,84,88] Lee and colleagues[84] demonstrated AL expression in 62% of metastatic breast carcinomas and emphasized its value in the workup of metastatic carcinomas of undetermined primary. AL is not a specific marker for breast carcinoma however, and may be expressed in a variety of other neoplasms, including ovarian epithelial neoplasms, colonic adenocarcinoma, skin appendage tumors, salivary gland tumors, and mesotheliomas.[43,84,88] Demonstration of a high ER content in a metastatic tumor may support a mammary origin of the neoplasm (Fig. 10.15; also see Plate 5.4H).[73,97] However, endometrial and ovarian adenocarcinomas, malignant carcinoid tumor, and some carcinomas of the lung and stomach may also contain ER.[22,75,77,81] In addition, it has been shown that some pulmonary carcinomas and pancreatic endocrine tumors contain progesterone receptors.[22,40] Therefore, there is currently no specific marker for breast carcinoma and the immunocytochemical results of sex steroid hormone receptor, and AL should be interpreted in the context of cytological and clinical findings.

Statistically, metastatic breast carcinomas are most likely to have tumor cells with intracytoplasmic lumens.[52,111] This feature is particularly useful in the ultrastructural investigation of aspiration biopsies of poorly differentiated carcinomas involving axillary lymph nodes. The variability of appearance of intracytoplasmic lumens and the range of adenocarcinomas in which they can be less frequently identified by electron microscopy are provided in the "Intracytoplasmic Lumens" section in Chapter 6. At the ultrastructural level, most infiltrating duct and lobular carcinomas (whether primary or metastatic) are relatively poorly differentiated adenocarcinomas. For diagnostic purposes, because breast carcinomas are frequently a consideration, this poor differentiation may successfully contrast with more specific differen-

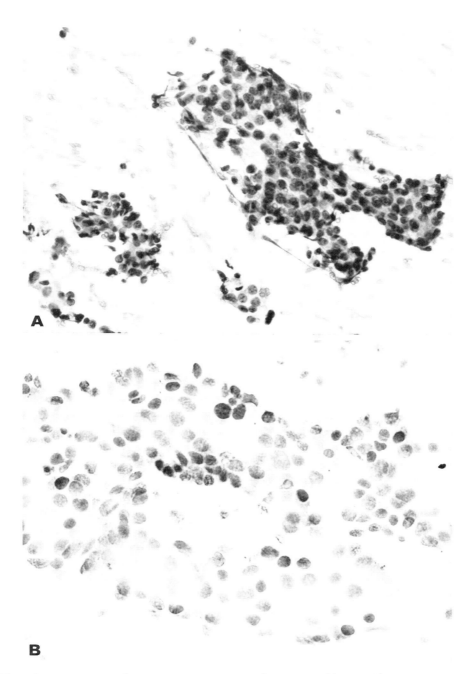

Fig. 10.15. Metastatic breast carcinoma, ABC of liver in a 64-year-old-woman with previously diagnosed breast carcinoma 8 years prior to the present admission. **A.** Smear preparation. Fragments of rather uniform tumor cells. × 260. **B.** Strong estrogen receptor staining in the majority of the nuclei supported a breast origin of this tumor. × 416.

tiation features that allow identification of some of the different adenocarcinomas detailed in other sections of this chapter.

# RENAL-CELL CARCINOMA VERSUS ADRENAL CORTICAL NEOPLASM

Radiologically guided NAB is a safe, minimally invasive, and reliable method for diagnosing renal and adrenal neoplasms.[71,169] Because of the close proximity of the adrenal glands to the upper pole of the kidneys, tumors of each organ may directly involve each other. We have described a case in which, both clinically and radiologically, a mass at the upper pole of the kidney was thought to be a renal-cell carcinoma, but following ABC, this tumor was proved to be of adrenal cortical origin.[170] Renal-cell carcinomas may also metastasize to the adrenal glands.[44,164]

On light microscopy in ABC or surgical pathology specimens, renal and adrenal cortical carcinomas may closely mimic each other, causing difficulty in interpretation.[31,130,164] These tumors should also be considered in the differential diagnosis of metastatic neoplasms with clear or oncocytic cytoplasm. Immunocytochemical and ultrastructural studies, in the context of light microscopic and clinical findings, may be helpful for accurate classification. Renal-cell carcinomas generally express cytokeratins, EMA, blood group isoantigens, and secretory component (Fig. 10.16), whereas adrenal cortical carcinomas are generally negative for these markers (Fig. 10.17).[31,59,164] Because both tumors usually express vimentin, this cytoplasmic filament does not contribute to the differential diagnosis.[31,41] Pheochromocytomas may rarely be encountered in abdominal ABCs.[148] Depending on the cytological presentation, they may mimic neuroendocrine neoplasms such as carcinoid tumor, carcinomas, and sarcomas. We encountered one case of pheochromocytoma that simulated adenocarcinoma. The aspirate was reported by the radiologist to be from the area of the "pancreas." Due to the pleomorphic nature of nuclei and the presence of occasional "acini," a diagnosis of malignancy was rendered and the possibility of adenocarcinoma was suggested. The patient later proved to have a large pheochromocytoma of the left adrenal gland. Electron microscopy (demonstrating neurosecretory granules) and immunocytochemistry (positive staining for neuron-specific enolase, chromogranin, and synaptophysin) can assist to confirm this diagnosis. Pheochromocytomas may also express a variety of polypeptides such as serotonin, somatostatin, and calcitonin.

The separation of renal-cell carcinoma and adrenal cortical tumors is an ideal application for electron microscopy of ABC specimens. Despite the fact that each may have a granular or clear-cell appearance on cytological assessment, the ultrastructural counterparts of renal-cell (Fig. 10.18B) and adrenal cortical (Fig. 10.20B,C) tumors are distinctive. In the minibiopsy specimens available through ABC specimens, even the growth patterns of these two tumors are different at the ultrastructural level. Renal adenocarcinomas display a glandular architecture, with the tumor cells arranged in compact nests or along thin, fibrovascular cores in the papillary variants. In renal-cell carcinomas, tumor cells are separated from the stroma, with its rich capillary network, by a basal lamina. This arrangement contrasts

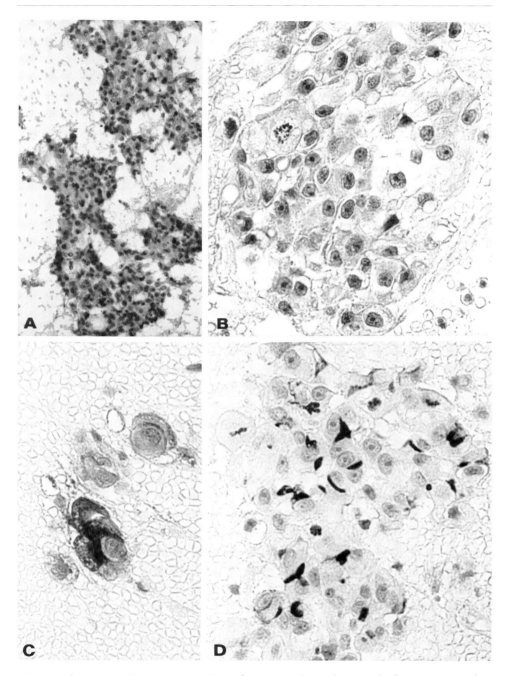

Fig. 10.16. Renal-cell carcinoma, ABC of a retroperitoneal mass. **A.** Smear preparation. Fragments of rather uniform tumor cells with round or oval nuclei. × 104. **B.** Cell block section. Note granularity of the cytoplasm. × 416. The differential diagnosis included renal-cell carcinoma and adrenal cortical tumor. The neoplastic cells expressed secretory component and epithelial membrane antigen (EMA), supporting the diagnosis of renal-cell carcinoma. **C.** Secretory component. × 416. **D.** EMA. The staining is predominantly membrane-based. × 416.

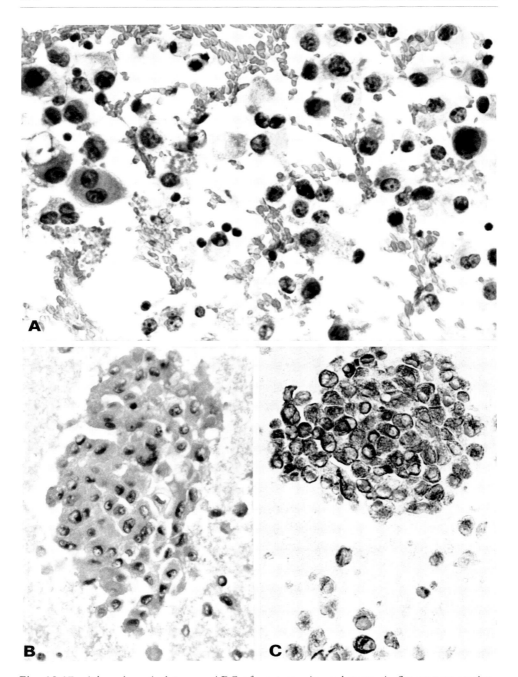

Fig. 10.17. Adrenal cortical tumor, ABC of a retroperitoneal mass. **A.** Smear preparation. Cellular aspirate composed of many isolated cells. Occasional binucleated tumor cells were seen. × 377. **B.** Cell block section. Note dense, somewhat granular, cytoplasm. × 260. **C.** The neoplastic cells strongly expressed vimentin. × 260. However, the stains for cytokeratins, epithelial membrane antigen, and secretory component were negative. The immunophenotype supported the diagnosis of adrenal cortical tumor over renal-cell carcinoma.

with adrenocortical lesions, whether benign or malignant, in which the growth pattern is generally sheet-like and there is no development of luminal structures or basal lamina.

In terms of cytoplasmic features, both tumors can have large amounts of lipid droplets, but glycogen accumulation is more a feature of renal adenocarcinomas. Microvilli are absent in adrenocortical lesions. It is essential to note, however, that microvilli may be sparse, absent, or aberrant in organization and form in some renal-cell carcinomas (Figs. 10.18B, 10.19A).[93] As emphasized in the "Microvilli" section in Chapter 6, many examples do not have the classic "brush border" microvilli of the normal renal tubular cells. Complex infoldings of the cell membrane at the base of tumor cells (Fig. 10.19B) can be seen in some renal adenocarcinomas (a reflection of this feature in the normal proximal renal tubular cell). This infolding is not evident in adrenal cortical neoplasms. Mitochondria can also be prominent in renal-cell and adrenal cortical tumors. Generally, mitochondria are either relatively smaller and darker staining (due to increased amounts of osmiophilic mitochondrial matrix) or have the characteristic tubulovesicular cristae in adrenal cortical neoplasms[139] (Fig. 10.20); this latter feature is not present in all such tumors. Prominent smooth endoplasmic reticulum and Golgi apparatus produce a much different general cytoplasmic appearance in adrenal cortical tumors compared with renal-cell carcinomas (compare Figures 10.18 and 10.20, which are ABC of a renal-cell carcinoma and an adrenal cortical tumor, respectively).

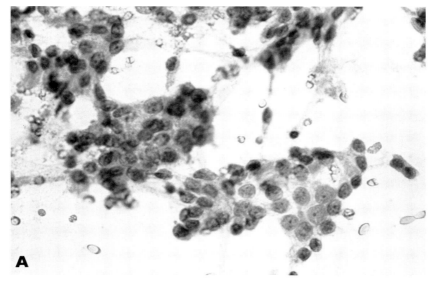

Fig. 10.18. Renal-cell carcinoma, ABC of a retroperitoneal mass. A. Smear preparation. Fragments of compact tumor cells with rather uniform round or oval nuclei are consistent with renal cell carcinoma. × 416.

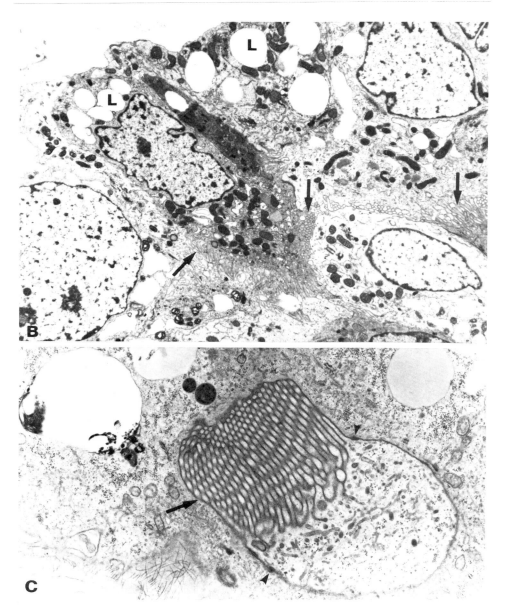

Fig. 10.18. B. Ultrastructurally, the tumor cells have darkly staining mitochondria, some prominent lipid droplets (L), and focally develop many microvilli (arrows) wedged between adjacent cells. × 2,800. C. A potential lumen (arrow) is crammed with well-defined microvilli with a more typical arrangement. Desmosomes (arrowheads) join the two adjacent cells. × 6,000.

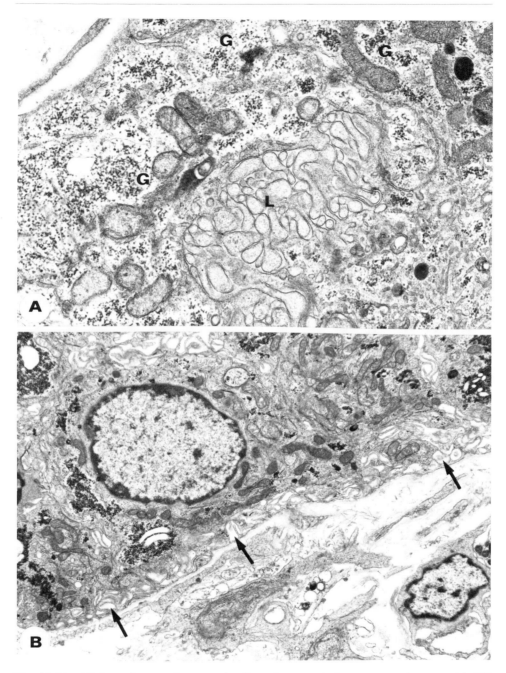

Fig. 10.19. ABC specimens of two renal-cell carcinomas. **A.** In one case, glycogen-rich (G) tumor cells form a putative lumen (L) occupied by abnormal microvilli. × 4,400. **B.** In another, basal aspects of tumor cells (arrows) have a band-like zone with many "clear" spaces formed as a result of repeated, complex infoldings of the cell membrane. × 4,400.

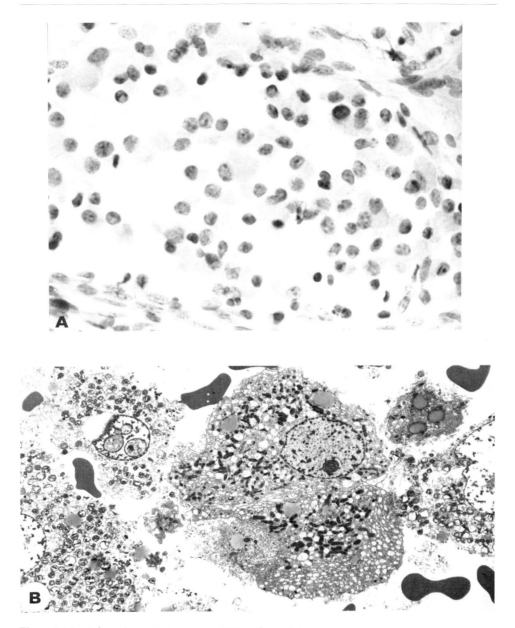

Fig. 10.20. Adrenal cortical tumor, ABC of an abdominal mass. **A.** Smear preparation. Cellular aspirate composed of many isolated cells with eccentric nuclei. × 260. **B.** Survey electron micrograph of this tumor with many darkly staining mitochondria, numerous small vesicular structures (representing smooth endoplasmic reticulum), and a few lipid droplets. × 1,800.

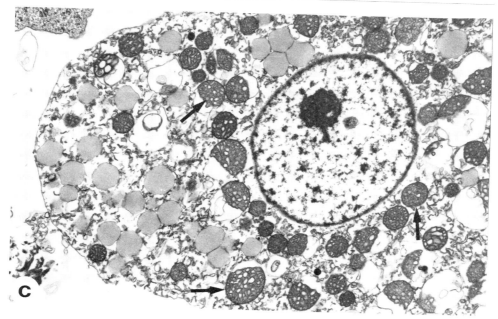

Fig. 10.20. C. Higher magnification reveals the characteristic vesicular cristae of the mitochondria (arrows) and many lipid droplets. This combination of features are diagnostic of adrenal cortical tumor. × 8,800.

# REFERENCES

1. Albores-Saavedra J, Nadji M, Civantos F, Morales AR: Thyroglobulin in carcinoma of the thyroid: An immunohistochemical study. *Hum Pathol* 14:62–66, 1983.
2. Albores-Saavedra J, Gorraez de la Mora T, de la Torre-Rendon F: Mixed medullary-papillary carcinoma of the thyroid: A previously unrecognized variant of thyroid carcinoma. *Hum Pathol* 21:1151–1155, 1990.
3. Allhoff EP, Proppe KH, Chapman CM, Lin C-W, Prout GR Jr: Evaluation of prostate specific acid phosphatase and prostate specific antigen in identification of prostatic cancer. *J Urol* 129:315–318, 1983.
4. Altmannsberger M, Dralle H, Weber K, Osborn M, Droese M: Intermediate filaments in cytological specimens of thyroid tumors. *Diagn Cytopathol* 3:210–214, 1987.
5. Amano S, Kataoka H, Hazama F, Nakatake M, Maki A: Alpha fetoprotein and carcinoembryonic antigen producing hepatocellular carcinoma: A case report studied by immunohistochemistry. *Acta Pathol Jpn* 35:969–974, 1985.
6. Bacus S, Flowers JL, Press MF, Bacus JW, McCarty KS Jr: The evaluation of estrogen receptor in primary breast carcinoma by computer-assisted image analysis. *Am J Clin Pathol* 90:233–239, 1988.
7. Bahu RM, Mangkornkanok-Mark M, Albertson D, et al: Detection of alpha-lactalbumin in breast lesions and relationship to estrogen receptors and serum prolactin. *Cancer* 46:1775–1780, 1980.
8. Bailey AJ, Sloane JP, Trickey BS, Ormerod MG: An immunocytochemical study of α-lactalbumin in human breast tissue. *J Pathol* 137:13–23, 1982.

9. Barry JD, Koch TJ, Cohen C, Brigati DJ, Sharkey FE: Correlation of immunohistochemical markers with patient prognosis in breast carcinoma: A quantitative study. *Am J Clin Pathol* 82:582–585, 1984.

10. Battifora H, Kopinski MI: Distinction of mesothelioma from adenocarcinoma; an immunohistochemical approach. *Cancer* 55:1679–1685, 1985.

11. Bedrossian CWM: Immunocytochemistry, in Astarita RW (ed): *Practical Cytopathology*. New York, Churchill Livingstone, 1990. Pp 403–457.

12. Bedrossian CWM, Davila RM, Merenda G: Immunocytochemical evaluation of liver fine-needle aspirations. *Arch Pathol Lab Med* 113:1225–1230, 1989.

13. Berman MA, Burnham JA, Sheahan DG: Fibrolamellar carcinoma of the liver: An immunohistochemical study of nineteen cases and a review of the literature. *Hum Pathol* 19:784–794, 1988.

14. Bolen JW, Hammar SP, McNutt MA: Reactive and neoplastic serosal tissue: A light-microscopic, ultrastructural, and immunocytochemical study. *Am J Surg Pathol* 10:34–47, 1986.

15. Bolen JW, Hammar SP, McNutt MA: Serosal tissue: Reactive tissue as a model for understanding mesotheliomas. *Ultrastruct Pathol* 11:251–262, 1987.

16. Bolen JW, Thorning D: Histogenetic classification of pulmonary carcinoma: Peripheral adenocarcinoma studied by light microscopy, histochemistry and electron microscopy. *Pathol Annu* 17:77–100, 1983.

17. Bollinger DJ, Wick MR, Dehner LP, et al: Peritoneal malignant mesothelioma versus serous papillary adenocarcinoma; a histochemical and immunohistochemical comparison. *Am J Surg Pathol* 13:659–670, 1989.

18. Bosman FT, Orenstein JM, Silverberg SG: Differential diagnosis of metastatic tumors, in Silverberg SG (ed): *Principles and Practice of Surgical Pathology*, ed 2. New York, Churchill Livingstone, 1990. Pp 119–144.

19. Brawer MK, Peehl DM, Stamey TA, Bostwick DG: Keratin immunoreactivity in the benign and neoplastic human prostate. *Cancer Res* 45:3663–3667, 1985.

20. Bruderman I, Cohen R, Leitner O, et al: Immunocytochemical characterization of lung tumors in fine-needle aspiration; the use of cytokeratin monoclonal antibodies for the differential diagnosis of squamous cell carcinoma and adenocarcinoma. *Cancer* 66:1817–1827, 1990.

21. Burns TR, Johnson EH, Cartwright J Jr, Greenberg SD: Desmosomes of epithelial malignant mesothelioma. *Ultrastruct Pathol* 12:385–388, 1988.

22. Cagle PT, Mody DR, Schwartz MR: Estrogen and progesterone receptors in bronchogenic carcinoma. *Cancer Res* 50:6632–6635, 1990.

23. Cagle PT, Truong LD, Roggli VL, Greenberg SD: Immunohistochemical differentiation of sarcomatoid mesotheliomas from other spindle cell neoplasms. *Am J Clin Pathol* 92:566–571, 1989.

24. Chacho MS, Greenbaum E, Moussouris HF, Schreiber K, Koss LG: Value of aspiration cytology of the thyroid in metastatic disease. *Acta Cytol* 31:705–712, 1987.

25. Chedid A, Chejfec G, Eichorst M, et al: Antigen markers of hepatocellular carcinoma. *Cancer* 65:84–87, 1990.

26. Choe B-K, Pontes EJ, Rose NR, Henderson MD: Expression of human prostatic acid phosphatase in a pancreatic islet cell carcinoma. *Invest Urol* 15:312–318, 1978.

27. Christensen WN, Boitnott JK, Kuhajda FP: Immunoperoxidase staining as a diagnostic aid for hepatocellular carcinoma. *Mod Pathol* 2:8–12, 1989.

28. Cibas ES, Corson JM, Pinkus GS: The distinction of adenocarcinoma from malignant mesothelioma in cell blocks of effusions: The role of routine mucin histochemistry and immunohistochemical assessment of carcinoembryonic antigen, keratin protein, epithelial membrane antigen, and milk fat globule-derived antigen. *Hum Pathol* 18: 67–74, 1987.

29. Clayton F, Ordóñez NG, Hanssen GM, Hanssen H: Immunoperoxidase localization of lactalbumin in malignant breast neoplasms. *Arch Pathol Lab Med* 106:268–270, 1982.

30. Cohen C, Bentz MS, Budgeon LR: Prostatic acid phosphatase in carcinoid and islet cell tumors. *Arch Pathol Lab Med* 107:277, 1983.

31. Cote RJ, Cordon-Cardo C, Reuter VE, Rosen PP: Immunopathology of adrenal and renal cortical tumors; coordinated change in antigen expression is associated with neoplastic conversion in the adrenal cortex. *Am J Pathol* 136:1077–1084, 1990.

32. Cotran RS, Kumar V, Robbins SL: *Robbins Pathologic Basis of Disease*, ed 4. Philadelphia, W. B. Saunders, 1989. Pp 958–962.

33. Cowen PN, Teasdale J, Jackson P, Reid BJ: Oestrogen receptor in breast cancer: Prognostic studies using a new immunohistochemical assay. *Histopathology* 17:319–325, 1990.

34. Dardick I: Diagnostic electron microscopy, in Gnepp DR (ed): *Pathology of the Head and Neck, Contemporary Issues in Surgical Pathology*, vol 10. New York, Churchill Livingstone, 1988. Pp 101–190.

35. Dardick I, Al-Jabi M, McCaughey WTE, et al: Ultrastructure of poorly differentiated diffuse epithelial mesotheliomas. *Ultrastruct Pathol* 7:151–160, 1984.

36. Dardick I, Jabi M, McCaughey WTE, et al: Diffuse epithelial mesothelioma: A review of the ultrastructural spectrum. *Ultrastruct Pathol* 11:503–533, 1987.

37. Davila RM, Bedrossian WM, Silverberg AB: Immunocytochemistry of the thyroid in surgical and cytologic specimens. *Arch Pathol Lab Med* 112:51–56, 1988.

38. Dekmezian R, Sniege N, Popok S, Ordóñez NG: Fine needle aspiration cytology of pediatric patients with primary hepatic tumors: a comparative study of two hepatoblastomas and a liver cell carcinoma. *Diagn Cytopathol* 4:162–168, 1988.

39. Devleeschouwer N, Faverly D, Kiss R, et al: Comparison of biochemical and immunoenzymatic macromethods and a new immunocytochemical micromethod for assaying estrogen receptors in human breast carcinoma. *Acta Cytol* 32:816–824, 1988.

40. Doglioni C, Gambacorta M, Zamboni G, Coggi G, Viale G: Immunocytochemical localization of progesterone receptors in endocrine cells of the human pancreas. *Am J Pathol* 137:999–1005, 1990.

41. Domagala W, Lasota J, Wolska H, et al: Diagnosis of metastatic renal cell and thyroid carcinomas by intermediate filament typing and cytology of tumor cells in fine needle aspirates. *Acta Cytol* 32:415–421, 1988.

42. Donna A, Betta P-G, Jones JSP: Verification of the histologic diagnosis of malignant mesothelioma in relation to the binding of an antimesothelial cell antibody. *Cancer* 63:1331–1336, 1989.

43. Doria MI, Adamec T, Talerman A: Alpha-lactalbumin in "common" epithelial tumors of the ovary; an immunohistochemical study. *Am J Clin Pathol* 87:752–756, 1987.

44. Duggan MA, Forestell CF, Hanley DA: Adrenal metastases of renal-cell carcinoma 19 years after nephrectomy; fine needle aspiration cytology of a case. *Acta Cytol* 31:512–516, 1987.

45. Ellis DW, Leffers S, Davies JS, Ng ABP: Multiple immunoperoxidase markers in benign hyperplasia and adenocarcinoma of the prostate. *Am J Clin Pathol* 81:279–284, 1984.
46. Epstein JI, Kuhajda FP, Lieberman PH: Prostate-specific acid phosphatase immunoreactivity in adenocarcinoma of the urinary bladder. *Hum Pathol* 17:939–942, 1986.
47. Epstein JI, Leiberman PH: Mucinous adenocarcinoma of the prostate gland. *Am J Surg Pathol* 9:299–308, 1985.
48. Feiner HD, Gonzalez R: Carcinoma of the prostate with atypical immunohistological features; clinical and histologic correlates. *Am J Surg Pathol* 10:765–770, 1986.
49. Fishleder A, Tubbs RR, Levin HS: An immunoperoxidase technique to aid in the differential diagnosis of prostate carcinoma. *Cleve Clin Q* 48:331–335, 1981.
50. Frable WJ: *Thin-Needle Aspiration Biopsy*. Philadelphia, W. B. Saunders, 1983. Pp 232–240.
51. Gal R, Aronof A, Gertzmann H, Kessler E: The potential value of the demonstration of thyroglobulin by immunoperoxidase techniques in fine needle aspiration cytology. *Acta Cytol* 31:713–716, 1987.
52. Ghadially FN: *Diagnostic Electron Microscopy of Tumours*, 2nd ed. London, Butterworths, 1985.
53. Goodman ZD, Ishak KG, Langloss JM, Sesterhenn IA, Rabin L: Combined hepatocellular-cholangiocarcinoma; a histologic and immunohistochemical study. *Cancer* 55:124–135, 1985.
54. Gown AM, Vogel AM: Monoclonal antibodies to human intermediate filament proteins. III. Analysis of tumors. *Am J Clin Pathol* 84:413–424, 1985.
55. Green C-A, Suen KC: Some cytologic features of hepatocellular carcinoma as seen in fine needle aspirates. *Acta Cytol* 28:713–718, 1984.
56. Hajdu SI, Ehya H, Frable WJ, et al: The value and limitations of aspiration cytology in the diagnosis of primary tumors; a symposium. *Acta Cytol* 33:741–790, 1989.
57. Hammar S: Adenocarcinoma and large cell undifferentiated carcinoma of the lung. *Ultrastruct Pathol* 11:263–291, 1987.
58. Hammar S, Bockus D, Remington F: Metastatic tumors of unknown origin: An ultrastructural analysis of 265 cases. *Ultrastruct Pathol* 11:209–250, 1987.
59. Harris JP, South MA: Secretory component; a glandular epithelial cell marker. *Am J Pathol* 105:47–53, 1981.
60. Heimann A, Moll U: Spinal metastasis of a thyroglobulin-rich Hürthle cell carcinoma detected by fine needle aspiration; light and electron microscopic study of an unusual case. *Acta Cytol* 33:639–644, 1989.
61. Herrera GA, Alexander B, De Moraes HP: Ultrastructural subtypes of pulmonary adenocarcinoma: A correlation with patient survival. *Chest* 84:581–586, 1983.
62. Holm R, Sobrinho-Simões M, Nesland JM, Sambade C, Johannessen JV: Medullary thyroid carcinoma with thyroglobin immunoreactivity; a special entity? *Lab Invest* 57:258–268, 1987.
63. Johnson DE, Herndier BG, Medeiros LJ, Warnke RA, Rouse RV: The diagnostic utility of keratin profiles of hepatocellular carcinoma and cholangiocarcinoma. *Am J Surg Pathol* 12:187–197, 1988.
64. Johnson TL, Lloyd RV, Burney RE, Thompson NW: Hurthle cell thyroid tumors; an immunohistochemical study. *Cancer* 59:107–112, 1987.

65. Johnston WW: Applications of monoclonal antibodies in clinical cytology as exemplified by studies with monoclonal antibody B72.3. *Acta Cytol* 31:537–556, 1987.
66. Jothy S, Brazinsky SA, Chin-A-Loy M, et al: Characterization of monoclonal antibodies to carcinoembryonic antigen with increased tumor specificity. *Lab Invest* 54:108–117, 1986.
67. Kagan AR, Steckel RJ: The limited role of radiologic imaging in patients with unknown tumor primary. *Semin Oncol* 18:170–173, 1991.
68. Kahn HJ, Thorner PS, Yeger H, Bailey D, Baumal R: Distinct keratin patterns demonstrated by immunoperoxidase staining of adenocarcinomas, carcinoids, and mesotheliomas using polyclonal and monoclonal anti-keratin antibodies. *Am J Clin Pathol* 86:566–574, 1986.
69. Kakizoe S, Kojiro M, Nakashima T: Hepatocellular carcinoma with sarcomatous change; clinicopathologic and immunohistochemical studies of 14 autopsy cases. *Cancer* 59:310–316, 1987.
70. Kakudo K, Miyauchi A, Katayama S, Watanabe K: Ultrastructural study of poorly differentiated medullary carcinoma of the thyroid. *Virchows Arch {A}* 410:455–460, 1987.
71. Katz RL, Patel S, Mackay B, Zornoza J: Fine needle aspiration cytology of the adrenal gland. *Acta Cytol* 28:269–282, 1984.
72. Katz RL, Raval P, Brooks TE, Ordóñez NG: Role of immunocytochemistry in diagnosis of prostatic neoplasm by fine needle aspiration biopsy. *Diagn Cytopathol* 1:28–32, 1985.
73. Keshgegian AA: Hormone receptors and other tumor markers, in Kline TS, Kline IK, (eds): *Guides to Clinical Aspiration Biopsy: Breast*. New York, Igaku-Shoin, 1989. Pp 215–234.
74. Keshgegian AA, Inverso K, Kline TS: Determination of estrogen receptor by monoclonal antireceptor antibody in aspiration biopsy cytology from breast carcinoma. *Am J Clin Pathol* 89:24–29, 1988.
75. Keshgegian AA, Wheeler JE: Estrogen receptor protein in malignant carcinoid tumor; a report of 2 cases. *Cancer* 45:293–296, 1980.
76. Khoury N, Raju U, Crissman JD, et al: A comparative immunohistochemical study of peritoneal and ovarian serous tumors, and mesotheliomas. *Hum Pathol* 21:811–819, 1990.
77. Kiang DT, Kennedy BJ: Estrogen receptor assay in the differential diagnosis of adenocarcinoma. *JAMA* 238:32–34, 1977.
78. Kirsten F, Chi CH, Leary JA, et al: Metastatic adeno or undifferentiated carcinoma from an unknown primary site—natural history and guidelines for identification of treatable subsets. *Q J Med* 62:143–161, 1987.
79. Kline TS, Kline IK: *Guides to Clinical Aspiration Biopsy: Breast*. New York, Igaku-Shoin, 1989.
80. Koelma IA, Nap M, Huitema S, Krom RAF, Houthoff HJ: Hepatocellular carcinoma, adenoma, and focal nodular hyperplasia; comparative histopathologic study with immunohistochemical parameters. *Arch Pathol Lab Med* 110:1035–1040, 1986.
81. Kojima O, Takahashi T, Kawakami S, Uehara Y, Matsui M: Localization of estrogen receptors in gastric cancer using immunohistochemical staining of monoclonal antibody. *Cancer* 67:2401–2406, 1991.
82. Lampertico P, Stagni F: Cytology and hormonal receptors in breast cancer. *Diagn Cytopathol* 2:17–23, 1986.

83. Leader M, Collins PM, Henry K: Anti-α1-antichymotrypsin staining of 194 sarcomas, 38 carcinomas, and 17 malignant melanomas, its lack of specificity as a tumor marker. Am J Surg Pathol 11:133–139, 1987.
84. Lee AK, DeLellis RA, Rosen PP, et al: Alpha-lactalbumin as an immunohistochemical marker for metastatic breast carcinomas. Am J Surg Pathol 8:93–100, 1984.
85. Lee AK, Rosen PP, DeLellis RA, et al: Tumor marker expression in breast carcinomas and relationship to prognosis: an immunohistochemical study. Am J Clin Pathol 84:687–696, 1985.
86. Leong AS-Y, Kan AE, Milios J: Small round cell tumors in childhood: Immunohistochemical studies in rhabdomyosarcoma, neuroblastoma, Ewing's sarcoma, and lymphoblastic lymphoma. Surg Pathol 2:5–17, 1989.
87. Leong AS-Y, Parkinson R, Milios J: "Thick" cell membranes revealed by immunocytochemical staining: A clue to the diagnosis of mesothelioma. Diagn Cytopathol 6:9–13, 1990.
88. Lloyd RV, Foley J, Judd WJ: Peanut lectin agglutinin and α-lactalbumin; binding and immunohistochemical localization in breast tissues. Arch Pathol Lab Med 108:392–395, 1984.
89. Logmans SC, Jobsis AC: Thyroid-associated antigens in routinely embedded carcinomas; possibilities and limitations studied in 116 cases. Cancer 54:274–279, 1984.
90. Lozowski M, Greene GL, Sadri D, et al: The use of fine needle aspiration of progesterone receptor content in breast cancer. Acta Cytol 34:27–30, 1990.
91. Lozowski MS, Mishriki Y, Chao S, et al: Estrogen receptor determination in fine needle aspirates of the breast; correlation with histologic grade and comparison with biochemical analysis. Acta Cytol 32:557–562, 1987.
92. Lundy J, Kline TS, Lozowski M, Chao S: Immunoperoxidase studies by monoclonal antibody B72.3 applied to breast aspirates: Diagnostic considerations. Diagn Cytopathol 4:95–98, 1988.
93. MacKay B, Ordóñez NG, Khoursand: The ultrastructure and immunocytochemistry of renal cell carcinoma. Ultrastruct Pathol 11:483–502, 1987.
94. Mansou EG, Hastert M, Park CH, Koehler KA, Petrelli M: Tissue and plasma carcinoembryonic antigen in early breast cancer; a prognostic factor. Cancer 51:1243–1248, 1983.
95. Marchetti E, Bagni A, Querzoli P, et al: Immunocytochemical detection of estrogen receptors by staining with monoclonal antibodies on cytologic specimens of human breast cancer. Acta Cytol 32:829–834, 1988.
96. Marshall RJ, Herbert A, Braye SG, Jones DB: Use of antibodies to carcinoembryonic antigen and human milk fat globule to distinguish carcinoma, mesothelioma, and reactive mesothelium. J Clin Pathol 37:1215–1221, 1984.
97. Masood S: Use of monoclonal antibody for assessment of estrogen receptor content in fine-needle aspiration biopsy specimen from patients with breast cancer. Arch Pathol Lab Med 113:26–30, 1989.
98. May EE, Perentes E: Anti-Leu 7 immunoreactivity with human tumours: Its value in the diagnosis of prostatic adenocarcinoma. Histopathology 11:295–304, 1987.
99. Mazoujian G, Bodian C, Haagensen DE Jr, Haagensen CD: Expression of GCDFP-15 in breast carcinomas; relationship to pathologic and clinical factors. Cancer 63:2156–2161, 1989.
100. McClellan RA, Berger U, Wilson P, et al: Presurgical determination of estrogen recep-

tor status using immunocytochemically stained fine needle aspirate smears in patients with breast cancer. *Cancer Res* 47:6118–6122, 1987.

101. McMillan JH, Levine E, Stephens RH: Computed tomography in the evaluation of metastatic adenocarcinoma from an unknown primary site—a retrospective study. *Radiology* 143:143–146, 1982.

102. Mendelsohn G, Bigner SH, Eggleston JC, Baylin SB, Wells SA Jr: Anaplastic variants of medullary thyroid carcinoma; a light-microscopic and immunohistochemical study. *Am J Surg Pathol* 4:333–341, 1980.

103. Michinami R, Oono Y: Carcinoembryonic antigen and lectin binding in the bile canalicular structures of hepatocellular carcinoma. *Virchows Arch {A}* 412:111–118, 1987.

104. Monteagudo C, Merino MJ, Laporte N, Neumann RD: Value of gross cystic disease fluid protein-15 in distinguishing metastatic breast carcinomas among poorly differentiated neoplasms involving the ovary. *Hum Pathol* 22:368–372, 1991.

105. Montgomery BT, Nativ O, Blute ML, et al: Stage B prostate adenocarcinoma; flow cytometric nuclear DNA ploidy analysis. *Arch Surg* 125:327–331, 1990.

106. Morohoshi T, Kanda M, Horie A, et al: Immunocytochemical markers of uncommon pancreatic tumors; acinar cell carcinoma, pancreatoblastoma, and solid cystic (papillary-cystic) tumor. *Cancer* 59:739–747, 1987.

107. Murthy L, Kapila K, Verma K: Immunoperoxidase detection of carcinoembryonic antigen in fine needle aspirates of breast carcinoma; correlation with studies in tissue sections. *Acta Cytol* 32:60–62, 1988.

108. Nadji M, Tabei SZ, Castro A, Chu TM, Morales AR: Prostatic origin of tumors; an immunohistochemical study. *Am J Clin Pathol* 73:735–739, 1980.

109. Nadji M, Tabei SZ, Castro A, et al: Prostatic-specific antigen: An immunohistologic marker for prostatic neoplasms. *Cancer* 48:1229–1232, 1981.

110. Naritoku WY, Taylor CR: Immunohistologic diagnosis of 2 cases of metastatic prostate cancer to breast. *J Urol* 130:365–367, 1983.

111. Nesland JM, Holm R: Diagnostic problems in breast pathology: The benefit of ultrastructural and immunocytochemical analysis. *Ultrastruct Pathol* 15:293–311, 1987.

112. Nuti M, Mottolese M, Viora M, et al: Use of monoclonal antibodies to human breast-tumor-associated antigens in fine-needle aspirate cytology. *Int J Cancer* 37:493–498, 1986.

113. Nystrom JS, Weiner JM, Wolf RM, Bateman JR, Viola MV: Indentifying the primary site in metastatic cancer of unknown origin; inadequacy of roentgenographic procedures. *JAMA* 241:381–383, 1979.

114. O'Hara CJ, Corson JM, Pinkus GS, Stahel RA: ME1; a monoclonal antibody that distinguishes epithelial-type malignant mesothelioma from pulmonary adenocarcinoma and extrapulmonary malignancies. *Am J Pathol* 136:421–428, 1990.

115. Ordóñez NG: The immunohistochemical diagnosis of mesothelioma; differentiation of mesothelioma and lung adenocarcinoma. *Am J Surg Pathol* 13:276–291, 1989.

116. Ostrzega N, Cheng L, Layfield LJ: Keratin immunoreactivity in fine-needle aspiration of the prostate: an aid in the differentiation of benign epithelium from well-differentiated adenocarcinoma. *Diagn Cytopathol* 4:38–41, 1988.

117. Otis CN, Carter D, Cole S, Battifora H: Immunohistochemical evaluation of pleural mesothelioma and pulmonary adenocarcinoma; a bi-institutional study of 47 cases. *Am J Surg Pathol* 11:445–456, 1987.

118. Ozzello L, DeRosa C, Habif DV, Greene GL: An immunohistochemical evaluation of

progesterone receptor in frozen sections, paraffin sections, and cytologic imprints of breast carcinomas. *Cancer* 67:455–462, 1991.

119. Papsidero LD, Croghan GA, Asirwatham J, et al: Immunohistochemical demonstration of prostate-specific antigen in metastases with the use of monoclonal antibody F5. *Am J Pathol* 121:451–454, 1985.

120. Pavelic ZP, Petrelli NJ, Herrera L, et al: D-14 monoclonal antibody to carcinoembryonic antigen: Immunohistochemical analysis of formalin-fixed, paraffin-embedded human colorectal carcinoma, tumors of non-colorectal origin and normal tissues. *J Cancer Res Clin Oncol* 116:51–56, 1990.

121. Pertschuk LP, Kim DS, Nayer K, et al: Immunocytochemical estrogen and progestin receptor assays in breast cancer with monoclonal antibodies; histopathologic, demographic, and biochemical correlations and relationship to endocrine response and survival. *Cancer* 66:1663–1670, 1990.

122. Pohoresky J, Yazdi HM: Value of B72.3 staining in "suspicious" fine needle aspiration biopsies of the breast. *Acta Cytol* 35:254–255, 1991.

123. Rastad J, Wilander E, Lindgren P-G, et al: Cytologic diagnosis of a medullary carcinoma of the thyroid by Sevier-Munger silver staining and calcitonin immunocytochemistry. *Acta Cytol* 31:45–47, 1987.

124. Reiner A, Neumeister B, Spona J, et al: Immunocytochemical localization of estrogen and progesterone receptor and prognosis in human primary breast cancer. *Cancer Res* 50:7057–7061, 1990.

125. Reiner A, Reiner G, Spona J, et al: Estrogen receptor immunocytochemistry for preoperative determination of estrogen receptor status on fine-needle aspirates of breast cancer. *Am J Clin Pathol* 88:399–404, 1987.

126. Reiner A, Spona J, Reiner G, et al: Estrogen receptor analysis on biopsies and fine-needle aspirates from human breast carcinoma; correlation of biochemical and immunohistochemical methods using monoclonal antireceptor antibodies. *Am J Pathol* 125:443–449, 1986.

127. Ro JY, El-Naggar A, Ayala AG, et al: Signet-ring-cell carcinoma of the prostate; electron-microscopic and immunohistochemical studies of eight cases. *Am J Surg Pathol* 12:453–460, 1988.

128. Ro JY, Grignon DJ, Ayala AG, et al: Mucinous adenocarcinoma of the prostate: Histochemical and immunohistochemical studies. *Hum Pathol* 21:593–600, 1990.

129. Ro JY, Têtu B, Ayala AG, Ordóñez NG: Small cell carcinoma of the prostate. II. Immunohistochemical and electron microscopic studies of 18 cases. *Cancer* 59:977–982, 1987.

130. Rosai J: *Ackerman's Surgical Pathology,* ed 7. St. Louis, CV Mosby, 1989.

131. Ruppert JM, Eggleston JC, deBustros A, Baylin SB: Disseminated calcitonin-poor medullary thyroid carcinoma in a patient with calcitonin-rich primary tumor. *Am J Surg Pathol* 10:513–518, 1986.

132. Schmid KW, Fischer-Colbrie R, Hagn C, et al: Chromogranin A and B and secretogranin II in medullary carcinomas of the thyroid. *Am J Surg Pathol* 11:551–556, 1987.

133. Schröder S, Klöppel G: Carcinoembryonic antigen and nonspecific cross-reacting antigen in thyroid cancer; an immunocytochemical study using polyclonal and monoclonal antibodies. *Am J Surg Pathol* 11:100–108, 1987.

134. Seymour L, Meyer K, Esser J, et al: Estimation of PR and ER by immunocytochemistry in breast cancer; comparison with radiological binding methods. *Am J Clin Pathol* 94(suppl 1):535–540, 1990.

135. Shah KD, Tabibzadeh SS, Gerber MA: Comparison of cytokeratin expression in primary and metastatic carcinomas; diagnostic application in surgical pathology. *Am J Clin Pathol* 87:708–715, 1987.
136. Shevchuk MM, Romas NA, Ng PY, Tannenbaum M, Olsson CA: Acid phosphatase localization in prostatic carcinoma; a comparison of monoclonal antibody to heteroantisera. *Cancer* 52:1642–1646, 1983.
137. Shousha S, Lyssiotis T, Godfrey VM, Scheuer PJ: Carcinoembryonic antigen in breast-cancer tissue: A useful prognostic indicator. *Br Med J* 1:777–779, 1979.
138. Sikri KL, Varndel IM, Hamid QA, et al: Medullary carcinoma of the thyroid; an immunocytochemical and histochemical study of 25 cases using eight separate markers. *Cancer* 56:2481–2491, 1985.
139. Silva EG, Mackay B, Samaan NA, et al: Adrenocortical carcinomas: An ultrastructural study of 22 cases. *Ultrastruct Pathol* 3:143–151, 1982.
140. Snee MP, Vyramuthu N: Metastatic carcinoma from unknown primary site: The experience of a large oncology centre. *Br J Radiol* 58:1091–1095, 1985.
141. Sobrinho-Simões M, Holm R, Johannessen JV: Diagnostic electron microscopy of thyroid tumours, in Johannessen JV, Gould V, Faria V, et al (eds): *Electron Microscopy in Diagnostic Pathology*. Lisbon, Portuguese Society of Pathology, 1984. Pp 75–99.
142. Srigley JR, Hartwick WJ, Edwards V, de Harven E: Selected ultrastructural aspects of urothelial and prostatic tumors. *Ultrastruct Pathol* 12:49–65, 1988.
143. Stanta G, Carcangiu ML, Rosai J: The biochemical and immunohistochemical profile of thyroid neoplasia, in Rosen PP, Fechner RE, (eds): *Pathology Annual (Part 1)*. Norwalk, Appleton & Lange, 1988. Pp 129–157.
144. Steckel RJ, Kagan AR: Metastatic tumors of unknown origin. *Cancer* 67:1242–1244, 1991.
145. Stein BS, Petersen RO, Vangore S, Kendall AR: Immunoperoxidase localization of prostate-specific antigen. *Am J Surg Pathol* 6:553–557, 1982.
146. Sterrett GF, Whitaker D, Shilkin KB, Walters MNI: Fine needle aspiration cytology of malignant mesothelioma. *Acta Cytol* 31:185–193, 1987.
147. Suen KC: Diagnosis of primary hepatic neoplasms by fine-needle aspiration cytology. *Diagn Cytopathol* 2:99–109, 1986.
148. Suen KC: *Guides to Clinical Aspiration Biopsy: Retroperitoneum and Intestine*. New York, Igaku-Shoin, 1987.
149. Swanson PE, Pettinato G, Lillemoe TJ, Wick MR: Gross cystic disease fluid protein-15 in salivary gland tumors. *Arch Pathol Lab Med* 115:158–163, 1991.
150. Szpak CA, Johnston WW, Roggli V, et al: The diagnostic distinction between malignant mesothelioma of the pleura and adenocarcinoma of the lung as defined by a monoclonal antibody (B72.3). *Am J Pathol* 122:252–260, 1986.
151. Taccagni G, Dell'Antonio G, Terreni MR, Cantaboni A: Heterogeneous subcellular morphology of lung adenocarcinoma cells: Identification of different cytotypes on cytological material. *Ultrastruct Pathol* 14:65–80, 1990.
152. Tahara E, Ito H, Taniyama K, Yokozaki H, Hata J: Alpha 1-antitrypsin, alpha 1-antichymotrypsin, and alpha 2-macroglobulin in human gastric carcinomas: A retrospective immunohistochemical study. *Hum Pathol* 15:957–964, 1984.
153. Tani E, Skoog L: Fine needle aspiration cytology and immunocytochemistry in the diagnosis of lymphoid lesions of the thyroid gland. *Acta Cytol* 33:48–52, 1989.
154. Tao L-C: Aspiration biopsy cytology of mesothelioma. *Diagn Cytolpathol* 5:14–21, 1989.

155. Tao L-C, Ho CS, McLoughlin MJ, Evans WK, Donat EE: Cytologic diagnosis of hepatocellular carcinoma by fine-needle aspiration biopsy. *Cancer* 53:547–552, 1984.

156. Taylor CR: *Immunomicroscopy: A Diagnostic Tool for the Surgical Pathologist.* Philadelphia, W. B. Saunders, 1986.

157. Tell DT, Khoury JM, Taylor HG, Veasey SP: Atypical metastasis from prostate cancer; clinical utility of the immunoperoxidase technique for prostate-specific antigen. *JAMA* 253:3574–3575, 1985.

158. Thung SN, Gerber MA, Sarno E, Hopper H: Distribution of five antigens in hepatocellular carcinoma. *Lab Invest* 41:101–105, 1979.

159. Tron V, Wright JL, Churg A: Carcinoembryonic antigen and milk-fat globule protein staining of malignant mesothelioma and adenocarcinoma of the lung. *Arch Pathol Lab Med* 111:291–293, 1987.

160. Uccini S, Monardo F, Paradiso P, et al: Synaptophysin in human breast carcinomas. *Histopathology* 18:271–273, 1991.

161. Visscher D, Zarbo RJ, Crissman JD: Application of monoclonal antibodies to high and low molecular weight cytokeratins in the differential diagnosis of human carcinomas. *Surg Pathol* 1:407–416, 1988.

162. Wakely Jr PE, Silverman JF, Geisinger KR, Frable WJ: Fine needle aspiration biopsy cytology of hepatoblastoma. *Mod Pathol* 3:688–693, 1990.

163. Weintraub J, Weintraub D, Redard M, Vassilakos P: Evaluation of estrogen receptors by immunocytochemistry on fine-needle aspiration biopsy specimens from breast tumors. *Cancer* 60:1163–1172, 1987.

164. Wick MR, Cherwitz DL, McGlennen RC, Dehner LP: Adrenocortical carcinoma; an immunohistochemical comparison with renal cell carcinoma. *Am J Pathol* 122:343–352, 1986.

165. Wick MR, Loy T, Mills SE, Legier JF, Manivel JC: Malignant epithelial pleural mesothelioma versus peripheral pulmonary adenocarcinoma: a histochemical, ultrastructural, and immunohistologic study of 103 cases. *Hum Pathol* 21:759–766, 1990.

166. Wirth PR, Legier J, Wright GL: Immunohistochemical evaluation of seven monoclonal antibodies for differentiation of pleural mesothelioma from lung adenocarcinoma. *Cancer* 67:655–662, 1991.

167. Wong MA, Yazdi HM: Hepatocellular carcinoma versus carcinoma metastatic to the liver; value of stains for carcinoembryonic antigen and naphthylamidase in fine needle aspiration biopsy material. *Acta Cytol* 34:192–196, 1990.

168. Yam LT, Janckila AJ, Lam WKW, Li C-Y: Immunohistochemistry of prostatic acid phosphatase. *Prostate* 2:97–107, 1981.

169. Yazdi HM: Genitourinary cytology, in Hajdu SI (ed): *Value and Limitations of Cytologic Evaluation, Clinics in Laboratory Medicine 11 (2)*. Philadelphia, W. B. Saunders, 1991. Pp 369–401.

170. Yazdi HM, Dardick I: What is the value of electron microscopy in fine-needle aspiration biopsy? *Diagn Cytopathol* 4:177–182, 1988.

# 11
# The Clinician's Viewpoint

## AN ONCOLOGIST'S PERSPECTIVE
William K. Evans, M.D.

In the last several decades, substantial progress has been made in the diagnosis and treatment of cancer. Orthovoltage machines have been replaced by megavoltage radiotherapy units, which not only deliver more radiation to deep-seated tumors, but also do so with far greater precision. Sophisticated radiotherapy treatment plans incorporate detailed information about a tumor's location and extent as revealed by modern imaging techniques, including computed tomographic (CT) and magnetic resonance scans. At the same time, the number and variety of cytotoxic agents that are effective against specific cancers have increased, and more effective combination regimens are now used in many advanced tumors with curative intent. A current trend is to combine several treatment methods to optimize the impact on the tumor both locally and systematically. Because of the complexity, expense, and toxicity of such strategies, oncologists require precise information regarding a tumor's pathological type and extent to guide treatment decisions and to provide valuable prognostic information to patients and their families. For these reasons, it has become very important for oncologists to eliminate as many uncertainties as possible through the judicious use of precise diagnostic aids.

Parallel to the development of more effective therapeutic strategies and staging procedures for cancer has been the possibility to biopsy organs with less need for operative intervention. Although the technique of aspiration biopsy has been practiced since the 1930s, its popularity only became established in the last two decades.[13] Until recently, there have been few pathologists with the necessary skill to interpret the aspirated material, and clinicians have been wary of the test's accuracy and fearful of spreading tumor cells along the needle tract. An increasing number of North American physicians have become expert in the use of needle aspiration biopsy (NAB), and a large body of medical literature has defined the usefulness, safety, and accuracy of NAB. NAB is also far less costly than any other alternative methods available. The use of NAB may obviate the need for hospitalization or, at the very least, abbreviate the length of hospital stays. One report based on the experience in an American community hospital demonstrated a potential cost reduction of 71 to 90% when NAB was used in lieu of conventional "operative" biopsies for the diagnosis of lesions of the breast, lung, and pancreas.[17] The demand for a prompt diagnosis with the least cost has helped accelerate the widespread acceptance of this technique. Improvements in the instrumentation necessary for aspiration biopsy and advances in radiological imaging techniques have made it possible to accurately direct biopsy needles to almost all sites in the body. Finally, there are increasing numbers of experienced cytopathologists who are able to accurately interpret aspiration biopsy cytology (ABC), which can be studied not only by light microscopy but by other sophisticated techniques as well. As detailed elsewhere in this book, immunocytochemical techniques, as well as the use of electron microscopy and genotypic analysis of DNA obtained through NAB, have expanded the usefulness of ABC. Oncologists should be fully knowledgeable of the potential uses of ABC and trained to perform NAB on superficial lesions. Internists or oncologists are dependent on radiology colleagues for biopsy of deep-seated lesions and on surgeons for intraoperative NAB.

# FINE-NEEDLE ASPIRATION BIOPSY TECHNIQUE FOR SUPERFICIAL LESIONS

The technique for NAB of superficial lesions can be readily learned and applied in the out-patient department of any hospital or clinic by internists, hematologists or oncologists. After minimal skin preparation, a local anesthetic may be used, although for palable lymph nodes and other superficial lesions a local anesthetic is generally not required and may obscure the location of a small lesion. A 22- to 25-gauge needle is attached to a 10-ml disposable syringe. The needle is introduced into the mass, which is immobilized by the opposite hand. Negative pressure is applied to the syringe and several sharp, quick strokes are made with the needle. These passes are made at different angles into the mass to sample representative areas of the mass. Before removing the needle, the negative pressure is released to keep the specimen in the needle when it is withdrawn from the skin. The specimen is then expressed onto a glass slide and a second glass slide is used to spread the aspirate in the same manner as a blood smear. Institutions may handle the aspirated material differently depending on local procedure and any anticipated special studies. In our institution, ABC specimens are handled so it is also possible to perform special studies (see Chapter 2). Whenever NAB is being performed to resolve a difficult diagnostic problem (e.g., unknown primary site, previous diagnosis of "undifferentiated malignancy"), it is desirable to discuss the case with a cytopathologist in advance so that arrangements can be made for the appropriate additional studies that may include a panel of immunocytochemical stains or electron microscopy. In either case, cytotechnologists or cytopathologists should be in attendance to ensure that there is adequate material to perform all the necessary studies and to ensure that the appropriate fixatives are used.

# POTENTIAL COMPLICATIONS OF NAB

NAB in the hands of an experienced operator is remarkably free of complications. For fine-needle aspirates of superficial lesions, bleeding is infrequent, is generally easily controlled, and stops within a few seconds with local pressure. NAB of deep lesions carries a greater risk of bleeding and other serious complications. Microscopic hematuria is the most common complication following renal biopsy; it occurs in 6.5% of patients.[20] Major complications of renal biopsy, including perirenal hemorrhage, pneumothorax, infection, arteriovenous fistula, and traumatic urinoma, occur infrequently. In a series of 5,674 renal aspirates performed in 84 institutions, the incidence of such complications was only 1.4%.[20] The frequency of complications relates both to the experience of the operator and to the intensity of the search to identify asymptomatic complications such as microscopic hematuria. In general, a low incidence of bleeding can be expected when a small-gauge needle is used with CT-scan guidance, which can direct the needle away from the renal hilum.

NAB of the lung can also be associated with bleeding. Ten percent of patients experience a short period of hemoptysis, and minor pleural hematomas occur in 1% of patients. A recent report[6] described two patients in whom bleeding into the pleural space was a major complication of the needle biopsy. Both had pleural effusion before biopsy, and the authors speculate that the pleural fluid may have prevented tamponade of the bleeding. The most common complication of NAB of the lung is pneumothorax, which occurs in 20 to 34% of biopsies,[2,16] and up to 10% of patients require a chest-tube insertion or other intervention.

NAB of the pancreas may result in elevation of amylase levels and abdominal pain. In most cases these side effects subside spontaneously over a number of hours, but there is one report in the literature of fatal necrotizing pancreatitis.[7] In this patient, there was no definite mass in the pancreas, which may have been important in the development of pancreatitis because a tumor would not contain functional tissue in communication with the pancreatic duct.

Ultrasound-guided NAB of bowel masses is now undertaken in those medically unfit individuals in whom the usual diagnostic investigations such as barium contrast studies and endoscopy cannot be performed. The inadvertent puncture of hollow viscera during abdominal percutaneous biopsies is common in this situation but does not usually cause complications of any clinical significance when a small-calibre needle is used.[29] The only relative contraindication to this procedure is bowel dilation. During the procedure, ultrasound is used to guide the needle directly into the thickened bowel wall without entering the lumen, thus reducing the possibility of adverse consequences. NAB of retroperitoneal lymph nodes for the diagnosis of malignant lymphoma, germ-cell tumors, and other malignancies is frequently undertaken with a transabdominal approach, which again must traverse hollow viscera. The frequency of untoward effects in this situation is also extremely rare.

It is a concern of many oncologists that needle aspiration may cause seeding of tumor cells along the needle tract. Although this complication is certainly possible and has been reported for a number of tumor sites,[9,35] the probability of this seeding occurring is diminished by the use of a fine-calibre needle and by maintaining negative pressure on the syringe until just before the needle exits the skin. Whenever possible, subsequent therapy should be designed to treat the needle tract, either by surgical excision or by inclusion in the radiotherapy treatment port. Review of the available medical literature and the experience at our own institution lead us to the conclusion that NAB is generally a very safe procedure.

## CLINICAL USES OF NAB

NAB is often used in situations where prompt institution of therapy is essential. In our institution, a report of the routine cytological interpretation from an ABC can generally be obtained within 30 minutes. If further studies are not required, a decision regarding therapy can often be made promptly. In clinical situations, such as superior vena caval obstruction, where there is an urgent need to institute therapy, the choice of approach (e.g., chemotherapy versus radiotherapy) can be determined rapidly based on the results of the ABC. NAB also avoids the postoperative recovery period that follows excisional biopsy, which frequently adds to the delay to onset of definitive therapy. Most radiation and medical oncologists are reluctant to commence treatment until at least seven to 10 days postoperatively, even if there have been no complications. Wound dehiscence or infection further delay the time to onset of treatment.

Frequently, oncologists at tertiary referral centers are referred patients who have already undergone a surgical biopsy. In many cases, the pathology report is not definitive, additional pathology material is unavailable for special studies, and a treatment decision cannot be made. Rather than submit such patients to a surgical biopsy, NAB with appropriate immunocytochemical stains and possibly electron microscopy examination may resolve the diagnostic problem with minimal risk of morbidity for the patient. Other patients may present with mass lesions and a past history of cancer. These patients, with presumed recurrent cancer, can be promptly diagnosed by NAB with little morbidity and quickly started on appropriate therapy. In other cases, ABC with special studies such as immunocytochemistry and electron microscopy may be helpful when there are no traces of cellular differentiation on light microscopy and a more specific diagnosis is required by the clinician for therapeutic and prognostic reasons. Chapter 8 details how the use of immunocytochemical or electron microscopy, or both, can assist in differentiating carcinomas from malignant lymphomas, carcinomas from malignant melanomas, or carcinomas from sarcomas. It also details how these techniques can contribute to the diagnosis of neuroendocrine tumors, germ-cell tumors, and small-cell neoplasms.

For these specialized techniques to be successful, however, it is first necessary to establish by light microscopy that malignant cells are present in the aspirate. With the exception of malignant lymphoma, immunophenotyping does not distinguish between neoplastic and nonneoplastic lesions of the same origin because they usually have the same complement of antigens. When confronted by a metastatic tumor of unknown origin, cytopathologists must translate the available light microscopy findings and the clinical data into specific questions that can be answered by immunophenotyping using a panel of well-defined antibodies. Consultation between clinicians and cytopathologists is therefore extremely important. On the basis of the clinical presentation and the cytological features of the aspirated material, including the immunophenotype, the primary site of the tumor may be indicated or the search for the primary site may be narrowed down to a few possibilities. Without such guidance the investigations patients may have to undergo to determine the site of origin of an undifferentiated neoplasm can be extremely extensive, time-consuming, and costly. It should be obvious that smears of malignant cells with distinct microscopic features of differentiated tumors do not require immunophenotyping, particularly in the hands of experienced cytopathologists. Immunocytochemical and electron microscopic studies are useful primarily in those cases where routine cytology does not provide an unequivocal diagnosis. It is important for clinicians to recognize that in these problematic cases, where special studies may be useful, there must be sufficient cytological material to set up the appropriate panel of antibodies or to perform electron microscopy. Opinions differ as to the most suitable method of processing cytological material. The technique used in our institution is described in Chapter 2 and is designed to enable special studies to be performed in all cases if required. Many cell markers may be demonstrated immunocytochemically on cell block sections or on smears fixed with alcohol. Fresh air-dried smears, however, allow better preservation of many more membrane-associated antigens and are more suitable for immunophenotyping of malignant lymphomas. Which antibodies are selected for study is dependent on knowledge of the characteristics of the antibodies, cytological findings of the aspirate, clinical presentation of the patient, and the number of cells or smears available for study. Therefore, consultation with cytopathologists is mandatory whenever hematologists or oncologists anticipate that immunocytochemical studies may be required. Some of the many antibodies available are outlined in Tables 5.6 and 9.2. Chapters 8, 9, and 10 provide detail on how these antibodies as well as electron microscopy can help differentiate tumors of different origins.

Although the small amounts of material available to light microscopists in aspirated material may make it difficult to appreciate the architectural arrangement, this is less of a problem for electron microscopists. In many instances, the relationships of tumor cells to one another and to the stroma can be determined from electron microscopy. The high resolution of electron microscopy permits the recognition of details such as small intercellular or intracellular lumina in adenocarcinoma. The identification of neurosecretory granules in small-cell lung cancer or premelanosomes in melanoma can direct clinicians to the correct diagnosis. It is estimated that electron microscopy yields additional information that is important to patient therapy in up to one-third of the aspirates examined.[37]

It has recently been demonstrated that it is feasible to perform molecular genetic analysis using DNA prepared from ABC specimens. The amount of DNA available in an aspirate is quite small; therefore, techniques have to be optimized. The small amount of DNA also limits the number of restriction enzyme analyses that can be performed. Once again, it is important to have a close liaison between clinicians and cytopathologists so that appropriate probes and restriction enzymes are selected. Southern blot analysis of gene rearrangements has proved to be the best method to prove clonality within lymphoid tissues.[23] Gene rearrangements precede the production of immunoglobulin or T-cell receptors; therefore, tumors at very early stages within the lymphoid lineage can be detected by these molecular genetic techniques. In cases where tumors contain an admixture of polyclonal lymphocytes, molecular genetic analysis can best differentiate nonlymphoid tumors from lymphomas. It is probable that molecular genetic analysis of gene rearrangements and chromosomal translocations will be used increasingly in the future for the diagnosis of lymphomas.

## Malignant Lymphoma

Hematologists and oncologists rely predominantly on the histopathological examination of surgically excised lymph nodes for decisions regarding therapy. In a number of situations, however, NAB may be useful in the initial evaluation of patients with lymphadenopathy. When the adenopathy is predominantly within the abdomen or chest and is not readily accessible for surgical biopsy, percutaneous NAB of an abdominal or mediastinal mass will at least distinguish metastatic carcinoma from malignant lymphoma. NAB can also be used to assess the extent of nodal involvement following a histological diagnosis without the need for further surgery. NAB is particularly useful in those patients who have previously been diagnosed and who show signs of relapse following treatment.

In many of these situations, the success rate of NAB is increased when CT scanning is used to localize the largest masses accessible for a safe biopsy. Immunophenotyping helps to classify lymphomas, which in turn influences the choice of therapy. Immunocytochemical studies are also of great help in separating nonneoplastic lymphoproliferative disorders such as lymphoid hyperplasia from low-grade lymphomas.[18]

NAB may prove particularly useful in confirming the diagnosis of malignant lymphoma in human immunodeficiency virus–positive patients in whom there may be reluctance to perform conventional lymph node biopsies.[38] By using ABC, only one individual is placed at minimal risk of a needle-stick injury. The specimen can be immediately placed in fixative, thereby minimizing the risk to laboratory staff.

NAB can also help distinguish benign or reactive lymphadenopathies from malignant lymphadenopathy. Oncologists must be mindful that not all lymphadenopathy, even in patients with known malignant disease, is necessarily due to metastatic tumor or malignant lymphoma. A variety of antigenic stimuli may result in lymphadenopathy and lymphadenitis, and some of these conditions occur more frequently in patients with malignancies. Mycobacteriosis, toxoplasmosis, infectious mononucleosis, cat-scratch disease, and dermatopathic lymphadenopathy may be suggested by the cell populations seen on ABC.[30] Appropriate serological tests or microbiological stains and cultures can then be used to establish the correct diagnosis. Despite the ease of performing NAB, it is important for clinicians to remember its limitations in the diagnosis of malignant lymphoma. Aspirated material does not allow for the distinction between follicular and diffuse lymphomas, which is extremely important when deciding on therapy.[1] It is also not possible to subclassify Hodgkin's disease based on aspirated material, and this informaton is of prognostic significance.

## Breast Cancer

It is essential that a diagnosis be established in any woman presenting with a dominant breast mass. Many physicians prefer to obtain mammography in advance of NAB to avoid the confusion created by a secondary hematoma if NAB is performed first. If there is a high probability of the mass being cystic, however, it can be aspirated first. Nonbloody fluid and complete resolution of the mass following aspiration give strong reassurance to physicians and patients alike that the mass is a benign cyst.

If the ABC of a breast mass is positive for malignant cells, surgeons can prepare patients psychologically for surgery. In the past, a one-step intraoperative procedure consisting of excisional biopsy followed by mastectomy if the biopsy was positive, left many women psychologically traumatized. It is now recognized that more conservative breast surgery is often possible and that therapeutic options need to be fully discussed with patients in advance of surgery.

In the follow-up of women who have had breast cancer, medical and radiation oncologists often encounter chest wall nodules or enlarged regional lymph nodes suggestive of tumor recurrence. NAB can confirm recurrent disease without the need for excisional biopsy. The speed of diagnosis and the absence of a surgical incision make it possible to change therapy almost immediately.

It is now possible through immunocytochemical means to determine whether aspirated tumor cells have estrogen and progesterone receptors.[25] This information is extremely valuable to oncology teams because it helps determine whether systemic management should be with hormones or chemotherapy. The recent introduction of mass breast screening programs will identify a large number of women with clinically occult lesions. Stereotaxic mammographic biopsies will be necessary to accurately diagnose these abnormalities if excisional biopsies are to be avoided in light of the associated physical scarring and psychological trauma. The sheer volume of cases requiring biopsy diagnosis could place enormous strain on surgical daycares, and the resultant delays in diagnosis could cause great stress for women. NAB again provides a better alternative as an accurate and quick means to establish a diagnosis.

## Lung and Mediastinal Lesions

Tumor histology and stage of disease are the two major determinants of therapy and prognosis in lung cancer. The diagnosis of lung cancer is most commonly established by bronchoscopy and mediastinoscopy. In patients with peripheral nodules, however, bronchial washings and brushing are frequently negative. Recent evidence indicates that the incidence of adenocarcinoma is increasing in North America and that adenocarcinomas are more commonly located peripherally in the lung as opposed to the central location of most squamous and small-cell tumors. Because NAB under fluoroscopy is most commonly used to diagnose peripheral lung shadows, it is likely that there will be increasing use of NAB in the initial diagnosis of lung cancer.

It is important to differentiate benign from malignant lesions and small-cell lung cancer from non–small-cell lung cancer in patients with an abnormal radiological shadow. This differentiation is generally accomplished by routine examination of the ABC. When the tumor appears "undifferentiated," electron microscopic study can often clarify whether the tumor is of squamous or adenocarcinoma differentiation as opposed to small-cell lung cancer. In addition, immunocytochemical studies of the ABC specimen often distinguish between neuroendocrine and nonneuroendocrine neoplasms. The importance of making the distinction between small-cell and non–small-cell carcinoma relates to the markedly different therapeutic strategies that are used for these two different types of lung cancer. For patients with small-cell lung cancer, the initial treatment is usually systemic chemotherapy, often followed by radiation therapy to the primary site and mediastinum. For patients with non–small-cell lung cancer, the standard approach is surgical resection for patients without mediastinal lymph node involvement. Patients with mediastinal disease (stage III) generally receive radiotherapy, although combined appoaches using chemotherapy and surgery or chemotherapy and radiotherapy are increasingly being employed. The complexity, cost, and toxicity of such therapies make it essential that the pathological diagnosis is secure and the stage of disease is precisely defined.

The lungs are a common site of metastatic disease, and metastatic nodules may be single or multiple. If solitary, a metastatic nodule can easily be confused with a primary bronchogenic carcinoma. It is therefore important to take a careful past medical history in the assessment of such patients. In the setting of multiple pulmonary nodules, metastatic disease should be suspected, although a number of benign conditions may produce a similar appreareance. In this case, NAB may differentiate a benign from a malignant disease process or suggest a primary tumor site if one is not evident clinically. In all of these situations, the alternative to NAB is an open surgical biopsy, which requires consultation, scheduling, anesthesia, and a return visit for wound inspection and suture removal. All of these components of the diagnostic procedure add substantially to cost, morbidity, and delay to onset of definitive therapy. The specificity of NAB in the diagnosis of pulmonary neoplasms is very high. In one large series of 2,421 biopsies, there were only eight false-positive results.[36]

Transthoracic NAB is also valuable in determining diagnosis in patients with anterior mediastinal masses. Differentiating among thymoma, lymphoma, germ-cell tumors, metastatic disease, and benign cysts is critically important because of the major differences in therapeutic strategy. NAB appears to be quite accurate in diagnosing metastatic disease, germ-cell tumors, and cysts, but routine cytological examinations alone is less accurate in the diagnosis of lymphoma and thymoma[14] because small, benign-appearing lymphocytes can be found in well-differentiated lymphocytic lymphomas, lymphocytic thymomas, and Hodgkin's disease. NAB in Hodgkin's disease will always have limited sensitivity because diagnostic malignant Reed Sternberg cells may be rare in the lesion and escape sampling. In addition, the epithelial nature of the neoplastic component of thymomas is often not recognizable if the cells are seen out of tissue context. Immunocytochemical studies and electron microscopy would be expected to improve the diagnostic accuracy of NAB in this clinical situation.

## Liver Cancer

NAB has significantly improved the accuracy of diagnosis of space-occupying liver lesions. Although standard "blind" needle biopsies yield diagnostic tissue in approximately 60% of cases.[26] NAB guided by ultrasound or CT almost always yields diagnostically useful tissue. In staging patients with a known primary malignancy, it is often important to confirm whether abnormalities identified on imaging studies are due to metastases or some other process. This confirmation is particularly important, for example, if the therapeutic decision is between "curative" surgical resection and palliative systemic chemotherapy. Unfortunately, it is sometimes assumed that an ultrasound- or CT-demonstrated abnormality is a metastasis. In several incidences, I have instituted systemic therapy based on radiological appearance, which was reported as positive for metastatic disease, only to find subsequently that the images were due to focal benign lesions. When a radioisotope or ultrasound scan is diffusely abnormal, NAB can be used to widely sample the liver with multiple passes. In a study of 40 patients with suspected malignant disease of the liver, NAB with radioisotope scintigraphic and fluoroscopic guidance was positive in 93% of those ultimately found to have malignancy, including 25 of 26 with liver metastases.[15] In 24 patients, conventional wide-bore needle biopsy was also performed. In the 16 patients who had a final diagnosis of hepatic malignancy, NAB was positive in 14 (88%), and conventional needle biopsy was positive in only 4 (25%). Therefore, in patients with suspected malignancy involving the liver, guided NAB appears to be the preferred diagnostic method. Similarly, lesions seen or palpated in the liver during abdominal or thoracic surgery should also be biopsied intraoperatively to verify what is assumed to be metastatic cancer.

## Thyroid Neoplasms

When evaluating patients with thyroid nodules, the overall aim of diagnostic tests is to remove all potentially lethal thyroid cancers and to undertake surgical removal of benign nodules as infrequently as possible. Solitary thyroid nodules represent the most common indication for thyroid surgery. Approaches to solitary nodules which by strict definition is a discrete mass 1 cm in diameter or larger arising in a gland of normal size and consistency, are widely divergent. At one extreme are those who advocate that all discrete nodules should be removed, whereas at the other extreme are those who advocate that only those lesions remaining after a course of hormone therapy should be surgically excised. Most clinicians attempt to use laboratory studies including NAB to identify those thyroid nodules at greatest risk of being malignant. This use of NAB had its opponents, who cite the danger of tumor implantation, difficulties in interpretation of the aspirated material, and the risk of sampling errors. The reports of tumor implants following needle biopsy are extremely rare. Crile[4] observed only one instance in over 2,000 thyroid needle biopsies. Walfish and colleagues[34] reported a correlation accuracy of 82% in 88 surgically treated cases of solitary nonfunctioning nodules in patients who had undergone NAB. There were no false-negative results, and most false-positives results occured in cystic lesions. Many others have also observed a low incidence of cytological misinterpretation of cold nodules with NAB.[8,22,31,34] Although it may be very difficult to differentiate follicular neoplasms from nonneoplasmic lesions on an aspiration biopsy, skilled cytopathologists can identify papillary carcinomas, medullary carcinomas, and anaplastic carcinomas with a high degree of accuracy.

NAB has been used both as a diagnostic and a therapeutic measure for thyroid cysts. Clinicians should remember that aspiration of a cystic lesion does not eliminate the possibility of neoplasm. Although malignacy is rare in purely cystic lesions, it may well be present in mixed cystic–solid lesions. Therefore, needle aspiration is best conducted in conjunction with ultrasonography.[34]

As outlined in Chapter 11, a panel of diagnostic antibodies to thyroglobulin, calcitonin, and chromogranin can be very useful in diagnosing medullary carcinoma as distinct from tumors of follicular origin.

Thyroid tumors, particularly papillary carcinomas, commonly metastasize to cervical lymph nodes, and a neck mass may be the first sign of thyroid malignancy. Thyroid cancer should be in the differential diagnosis of cervical lymphadenopathy. ABC with special stains for thyroglobulin, which is the hallmark of thyroid cancer, can lead to a prompt diagnosis. The panel of antibodies should also contain antibodies to leukocyte common antigen, cytokeratin, and calcitonin to exclude malignant lymphoma, carcinoma of the head and neck, and medullary carcinoma of the thyroid, which are all important diagnostic possibilities in patients presenting with cervical adenopathy.

### Prostate Cancer

Prostate cancer has become the second most common malignancy among men in North America. It is usually diagnosed either after patients present with urinary tract obstruction or are found on routine rectal examination to have an abnormal consistency or configuration to the prostate gland. Although detection of prostatic abnormalities on rectal examination is the most practical means of detecting prostatic cancer, abnormalities of consistency of the prostate may also be due to prostatitis, nodular hyperplasia, infarct, tuberculosis, or lithiasis. Transrectal ultrasound of the prostate will demonstrate hypoechogenic areas in most cases of prostatic cancer but not in all. To obtain confirmation of the diagnosis, urologists commonly obtain tissue either by transrectal or transperineal core biopsies, open perineal biopsies, or transurethral resection. These procedures require hospitalization and are accompanied by significant morbidity. Alternatively, NAB has proved to be an accurate method of diagnosis for prostate cancer in experienced hands.[19] One difficulty is that well-differentiated tumors are hard to diagnose definitively on NAB material.[13] Histological grading of prostatic adenocarcinomas is useful in conjunction with clinical staging to determine therapy and prognosis. Although cytological grading can be done, the commonly used Gleason grading system cannot be applied to ABC material. The available data indicate that a definitive diagnosis of moderately or poorly differentiated adenocarcinoma can be made with a high degree of sensitivity and specificity, and in the hands of experienced cytopathologists has an accuracy at least as good as a core biopsy. One potential advantage of NAB is that it allows assessment of several areas of the prostate that core biopsies do not. NAB can accurately diagnose benign prostatic hyperplasia, but confusion can result from the inflammatory atypia induced by prostatitis or therapy-induced changes following irradiation or chemotherapy. Complications of NAB of the prostate are rare. A few cases of sepsis have followed this procedure, and those caring for these patients should be aware of this potential complication of the procedure.

## SUMMARY

An oncologist's need for prompt and precise information to establish a diagnosis, to institute therapy, and to avoid as much psychological distress for patients as possible is well met by NAB. Unfortunately, although the technique of fine-needle aspiration for superficial lesions is simple to learn, most oncologists have not been trained in its use and do not use it in their practice. As a result, needless excisional biopsies and other surgical procedures continue to be undertaken, which incur unnecessary physical and mental discomfort to patients, delays in the onset of appropriate therapy, and added health care costs. It is imperative that physicians caring for cancer patients become familiar with the technique of NAB and proficient in the aspiration of superficial lesions. The effectiveness of NAB has been increased substantially by the introduction of immunocytochemical staining, the use of electron microscopy, and the ability to perform molecular genetic analysis on aspirated material. Clinicians should be aware of these advances to make optimal use of the cytopathology laboratory's full potential. As a result, accurate diagnoses can be established for patients while minimizing their physical and emotional distress and generating the least possible cost for the healthcare system.

# A RADIOLOGIST'S PERSPECTIVE
Hardy H. Tao, M.D.

Until recently, diagnostic radiology was considered a minor specialty, dealing with only shadows or summation of shadows. With experience and intuitive interpretation, radiologists were usually able to form diagnostic opinions or provide a long list of differential diagnoses. The final decision was always left to clinicians, whether general practitioners, oncologists, or surgeons based on a radiologist's differential diagnoses in concert with the clinical situation. Radiologists were usually not in a position to enter into the actual decision-making process. With the advantages gained by contrast examinations, such as excretory urography, percutaneous transhepatic cholangiography, angiography, and contrast studies of the gastrointestinal tract, radiologists increased their capability of providing a definitive diagnosis. The introduction of diagnostic ultrasound, CT, and magnetic resonance, by displaying superb anatomical detail and providing information not readily obtained through routine physical examinations, strengthened the role of radiologists. Nevertheless, many lesions are still difficult to differentiate no matter how detailed radiological examinations are; radiologists are still dealing with shadows or summation of shadows. ABC ushered in a new era for radiologists. With assistance from cytopathologists, radiologists' long list of differential diagnoses can often become a definite diagnosis within minutes. The use of electron microscopy and immunocytochemistry has further enhanced the potential of ABC through increased diagnostic accuracy and better tumor classification. From the point of view of radiologists, ABC has been a major advance.

Because ultrasound or CT can demonstrate a mass lesion anywhere in the body, many diagnostic problems that previously required surgical exploratory laparotomy or incisional biopsy can now be easily solved by a simpler, safer, and much less expensive method. The discomfort, expense, and risk of exploratory surgery (often without significant therapeutic benefit) have been largely eliminated. The ability to make a definite diagnosis reduces patients anxiety and physician hesitation. An accurate cytopathological diagnosis enables both appropriate preoperative staging

TABLE 11.1 Responsibilities of Clinicians to Radiologists

| |
|---|
| Be familiar with limitations and indications of ABC |
| Provide patients' past and recent history |
| Provide a tentative diagnosis before ABC |
| Alert radiologists to possible risk factors |
| Inform patients of the necessity for the procedure |
| Inform patients of the ABC results and correlate this with other clinical findings |
| Be familiar with possible complications and how to manage them |

procedures and multidisciplinary consultation. Surgeons can proceed with a definite treatment plan if the tumor is operable. Even with unresectable tumors, there are benefits for patients. A positive cytology eliminates the need for surgical exploration and allows consideration of chemotherapy or radiotherapy.

As with all other diagnostic parameters, the reliability of ABC depends on (1) the experience and skill of the individual obtaining the samples, (2) the proper preparation of the specimen by cytotechnologists, and (3) the knowledge and experience of the cytopathologists. Although ABC results can be reported independently, a basic principle of the practice of medicine, the team approach, has many advantages. The team consists of a referring clinician, a radiologist, and a cytopathologist. The team members must communicate with each other to achieve the full advantage of ABC. Radiologists are usually the link between clinicians and cytopathologists. Radiologists usually do not have detailed knowledge of a patient's past history and are seldom involved in the eventual treatment and follow-up. Radiologists are entitled to expect certain responsibilities from clinicians (Table 11.1).

Needless to say, clinicians should be familiar with the indications and limitations of ABC. It is impossible and impractical to aspirate all abnormal shadows seen in radiology. Some clinical judgment is needed to decide whether a biopsy is necessary. For example, it is not necessary to aspirate bilateral large adrenal masses in a proved case of lung cancer. It is also an unreasonable request to aspirate some abnormal soft-tissue shadows demonstrated by CT in a patient who is still in the postoperative period following recent surgery.

Who should inform patients of a positive cytopathological diagnosis? This is not the role of radiologists because they usually lack details of the clinical context, past history, and other investigations. It is the responsibility of attending clinicians to correlate the cytopathological diagnosis and the clinical findings to ensure that the ABC diagnosis satisfactorily explains the total clinical picture. If the ABC is negative for malignancy, clinicians should reevaluate the results in light of other investigations and decide the probability of a false-negative diagnosis. If deemed necessary, clinicians should request a repeat ABC.

All physicians should learn the principles underlying NAB and practice this technique. Clinically palpable masses are easily aspirated, and surgeons should be encouraged to perform NAB under direct vision during surgery. Percutaneous NAB of deep lesions is a rather different matter. Radiologists have the clear advantage of lesion localization because the true size and exact location of the lesion often can be accurately assessed only by the ultrasound or CT techniques that guide the NAB. For lung lesions, NAB is done under fluoroscopic control after CT. For

TABLE 11.2. Major Reasons for False-Negative ABC Results

| |
|---|
| Sampling error |
| Contaminated, diluted, necrotic, and air-dried specimens |
| Limitation of cytomorphological criteria |
| False interpretation |

percutaneous breast NAB of very small or nonpalpable lesions, the mammographic stereotaxic method is the most reliable and accurate.

Failure to diagnose a cancer is unacceptable. The main reasons for a false-negative diagnosis are listed in Table 11.2. Most of the false-negative diagnoses of cancer are not the fault of cytopathologists, but result from a nonpresentative specimen. False-negative interpretations by skilled and experienced cytopathologists should be negligible because of the strict cytomorphological criteria and the additional information obtained from electron microscopy and immunocytochemistry. Sampling error then becomes a major part in determining sensitivity. Radiologists should be familiar with various localization methods, use considerable care pinpointing the exact location of the lesion, and ensure that the needle is in the lesion before aspiration. For a typical procedure, 60% of the time is spent localizing the lesion and the rest explaining the procedure to the patient prior to preparing and performing the actual aspiration. If the specimen is truly representative, a negative report will emphasize the benign nature of the lesion. Because further assessments are unnecessary, the cost savings are considerable. Relief of patient anxiety is also an immeasurable benefit.

Problems of diagnosis occur because of contaminated specimens, necrotic material, diluted specimens, and air-dried samples. All are avoidable. Contaminated specimens can be avoided by precise needle localization, and careful aspiration ensures that the needle is always within the lesion. Many tumors have necrotic centers, containing few viable cells; the nonnecrotic parts of the tumor, guided by ultrasound or CT, must therefore be aspirated. Air-dried specimens can be avoided by speedy and careful preparation of the slide on site by a cytotechnologist.

Because radiologists have the advantage of lesion localization, they should be prepared to do most of the nonsurgical biopsies. Additional responsibilities of radiologists performing NAB are summarized in Table 11.3. Good communication must occur between radiologists and cytopathologists. Cytopathologists should be

TABLE 11.3. Responsibilities of Radiologists

| |
|---|
| Be familiar with the limitations and indications of ABC |
| Be aware of possible complications and how to avoid them |
| Inform clinicians of significant complications |
| Provide cytopathologists with necessary information |
| Obtain patients' consent to the procedure |
| Perform precise localization of the lesion |
| Be familiar with the art of NAB |
| Be ready to repeat the biopsy if required |

provided not only with the site of the lesion sampled, but also with the pertinent clinical details, particularly when there is a history of previous or concurrent pathology in other parts of the body.

Although NAB is considered safe, the procedure is not entirely free of complications, including severe and even fatal ones. Theoretically, the risk from needle aspiration increases as the diameter of the needle increases. In addition, there is a significant increase in risk when a cutting needle is employed. Over the years, I have used a wide range of needles of different shapes, sizes, and designs. I now only use two needles: the classic 22-gauge Chiba needle (Cook, Bloomington, IN) and the 22-gauge spinal needle, both without a cutting edge or slot. The Chiba needle is the more flexible of the two and it usually provides more adequate samples. Spinal needles are available in different lengths, and their rigidity provides better path control.

The main complication of NAB is bleeding. Most cases require simple observation; significant bleeding requiring transfusion occurs infrequently. Intraabdominal bleeding is more likely to occur if the spleen is punctured. Other factors that influence bleeding are highly vascular lesions and abnormal prothrombin time and partial thromboplastin time. Following the NAB, we usually do not perform a physical examination during the 30-minute monitoring period unless indications warrant it. The most common complication arising from a percutaneous lung biopsy is pneumothorax, especially in patients with chronic obstructive pulmonary disease. We always perform a chest radiograph after the examination as a precautionary measure against pneumothorax. Radiologists should know the indications for treatment of pneumothorax and be prepared to initiate it.

NAB should not be performed without a quality service provided by cytopathology. The responsibilities of cytotechnologists and cytopathologists are summarized in Table 11.4. At the Ottawa Civic Hospital, a cytotechnologist is available to be in the Radiology Department within five minutes to prepare the smears and needle rinse. As stated previously, an improperly prepared smear will make interpretation very difficult or impossible. It is very important to provide a speedy initial verbal diagnosis so that a suitable course of action can be planned and later implemented. It is also important to know whether a repeat biopsy is necessary. We ask patients to stay close at hand for 30 minutes, which allows us to monitor any immediate complications and, if necessary, to do a repeat biopsy without another appointment.

Most of the diagnoses in ABC cases are based on subjective morphological appearance of the cells. This principle is likely to become the mainstream in years to come. To have clinicians and radiologists accept the positive cytological diagnosis as the final diagnosis, the ABC diagnosis should be accurate and specific. There should be a quality control program to constantly monitor cases with a positive

TABLE 11.4. Responsibilities of Cytotechnologists and Cytopathologists

| |
|---|
| Take full responsibility for the aspirated specimen |
| Provide an initial cytomorphological diagnosis within 30 minutes |
| Avoid a false-positive diagnosis |
| Indicate whether a repeat NAB is necessary |
| Always prepare for electron microscopy and immunocytochemistry |

cancer diagnosis. As clearly illustrated by Yazdi and Dardick in the previous chapters, electron microscopy or immunocytochemistry, or the combination of both, can offer more specific diagnoses and better tumor classification.

# THORACIC SURGEONS AND FINE-NEEDLE ASPIRATION BIOPSY

Farid M. Shamji, M.B.

Percutaneous NAB has established itself in the last 10 years as a valuable and acceptable technique for diagnosis of intrathoracic lesions.[28,33] NAB is a method whereby a very small quantity of tissue, fluid, and cells is aspirated from a lesion for cytological examination.[36] In every instance, the specimen should be prepared for cytomorphological diagnosis and, in difficult cases, for immunocytochemistry and electron microscopy as well.

The principal advantages for this diagnostic procedure are as follows:

1. The equipment required (biplane fluoroscopy) is simple and readily available. Ultrasound and CT scans may be needed if the lesion cannot be localized on fluoroscopy.
2. The technique is simple, safe, and inexpensive.
3. The technique is applicable to out-patients, which has the added advantage of reducing healthcare costs and still maintaining high diagnostic accuracy and patient safety.
4. It is not time-consuming, requiring on the average 20 minutes to perform. With proper scheduling it is not disruptive to the normal running of the two departments involved—Radiology and Pathology.
5. Familiarity with cytological techniques, immediate preparation of the smears by cytotechnologists, and availability of cytopathologists to provide an initial cytomorphological diagnosis within 30 minutes is very beneficial to expeditious patient care and investigation. It is important to know, as soon as possible, whether a repeat NAB is necessary while the patient is still in the Radiology Department
6. With respect to patient care, NAB has had an important role in helping to reduce the number of patients who have to be managed with an ill-advised "wait and see" policy on the one hand (based on radiological assessment of the growth behavior of the lesion) and ill-advised thoracotomy on the other.
7. With experience, cytopathologists are able to specify the cell type in primary lung cancer. Typing of tumor is important in primary bronchogenic carcinoma because it influences the treatment plan and determines prognosis. Several reports have shown a good correlation between cytological diagnosis and subsequent pathological findings,[33,36] which can be further improved by immunocytochemical and electron microscopic studies. A high cytological accuracy in the range of 70 to 80% is possible for squamous-cell carcinoma and adenocarci-

noma; for bronchioloalveolar carcinoma it reaches 84%. The correlation for poorly differentiated large-cell carcinoma is low (18%). Furthermore, many reports have indicated that cytological diagnostic accuracy in small-cell anaplastic carcinoma is high (close to 82%). This finding has significant impact on patient management because small-cell lung cancers are treated with a multimethod approach of neoadjuvant chemotherapy and subsequent pulmonary resection for limited disease. Furthermore, cytological examination is reliable in differentiating between primary lung cancer and metastatic tumors in the lung.

The NAB technique for intrathoracic lesions, which has gained increased acceptance, consists of the use of a fluoroscopically guided fine-needle aspiration (22-gauge Chiba needle).[36] Almost all lung lesions, regardless of their size and location, can be sampled for biopsy using biplane fluoroscopy. Rarely, CT-guided biopsy will be necessary for nodules that are inaccessible because of their size and location.

The accuracy of cytological diagnosis is related to proper sampling of lesions by chest radiologists, immediate proper preparation of specimens, and expertise of cytopathologists. Cytopathologists should, in every instance, attempt to apply strict criteria to their assessment to avoid, at all cost, false-positive diagnosis of malignancy. This may result in a number of cases being diagnosed as suspicious of malignancy; this category must be minimized if the technique is to be of practical use. With NAB, positive diagnoses are obtained in 90 to 95% of lung cancer. The false-negative rate varies from 5 to 20%; the higher rate remains a concern and indicates a need for sound clinical judgment in deciding whether a nonmalignant aspirate is truly representative. Whenever an NAB of lung lesion yields a benign diagnosis (i.e., definite cytological evidence of abnormal cells possessing no malignant characteristics) or is unsatisfactory (i.e., material was scant, poorly prepared, or shows normal cellular elements of the lung), there should be discussion between the clinician, the chest radiologist, and the cytopathologist. Whereas the diagnosis of benign neoplasm appears to allow for greater confidence, the acceptance of other nonspecific benign diagnoses warrants consideration of risk factors for lung cancer, thorough evaluation of all the available chest radiographs to assess growth behavior, clinical history, and the operative risk of thoracotomy. In the report by Sagel and colleagues,[28] although the sensitivity in diagnosing malignant pulmonary neoplasm by NAB was extremely high (99.8%, or 858 of 860 patients), the specificity was only 87% (255 of 293). The authors emphasized that a finding of "no malignant cells" should not be accepted as conclusive proof of the absence of malignancy. In the same report, of the 858 patients with pulmonary lesions for whom a positive diagnosis of malignancy was obtained, 75 patients required a second biopsy for diagnosis, indicating that "one negative procedure is considered an incomplete diagnostic test in the cancer suspect." A very low false-positive rate of 1 to 2% has been reported in several large series of pulmonary NABs.[11,28,33]

There are several specific indications for NAB.

1. For evaluation of solitary pulmonary nodules in patients with normal previous chest radiographs and no past history of cancer. In this situation, a positive preoperative diagnosis of malignancy may expedite and simplify the operative approach. It allows surgeons to fully inform patients about the nature of the disease and about the nature of the anticipated surgical procedure. It shortens the operative time by avoiding intraoperative diagnostic maneuvers such as

wedge resection with quick section analysis. In patients who are poor operative risks, NAB spares them from a thoracotomy for diagnosis, which has a definable postoperative morbidity and mortality.[10]

2. For evaluation of solitary pulmonary nodules in patients with past history of cancer—either a primary lung or a nonthoracic malignancy. There is a definitive advantage in establishing a precise diagnosis before resorting to thoracotomy. It is important to know whether (1) the lung lesion is benign, in which case there is no need for thoracotomy; (2) the lung lesion is a metastasis, in which case a complete metastatic work-up is necessary before resecting the pulmonary lesion; and (3) the lung lesion is a second primary lung cancer, in which case patients should be assessed thoroughly for both extrathoracic and intrathoracic spread as well as for cardiopulmonary status before recommending thoracotomy.

3. For evaluation of multiple pulmonary nodules with or without past history of cancer. Patients may have multicentric bronchioloalveolar carcinomas or pseudolymphomas, for which thoracotomy should not be performed for diagnosis if possible.

4. For evaluation of two undiagnosed lung lesions in different lobes or lungs. This diagnosis is clearly of considerable importance to patient care. An unnecessary thoracotomy can be avoided if both lung lesions prove to be benign; or appropriate pulmonary resection can be performed, in the absence of contraindication to thoracotomy, if the diagnosis is synchronous primary lung cancers or pulmonary metastases from previously treated nonthoracic malignancy such as renal-cell carcinoma, colon carcinoma, or sarcoma, provided the lungs are the only site of spread.

5. For evaluation of pulmonary lesions that are clearly unresectable, (e.g., presence of associated superior vena cava obstruction, recurrent laryngeal nerve palsy, distant metastases, and poor cardiopulmonary reserve). In these situations, NAB helps avoid an unnecessary thoracotomy for diagnosis and its attendant risks.

6. For evaluation of patients with suspected lung cancer who cannot tolerate major surgery. In this situation, definitive cytological diagnosis is required before radiotherapy or chemotherapy can be instituted.

7. For evaluation of centrally located undiagnosed pulmonary lesions that are likely to require pneumonectomy for resection. This situation has important medical/legal implications because confirmation of diagnosis by lobectomy is acceptable for a benign lesion; therefore, pneumonectomy is definitely out of the question.

8. For evaluation of patients presenting with central nervous system symptoms and radiological evidence of a concomitant lung lesion.

Percutaneous NAB is used widely for both diffuse and nodular lung disease in centers where there is access to good cytopathology.[3,12] It is of particular benefit in patients with lymphangitic carcinoma.

Although NAB has proved to be a useful technique for the diagnosis of lung lesion, it has not obtained the same popularity in the study of mediastinal disease. One reason is reluctance on the part of chest radiologists to introduce the needle in areas close to major vascular structures or to the pericardium for fear of producing cardiac tamponade. Another reason is the difficulty with diagnostic interpretation of aspirated material from anterior mediastinal tumors and the considerable experience required. According to Herman and colleagues,[14] reporting on 143 transthoracic needle biopsies in 126 patients with anterior mediastinal masses, the technique proved useful in the patients with metastatic disease to the anterior mediastinum; there were no false-positive results in this group and the specificity was 100%. In the same report, in the 11 patients with germ-cell tumors of the anterior mediastinum, sensitivity was 91% and specificity was 98%. Unfortunately, lymphoma and thymoma were less reliably diagnosed by this method, unless electron microscopy and immunocytochemistry methods were used in addition to routine cytomorphological assessment. Differentation among thymoma, lymphoma, and germ-cell tumors is of considerable importance because of major differences in the treatment of these tumors. Therefore, it is often necessary to obtain adequate incisional biopsy specimens for accurate histological diagnosis when dealing with anterior mediastinal masses.

Contraindications to percutaneous fine-needle aspiration biopsy include bleeding diathesis, pulmonary hypertension, suspected pulmonary arteriovenous malformations, suspected echinococcal cysts, and severe pulmonary emphysema.

Fatal complications from NAB are extremely rare. A review of the literature up to 1981 indicates that there have been nine deaths following NAB. These deaths were attributed to cardiac arrest (3), air embolism (2), pneumothorax (1), and hemorrhage (3). All but two of these deaths occurred prior to 1974. Todd and colleagues[33] and Sagel and colleagues[28] reported no deaths in 2,114 and 1,211 biopsies, respectively.

Pneumothorax, by far the most common complication of NAB, is easily managed. The incidence is approximately 20 to 30%, and most pneumothoraces are small and apical.[33] Intercostal chest tube drainage is required in approximately 10 to 20% of patients and usually for a period of 24 to 48 hours. Significant pneumothoraces rarely develop after four hours, and tension pneumothoraces, if they occur, do so within minutes after the procedure. It is likely that the rate of pneumothoraces can be reduced by the use of thin needles.[27] Transient minor hemoptysis occurs in approximately 6% of patients. Hemothorax rarely occurs. There are rare reports of tumor implantation in needle tracts.[21]

In summary, the diagnostic yield of percutaneous fine-needle aspiration biopsy of intrathoracic lesions greatly depends on the procedure being performed correctly and the material being prepared and examined properly. By adding electron microscopy and immunocytochemistry to the standard cytological examination, not only does the diagnostic yield increase but also a more specific cell type diagnosis can be obtained. The value of NAB lies in the rapid diagnosis of undiagnosed pulmonary lesions, becauase 84% of malignancies will be identified promptly. The morbidity of NAB is very low, which makes the technique attractive. Several excellent reports indicate high sensitivity and specificity.

# REFERENCES

1. Cafferty LL, Katz RL, Ordonez NG, Carrasco CH, Cabanillas FR: Fine needle aspiration diagnosis of intraabdominal and retroperitoneal lymphoma by a morphologic and immunocytochemical approach. *Cancer* 65:72–77, 1990.

2. Chin WS, Eyee IST: Percutaneous aspiration biopsy of malignant lung lesions using the CHIBA needle. *Clin Radiol* 29:617–619, 1978.

3. Conces DJ Jr, Clark SA, Tarver RD, Schwenk RG: Transthoracic aspiration needle biopsy: Value in the diagnosis of pulmonary infections. *AJR* 152:31–34, 1989.

4. Crile G: Needle biopsy of thyroid nodules, in Varco RL, Delaney JP (eds): *Controversy in Surgery*. Philadelphia, W. B. Saunders, 1976, pp 159–162.

5. Crile G, Vickery AL: Special uses of the Silverman needle biopsy in office practice and at operation. *Am J Surg* 83:83–85, 1972.

6. Doyle T, Mullerworth M: Bleeding as a complication of fine needle lung biopsy. *Thorax* 43:1013–1014, 1988.

7. Evans WK, Ho CS, Mcloughlin MJ, Tao LC: Fatal necrotizing pancreatitis following fine-needle aspiration biopsy of the pancreas. *Radiology* 141:61–62, 1981.

8. Friedman M, Shimaoka K, Getaz P: Needle aspiration of thyroid lesions. *Acta Cytol* 23:194–203, 1979.

9. Gibbons RP, Bush WH Jr, Burnett LL: Needle tract seeding following aspiration of renal cell carcinoma. *J Urol* 118:865–867, 1977.

10. Ginsberg RJ, Hill LD, Eagan RT, et al: Modern thirty-day operative mortality for surgical resections in lung cancer. *J Thorac Cardiovasc Surg* 86:654–658, 1983.

11. Greene RE: Transthoracic needle aspiration biopsy, in Athanasoulis CA, Pfister RC, Greene RE, Robertson GH (eds): *Interventional Radiology*. Philadelphia, W. B. Saunders, 1982. pp 587–634.

12. Greenman RL, Goodall PT, King D: Lung biopsy in immunocompromised hosts. *Am J Med* 59:488–496, 1975.

13. Hajdu SI, Ehya H, Frable WJ et al: The value and limitations of aspiration cytology in the diagnosis of primary tumors; a symposium. *Acta Cytol* 33:741–790, 1989.

14. Herman SJ, Holub RV, Weisbrod GL, Chamberlain DW: Anterior mediastinal masses: Utility of transthoracic needle biopsy. *Radiology* 180:167–170, 1991.

15. Ho CS, McLoughlin MJ, Tao LC, Blendis L, Evans WK: Guided percutaneous fine-needle aspiration biopsy of the liver. *Cancer* 47:1781–1785, 1981.

16. Johnson RD, Gobien RB, Valicenti JF: Current status of radiologically directed pulmonary thin needle aspiration biopsy. *Ann Clin Lab Sci* 13:225–239, 1983.

17. Kaminsky DB: Aspiration biopsy in the context of the new Medicare fiscal policy.*Acta Cytol* 28:333–336, 1984.

18. Kern WH: Exfoliative and aspiration cytology of malignant lymphomas. *Semin Diagn Pathol* 3:211–218, 1986.

19. Kline TS: *Guides to Clinical Aspiration Biopsy*. Prostate. New York, Igaku-Shoin, 1985, p 2.

20. Lang EK: Renal cyst puncture and aspiration. A survey of complications. *AJR* 128:723–727, 1977.

21. Lauby VW, Burnett WE, Rosemond GP, et al: Value and risk of biopsy of pulmonary lesions by needle aspiration. Twenty-one years experience. *J Thorac Cardiovasc Surg* 49:159–172, 1965.

22. Lowhagen T, Granberg PO, Lundell G, et al: Aspiration biopsy cytology (ABC) in nodules of the thyroid gland suspected to be malignant. *Surg Clin North Am* 59:3–18, 1979.

23. Lubinski J, Chosia M, Kotanska K, Huebner K: Genotypic analysis of DNA isolated from fine needle aspiration biopsies. *Anal Quant Cytol Histol* 10:383–390, 1988.

24. Miller JM, Hamburger JI, Kini S: Diagnosis of thyroid nodules. Use of fine-needle aspiration and needle biopsy. *JAMA* 241:481–484, 1979.
25. McClelland RA, Berger U, Wilson P, et al: Presurgical determination of estrogen receptor status using immunocytochemically stained fine needle aspiration smears in patients with breast cancer. *Cancer Res* 47:6118–6122, 1987.
26. Parets AD: Detection of intrahepatic metastases by blind needle liver biopsy. *Am J Med Soc* 237:335–340, 1959.
27. Pinsker KL, Kamholz SL, Johnson J, Schreiber K: Fine needle aspiration biopsy of intrathoracic lesions. *Chest* 78:3, 1980.
28. Sagel SS, Ferguson TB, Forrest JV, et al: Percutaneous transthoracic aspiration needle biopsy. *Ann Thorac Surg* 26:399–405, 1978.
29. Solbiati L, Montali G, Croce F, et al: Fine-needle aspiration biopsy of bowel lesions under ultrasound guidance: Indications and results. *Gastrointest Radiol* 11:172–176, 1986.
30. Stani J: Cytologic diagnosis of reactive lymphadenopathy in fine needle aspiration biopsy specimens. *Acta Cytol* 31:8–13, 1987.
31. Stavric GD, Karanfilski BT, Kalamaras AK, et al: Early diagnosis and detection of clinically non-suspected thyroid neoplasia by the cytologic method: A critical review of 1536 aspiration biopsies. *Cancer* 45:340–344, 1980.
32. Tao LC, Pearson FG, Delarue NC, et al: Percutaneous fine-needle aspiration biopsy. It's value in clinical practice. *Cancer* 45:1480–1485, 1980.
33. Todd TR, Wiesbrod GL, Tao LC, et al: Aspiration needle biopsy of thoracic lesions. *Ann Thorac Surg* 32:154–161, 1981.
34. Walfish PG, Hazani E, Strawbridge HT, et al: A prospective study of combined ultrasonography and needle aspiration biopsy in the assessment of the hypofunctioning thyroid nodule. *Surgery* 82:474–482, 1977.
35. Wang CA, Vickery AL Jr, Maloof F: Needle biopsy of the thyroid. *Surg Gynecol Obstet* 142:365–368, 1976.
36. Weisbrod GL: Transthoracic percutaneous lung biopsy. *Radiol Clin North Am* 28:647–655, 1990.
37. Wills EJ, Carr S, Philips J: Electron microscopy in the diagnosis of percutaneous fine needle aspiration specimens. *Ultrastruct Pathol* 11:361–387, 1987.
38. Wotherspoon AC, Norton AJ, Lees WR, Shaw P, Isaacson PG: Diagnostic fine needle core biopsy of deep lymph nodes for the diagnosis of lymphoma in patients unfit for surgery. *J Pathol* 158:115–121, 1989.

# 12
# Developing and Potential Applications

The many advantages of fine-needle aspiration biopsy in diagnostic pathology, both for patients and pathologists, have lead to technological trials of a wide variety of types on aspiration samples. Certainly some of these studies will be of practical value as further experience is gained and other applications are sought. A few examples from the recently published literature are illustrative.

DNA content analysis, including assessment of ploidy and S-phase fractions, may be of value for determining the biological behavior of certain tumors and for the planning of treatment. Suitable cellular specimens can be readily obtained by needle aspiration biopsy (NAB),[5,10,18] obviating the need for surgical intervention in certain situations. Quantitative assessment by image analysis of immunocytochemical stains such as estrogen receptor provides better reproducibility than the usual subjective evaluation. Developments in flow cytometry and image analysis no doubt will expand the diagnostic applications to other quantitative analyses of aspiration biopsy cytology (ABC) specimens. The progress in clinical quantitative cytology is well summarized in a recent publication by Herman and colleagues.[9]

Genotyping is a very specific technique for establishing clonality of lymphoid proliferations, distinguishing lymphoma from nonlymphomatous neoplasms, and detecting the presence of oncogene rearrangements. Lubiński and colleagues[14] demonstrated the feasibility of performing molecular genetic analysis of gene rearrangements and translocations using DNA prepared from ABC specimens. Williams and colleagues[19] studied the utility of Southern blot analysis in ABC samples of 27 cases of known or suspected non-Hodgkin's lymphomas with promising results. They analyzed immunoglobulin, T-cell receptor, *c-myc,* and *bcl-2* gene rearrangements. These studies are very encouraging and further expand the diagnostic value of NAB.

*In situ* hybridization (ISH) techniques alone or in combination with immunocytochemical studies are now widely used in diagnostic pathology for localization of viruses and gene products for specific hormones, amplified oncogenes, and growth factors; this technique can be performed on tissue sections and cytological preparations.[12] Combined ISH and immunocytochemical studies allow the detection of viral- or tumor-encoding messenger RNA and their proteins; however, further technical improvements and demonstration of cost-effectiveness are required before these studies can be recommended on a routine basis for ABC specimens.

Amplification of certain oncogenes is associated with the development of malignancy. For example, it has been shown that amplification of the c-erbB-2 (also called HER2 and *neu*) oncogene in breast carcinomas is associated with tumor aggressiveness.[13,16] Soomro and colleagues[16] demonstrated a good correlation between the amplification of c-erbB-2 and immunoreactivity for its protein products. Wright and colleagues[20] stained sections of formalin-fixed, paraffin-embedded tissue from 185 primary breast carcinomas with a polyclonal antibody reactive with the c-erbB-2 oncoprotein. Positive staining was associated with earlier relapse, shorter post-relapse survival, and shorter overall survival.[20] The specimen for hybridization or immunocytochemical studies can be obtained by NAB,[13] which allows repeated sampling from primary and metastatic tumors during progression of cancer. Elucidation of other oncodevelopmental markers and antibodies labeling oncogene protein products would certainly increase the usefulness of NAB.

Specific DNA amplification is now possible by polymerase chain reaction (PCR) performed on DNA obtained from biopsy specimens, mononuclear cells from peripheral blood, and bone marrow aspirates.[2] PCR technology can be used for specific amplification of DNA allowing detection of minimal number of neoplastic cells. Crescenzi and colleagues,[2] using DNA polymerase chain amplification of t(14;18) chromosome breakpoints, were able to detect one lymphoma cell in $10^6$ normal cells. The sensitivity of the PCR technique far exceeds that obtainable by routine morphology, immunophenotyping, and genotyping techniques.[17] We expect increasing use of the PCR technique on material obtained by NAB in the future.

Immunocytochemical studies on ABC specimens are not only useful for diagnostic purposes, but also may have an increasing prognostic role. Expression of cathepsin D (an estrogen-regulated lysosomal protease), for example, may prove to be valuable in assessing the prognosis in breast cancer.[8] It has been suggested that vimentin expression may be a useful prognostic index in renal-cell carcinomas and node-negative breast carcinomas.[3,4] Castronovo and colleagues[1] evaluated the expression of laminin receptor using the immunoperoxidase technique in 81 breast ABCs. In contrast to benign lesions, which contained a low level of laminin receptor antigen, 71% of malignant tumors contained cells that were strongly stained by antilaminin antibody. These investigators suggested that immunodetection of laminin receptor could be a valuable adjunct in assessment of prognosis of breast lesions.[1] These and other prognostic indicators have the potential to serve as discriminatory factors in treatment decisions.

Determination of tumor proliferation rate using monoclonal antibodies such as Ki-67 and PC10 may be valuable in assessing prognosis.[4,7,11] Monoclonal antibody Ki-67 recognizes a nuclear antigen in proliferating cells but only works on cryostat sections of snap frozen material and cytological smears. Monoclonal antibody PC10 recognizes a proliferation-cell nuclear antigen (PCNA) that is also known as cyclin. In lymphoid neoplasms, PCNA expression can be used as an index of cell proliferation (Fig. 12.1).[7] This antibody is reactive on paraffin-embedded cells and tissues. Further development of these and other similar antibodies will allow study of the growth component of different neoplasms.

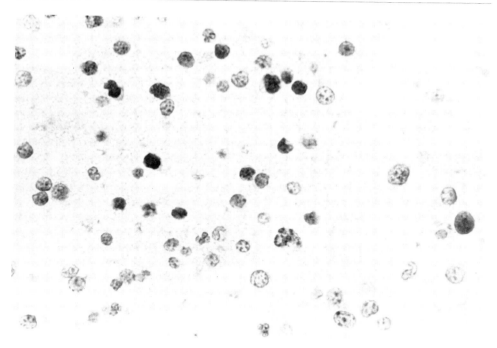

Fig. 12.1. ABC of a high-grade non-Hodgkin's lymphoma. Note strong expression of proliferating cell nuclear antigen (PCNA) in a proportion of tumor cells. Monoclonal antibody PC10, × 556.

Further development and commercial availability of sensitive and specific monoclonal antibodies to cell surface, cytoplasmic, and nuclear antigens may substantially improve the value of immunocytochemistry in cytopathology. Introduction of additional monoclonal antibodies for immunophenotyping of malignant lymphoma in paraffin-embedded ABC specimens using routine techniques would certainly facilitate lineage assignment of non-Hodgkin's lymphomas in more diagnostic laboratories.

There is considerable potential for expansion of ultrastructural techniques to the analysis of needle aspiration specimens for diagnostic purposes. An obvious application is postembedded immunogold methods[15] to epoxy resin–embedded tumor tissue. Pulmonary aspirates are commonly used to distinguish small-cell from nonsmall–cell carcinoma. For this purpose, identification of neurosecretory granules at the electron microscopic level can be crucial but tedious. If only a few, somewhat atypical, small membrane-bound granules are detected, how diagnostic are they? Detection of chromogranin within such granules using immunoelectron microscopy can eliminate diagnostic doubts. As illustrated in Figure 12.2, immunoelectron microscopy is readily applied to ABC specimens of lung tumors. Given the minibiopsy nature of many ABC specimens, other diagnostic techniques now in use for ultrastructural studies should be applicable. ABC specimens could be embedded in nonepoxy (acrylic) resins. Their superior preservation and access to tissue antigens allows application of a wider variety of diagnostic antibodies than is possible with the epoxy type. This approach and techniques such as cryoultramicro-

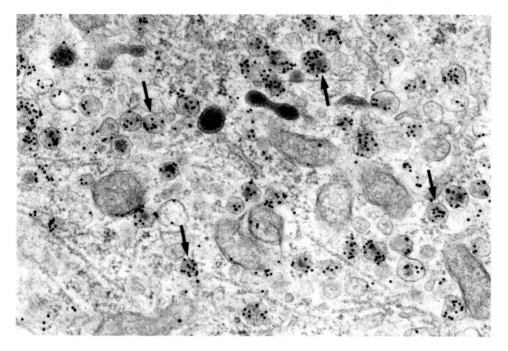

**Fig. 12.2.** Immunoelectron micrograph of an NAB of a carcinoid tumor of the lung. The feasibility of identifying neurosecretory granules is demonstrated using an antichromogranin antibody/protein A-gold technique on epoxy resin–embedded tissue. Colloidal gold particles are well localized to most of the secretory granules (arrows). × 49,500.

tomy could improve the efficacy of immunogold technologies as diagnostic aids. The diagnostic specificities of *in situ* hybridization might also become available to ABC specimens with combinations of such techniques. Because light microscope immunocytochemistry classifies tumors by attempting to detect functional markers, increasing the sensitivity by ultrastructural methods has considerable diagnostic potential in commonly encountered "poorly or undifferentiated" tumors.

Perhaps an editorial written by Goellner[6] entitled "The Maturation of Aspiration Cytology" serves as a fitting conclusion to this chapter. This editorial notes the effects on data interpretation when applying some of the new technologies to cytopathology. What this cautionary editorial emphasizes is that the discipline of cytopathology is far from mature but is, in fact, still in an adolescent state. No negative implication is inherent in such an interpretation; rather, it highlights the immense and still untapped potential for ABC. The diagnostic techniques forming the basis for this book are hopefully one barometer of the utility of cytopathology.

# REFERENCES

1. Castronovo V, Colin C, Claysmith AP, et al: Immunodetection of the metastasis-associated laminin receptor in human breast cancer cells obtained by fine-needle aspiration biopsy. *Am J Pathol* 137:1373–1381, 1990.

2. Crescenzi M, Seto M, Herzig GP, et al: Thermostable DNA polymerase chain amplification of t(14;18) chromosome breakpoints and detection of minimal residual disease. *Proc Natl Acad Sci USA* 85:4869–4873, 1988.
3. Dierick A-M, Praet M, Roels H, et al: Vimentin expression of renal cell carcinoma in relatiaon to DNA content and histological grading: A combined light microscopic, immunocytochemical and cytophotometrical analysis. *Histopathology* 18:315–322, 1991.
4. Domagala W, Lasota J, Dukowicz A, et al: Vimentin expression appears to be associated with poor prognosis in node-negative ductal NOS breast carcinomas. *Am J Pathol* 137:1299–1304, 1990.
5. Eliasen CA, Opitz LM, Vamvakas EC, et al: Flow cytometric analysis of DNA ploidy and S-phase fraction in breast cancer using cells obtained by ex vivo fine-needle aspiration: An optimal method for sample collection. *Mod Pathol* 4:196–200, 1991.
6. Goellner JR: The maturation of aspiration cytology. *Am J Clin Pathol* 94:793–794, 1990.
7. Hall PA, Levison DA, Woods AL, et al: Proliferating cell nuclear antigen (PCNA) immunolocalization in paraffin sections: An index of cell proliferation with evidence of deregulated expression in some neoplasms. *J Pathol* 162:285–294, 1990.
8. Henry JA, McCarthy AL, Angus B, et al: Prognostic significance of the estrogen-regulated protein, cathespin D, in breast cancer; an immunohistochemical study. *Cancer* 65:265–271, 1990.
9. Herman CJ, McGraw TP, Marder RJ, Bauer KD: Recent progress in clinical quantitative cytology. *Arch Pathol Lab Med* 111:505–512, 1987.
10. Koha M, Caspersson TO, Wikström B, Brismar B: Heterogeneity of DNA distribution pattern in colorectal carcinoma; a microspectrophotometric study of fine needle aspirates. *Anal Quant Cytol Histol* 12:348–351, 1990.
11. Kuenen-Boumeester V, Van Der Kwast ThH, Van Laarhoven HAJ, Henzen-Logmans SC: Ki-67 staining in histological subtypes of breast carcinoma and fine needle aspiration smears. *J Clin Pathol* 44:208–210, 1991.
12. Lloyd RV: Use of molecular probes in the study of endocrine diseases. *Hum Pathol* 18:1199–1211, 1987.
13. Lönn U, Lönn S, Nylen U, Winblad G, Stenkvist B: Amplification of oncogenes in mammary carcinoma shown by fine-needle biopsy. *Cancer* 67:1396–1400, 1991.
14. Lubiński J, Chosia M, Katańska K, Huebner K: Genotypic analysis of DNA isolated from fine needle aspiration biopsies. *Anal Quant Cytol Histol* 10:383–390, 1988.
15. Silver MM, Hearn SA: Postembedding immunoelectron microscopy using protein A-gold. *Ultrastruct Pathol* 11:693–703, 1987.
16. Soomro S, Shousha S, Taylor P, Shepard HM, Feldmann M: c-*erb*B-2 expression in different histological types of invasive breast carcinoma. *J Clin Pathol* 44:211–214, 1991.
17. Stetler-Stevenson M, Raffeld M, Cohen P, Cossman J: Detection of occult follicular lymphoma by specific DNA amplification. *Blood* 72:1822–1825, 1988.
18. Weger AR, Glaser KS, Schwab G, et al: Quantitative nuclear DBA content in fine needle aspirates of pancreatic cancer. *Gut* 32:325–328, 1991.
19. Williams ME, Frierson HF Jr, Tabbarah S, Ennis PS: Fine needle aspiration of non-Hodgkin's lymphoma; Southern blot analysis for antigen receptor, *bcl*-2, and *c-myc* gene rearrangements. *Am J Clin Pathol* 93:754–759, 1990.
20. Wright C, Angus B, Nicholson S, et al: Expression of c-*erb*B-2 oncoprotein: A prognostic indicator in human breast cancer. *Cancer Res* 49:2087–2090, 1989.

# Index

**A**
A1-ACT, *see* Alpha-1-antichymotrypsin
A1-AT, *see* Alpha-1-antitrypsin
ABC, *see* Aspiration biopsy cytology (ABC)
*ABC, see* Avidin-biotin-peroxidase complex (*ABC*)
Actin
　as antibody, 54
　microvilli and, 103–105
Actin/myosin complexes, as cytoplasmic filament, 131, 133
Adenocarcinoma(s), *see also specific types of*
　gastrointestinal, primary site and, 243–246
　liver, versus hepatocellular carcinoma, primary site and, 247–257
　versus mesothelioma
　　aspiration biopsy cytology and, 257–260
　　electron microscopy and, 260
　　primary site and, 257–265
　prostatic, primary site and, 240–243
　renal, 274
　versus squamous-cell carcinoma, primary site and, 236–240
Adrenal cortical neoplasm
　needle aspiration biopsy and, 274–281
　versus renal-cell carcinoma, primary site and, 274–281
AEC, *see* Aminoethylcarbazole
AFP, *see* Alpha-fetoprotein
Alcohol-based fixatives, 11
Alcohol-fixed unstained smears, as cytological preparations for immunocytochemistry, 38–39
Alpha-1-antichymotrypsin (A1-ACT), 55
Alpha-1-antitrypsin (A1-AT), 55
Alpha-fetoprotein (AFP), 54
Alpha lactalbumin, 271–272
Aminoethylcarbazole (AEC), 34
Amplification, of oncogenes, 312
Anaplastic carcinomas, as pleomorphic-cell neoplasm, 157
Anemone cell tumors, 109
Angiosarcoma, 194
Antibody control, as method of immunocytochemistry, 35–36
Antibody(ies), 53–76
　to actin, 54
　to alpha-1-antichymotrypsin (A1-ACT), 55
　to alpha-1-antitrypsin (A1-AT), 55
　to alpha-fetoprotein (AFP), 54

　B72.3 as, 55, 57
　to calcitonin, 57–58
　to carcinoembryonic antigen (CEA), 58–59
　to chromogranins (CGs), 59
　to epithelial membrane antigen (EMA), 60
　to estrogen receptor (ER), *see* Breast carcinoma
　to human chorionic gonadotropin (hCG), 60–61
　immunocytochemistry and, 29–31, 43–45
　to intermediate filaments, 61–67
　melanoma-specific antibody (HMB-45) as, 53, 68–69
　monoclonal, 29–31
　　for prognosis, 312–313
　to muscle-specific actin (HHF35), 53, 54
　neoplasms and, 43–45
　to neurofilament proteins (NFP), *see* Intermediate filament(s)
　to neuron-specific enolase (NSE), 69
　NKI/C3 as, 69
　NKK/Bteb as, 69
　to placental alkaline phosphatase (PLAP), 69–70
　polyclonal, 29–31
　to progesterone receptor (PR), *see* Breast, carcinoma
　to prostate-specific antigen (PSA), 70–71
　to prostatic acid phosphatase (PrAP), 70
　selection of, 43–45, 53–54
　to S-100 protein, 71–74
　to synaptophysin (SYN), 74
　to thyroglobulin (TG), 74, 76
　titers of, 31
Architectural features, for diagnosis, in electron microscopy, 90–93
Artifacts, in immunocytochemistry, 46, 49–50
Aspiration biopsy cytology (ABC)
　breast carcinoma and, 269–270
　choice of fixative for, 11–13
　determining primary site of metastatic neoplasm and, 235–236
　diagnosis of neoplasms by, 3
　diagnostic role of, 1
　differential diagnoses by, 43
　electron microscopy and, 89, 143–144
　hepatocellular carcinoma and, 247–252
　in Hodgkin's disease, 217, 229
　immunocytochemistry and, for prognosis, 312
　liver neoplasms and, 247–252
　malignant lymphoma and, 217

317

Aspiration biopsy cytology (ABC) (contd.)
  mesothelioma versus adenocarcinoma and, 257–260
  minibiopsy nature of, 2
  recognition of neuroendocrine lesions in, 118, 120
  role of ancillary techniques on, 3–4
Aspiration biopsy cytology (ABC) specimens
  electron microscopy techniques and, 2, 13–14, 18–26
  immunocytochemistry and, 13–14, 17–18
  molecular genetic analysis of, 295–296
  processing techniques for specimens and, 13–14
Automation systems, as technique for immunocytochemistry, 36–37
Avidin-biotin-peroxidase complex (ABC), 29
  as detection system, for immunocytochemistry, 31–33

## B

B72.3, 55, 57
Background staining, in immunocytochemistry, 33–34, 46
B-cell markers, as lymphoid marker, 219
Benign lesions, in electron microscopy, 135, 137
BerH$_2$, as monoclonal antibody, 219
Blocking methods, in immunocytochemistry, 33–34
Bowel, complications of needle aspiration biopsy in, 293
Breast carcinoma
  aspiration biopsy cytology and, 269–270
  classification of with immunocytochemistry, 270–271
  primary site and, 269–274
  prognosis of
    with estrogen receptor, 271–272
    with progesterone receptor, 271–272
Breast neoplasm, needle aspiration biopsy for, 296–297
Bronchioloalveolar carcinoma, electron microscopy of, 263
"Built-in" controls, in immunocytochemistry, 51–52

## C

Calcitonin, 57–58
Cancer, see Neoplasm(s); specific neoplasms
Carcinoembryonic antigen (CEA), 58–59
Carcinoma(s)
  anaplastic, as pleomorphic-cell neoplasm, 157
  breast
    aspiration biopsy cytology and, 269–270
    classification of with immunocytochemistry and, 270–271
    primary site and, 269–274
    prognosis of
      with estrogen receptor, 271–272
      with progesterone receptor, 271–272
    bronchioloalveolar, electron microscopy of, 263

  embryonal, as germ-cell neoplasms, 201
  gastrointestinal, 243–246
  hepatocellular, aspiration biopsy cytology and, 247–252
  medullary, of thyroid gland, diagnosis of, 266
  papillary, 269, 300
  prostatic, 240–243
  renal-cell, versus adrenal cortical neoplasm, primary site and, 274–281
  sarcomatoid, as spindle-cell neoplasm, 157–159
  small-cell, as small-cell neoplasm, 159
  squamous-cell
    versus adenocarcinoma, primary site and, 236–240
    electron microscopy of, 237–240
    features of, 237
  undifferentiated, as large/polygonal neoplasm, 155–156
Cartilaginous neoplasms, 198–199
Cathepsin D, 312
CEA, see Carcinoembryonic antigen
Cell blocks
  as cytological preparations for immunocytochemistry, 38
  preparation of, 2
  processing of, for light microscopy, 16–17
  sections of, interpretation of, 43
Cell control, in immunocytochemistry, 34–35, 51–52
Cells
  diagnostic features of, in electron microscopy, 93
  malignant, in immunocytochemistry, 45–46
  nonneoplastic, in immunocytochemistry, 45–46
  relationships of, in electron microscopy, 90–93
C-erb B-2 oncogene, 312
CG, see Chromogranin
Chondrosarcoma, 198–199
Chordoma, 199
Chromogens, 34
Chromogranin (CG)
  as antibody, 59
  for neuroendocrine neoplasms, 166
CK, see Cytokeratin
Clear-cell sarcomas, as large/polygonal neoplasm, 155–156
Control(s), in immunocytochemistry, 34–36, 51–52
Cost-effectiveness
  of electron microscopy, 5–7
  of immunocytochemistry, 5–7
Cytokeratin (CK)
  as cytoplasmic filament, 129–131
  as intermediate filament, 62–65
Cytological preparations, for immunocytochemistry, 38–39
Cytopathology, diagnostic role of, 1
Cytoplasm
  apical, 108
  microvillous-like extensions of, 109
Cytoplasmic filaments, microvilli and, 128–129

Cytospin preparations, as cytological preparations for immunocytochemistry, 39

## D

DAB, *see* Diaminobenzidine
Desmin
  as cytoplasmic filament, 131
  as intermediate filament, 65, 67
Desmoplastic mesotheliomas, as spindle-cell neoplasm, 157–159
Desmosomes, intercellular junctions and, 97, 99
Detection systems, as method of immunocytochemistry, 31–33
Diaminobenzidine (DAB), 34
DNA content analysis, potential applications for, 311

## E

Electron microscopy
  adenocarcinoma versus squamous cell carcinoma and 236–240
  advantages of, 89–90
  as ancillary role in aspiration biopsy cytology, 3
  aspiration biopsy cytology and, 89, 143–144
  aspiration biopsy cytology specimens and, 2, 13–14, 20–25
  basal lamina in, 127–128
  benign lesions in, 135, 137
  breast carcinoma and, 269–272, 274
  of bronchioloalveolar carcinoma, 263
  clinician's viewpoint of, 291–308
  cost-effectiveness of, 5–7
  cytoplasmic filaments in, 128
    actin/myosin complexes as, 131
    cytokeratin as, 129
    desmin as, 131
    vimetin as, 131
  developing applications for, 311–314
  for diagnoses, 143–144
  diagnostic architectural features in, 90–93
  diagnostic cellular features in, 93
  gastrointestinal adenocarcinoma and, 243
  germ cell neoplasms as, 200–201, 205
  glycocalyx and glycocalyceal bodies in, 110–111
  glycogen in, 115
  glycogen preservations and, 25–26
  hepatocellular carcinoma versus adenocarcinoma and, 250–257
  of Hodgkin's disease, 230–231
  indications for, 4–5
  intercellular junctions in, 93, 97–101
  intracytoplasmic lumens in, 133–135
  limitations of, 146–151
  lipid in, 120
  of malignant lymphoma, 230–231
  malignant melanoma and, 159
  melanosomes in, 120, 122, 127
  mesenchymal neoplasms and, 178–179, 182
    angiosarcoma as, 195
    cartilaginous neoplasms as, 198–199
    chordoma as, 199
    liposarcoma as, 192–193
    malignant fibrous histiocytoma as, 192
    myogenic sarcoma as, 182–183, 185, 188
    neurogenic neoplasms as, 188, 192
    osteogenic neoplasms as, 195, 198
    synovial sarcoma as, 193
  mesothelioma versus adenocarcinoma and, 257, 259–260, 263–264
  microvilli in, 101, 103–110, 133–135
  microvillous core rootlets in, 105, 108–109
  mucin in, 111, 115
  neuroendocrine neoplasms and, 164, 166–167, 169, 171, 173
  neurosecretory granules in, 115–118, 120
  oncologist's perspective of, 291–292
  potential applications for, 311–314
  practical application of, 5–7
  prostatic adenocarcinoma and, 240, 242–243
  principles of, 89–137
  quantification of, 143–145
  radiologist's perspective of, 301–305
  renal cell carcinoma versus adrenal cortical neoplasm and, 274, 277
  small cell neoplasms and, 173–174
  thyroid neoplasms and, 266, 269
  specimen processing for, 13–14, 18–26
  squamous-cell carcinoma versus adenocarcinoma and, 236–240
  for surgical pathology, 144–145
  tumor cells in, 89
  value of, 143–145
EMA, *see* Epithelial membrane antigen
Embryonal carcinoma, 200–205
Epithelial membrane antigen (EMA), 60
Epithelial tumors, intercellular junctions and, 99, 101
Epithelioid leiomyosarcomas, as large/polygonal neoplasm, 155–156
Epithelioid sarcoma
  as large/polygonal neoplasm, 155–156
  as mesenchymal neoplasm, 179, 200
Estrogen receptor (ER)
  as antibody, *see* Breast carcinoma
  in prognosis of breast carcinoma, 271–272
Ewing's sarcoma, 159, 177–178
Extragonadal germ-cell tumors, 200

## F

Factor VIII-related antigen, 195
Fibrosarcoma, 200
Fibrous histiocytoma, malignant
  as mesenchymal neoplasm, 192
  as pleomorphic-cell neoplasm, 156–157
Filaments, *see* Cytoplasmic filaments; Intermediate filament(s)
Filipodia, 109
Fixation, as processing technique for specimens, 11–13
Fixative(s)
  alcohol-based, 11
  choice of, for aspiration biopsy cytology, 11–13
  formalin, 11–13

Follicular neoplasms, diagnosis of, 266
Formalin fixatives, 11–13

## G

Gastrointestinal adenocarcinoma, 243–246
Gastrointestinal tract adenocarcinoma, microvilli of, 103
GCDFP-15, *see* Gross cystic disease fluid protein
Genotyping, potential applications for, 311
Germ-cell neoplasms, classification of, 200–205
Germ-cell neoplasms, 200–205
    as large/polygonal neoplasm, 155–156
    as pleomorphic-cell neoplasm, 157
GFAP, *see* Glial fibrillary acidic protein
Glial fibrillary acidic protein (GFAP), as intermediate filament, 67
Glycocalyceal bodies, as secretory product, 111
Glycocalyx, as secretory product, 110–111
Glycogen
    preservation of, electron microscopy and, 25–26
    as secretory product, 115
Gross cystic disease fluid protein (GCDFP-15), 270–271

## H

hCG, *see* Human chorionic gonadotropin
Hepatocellular carcinoma
    aspiration biopsy cytology and, 247–252
    electron microscopy of, 250, 252–253
    versus adenocarcinoma and, 247–257
HER2 oncogene, 312
HHF35, *see* Muscle-specific actin
Histiocytoma, malignant fibrous, as pleomorphic-cell neoplasm, 156–157
HMB-45 *see Melanoma-specific antibody*
Hodgkin's disease
    aspiration biopsy cytology of, 229
Human chorionic gonadotropin (hCG), 60–61
Hybridization techniques, *in situ*, potential applications for, 311–312
Hyperplasia, lymphoid, 229–230

## I

If, *see* Intermediate filament(s)
Immunocytochemistry
    antibodies and, 29–31, 43–45
    artifacts in, 46, 49–50
    aspiration biopsy cytology and ancillary role in, 3
    for prognosis, 312
    aspiration biopsy specimens in, 13–14
    automation systems in, 36–37
    background staining in, 46
    clinician's viewpoint of, 291–308
    control in, 51–52
    cost-effectiveness of, 5–7
    cytological preparations for, 38–39
    developing applications for, 311–314
    history of, 29
    immunophenotyping of malignant lymphoma and, 223
    indications for, 4–5
    interpretation of, 43–52
    limitations of, 43–52
    malignant cells in, 45–46
    of malignant lymphoma, 221–230
        versus nonlymphoma neoplasms, 223
        processing of, 221–223
    methods of, 31–38
        blocking methods as, 33–34
        chromogens as, 34
        controls as, 34–36
        detection systems as, 31–33
    nonneoplastic cells in, 45–46
    oncologist's perspective of, 291–292
    patterns of staining in, 45
    potential applications for, 311–314
    practical applications of, 5–7
    principles of, 29–39
    quality assurance in, 37–38
    radiologist's perspective of, 301–305
    specimen processing for, 13–14, 17–18
Immunoelectron microscopy, 313–314
Immunoglobulins, as lymphoid marker, 219
Immunoperoxidase methods, as detection system, for immunocytochemistry, 31
Immunophenotyping
    of malignant lymphoma, 225–230
    of neoplasms, 43–45
*In situ* hybridization (ISH) techniques, potential applications for, 311–312
Intercellular junctions, in electron microscopy, 93, 97–101
Intermediate filament(s)
    antibodies to, 61–67
    cytokeratin (CK) as, 62–65
    desmin as, 65, 67
    glial fibrillary acidic protein (GFAP) as, 67
    neurofilament proteins (NFPs) as, 67
    vimentin as, 65
Internal controls, in immunocytochemistry, 51–52
Intracytoplasmic lumens, microvilli in, 133–135

## J

Junctional complex, intercellular junctions and, 101
Junctions, intercellular, in electron microscopy, 93, 97–101

## K

Keratin, *see* Cytokeratin
Ki-1 antigen, as lymphoid marker, 219
Ki-67, 312

## L

L26, as lymphoid marker, 219
LAB, *see* Peroxidase-labeled avidin-biotin
Lamina, basal, in electron microscopy, 127–128
Large-cell malignant lymphoma, as large/polygonal neoplasm, 155–156
Large-cell neoplasms, as large/polygonal neoplasm, 155–156

LCA, *see* Leukocyte common antigen
Leiomyosarcoma
  epithelioid, as large/polygonal neoplasm, 155–156
  as mesenchymal neoplasm, 179
  as myogenic sarcoma, 183
  as pleomorphic-cell neoplasm, 156–157
Leishman stain, for malignant lymphoma, 221–222
Lesions, *see also specific lesions*
  benign, in electron microscopy, 135, 137
  intrathoracic, needle aspiration biopsy and, 306–308
  superficial, needle aspiration biopsy for, 292
Leu-7, as lymphoid marker, 221
Leukocyte common antigen (LCA), as lymphoid marker, 218–219
Leu-M1, as lymphoid marker, 219–221
Light microscopy
  cell block processing for, 16–17
  needle rinse processing for, 16
  smear processing for, 14–15
  specimen processing for, 14–16
Lipid, as secretory product, 118
Liposarcoma
  as mesenchymal neoplasm, 192–193
  as pleomorphic-cell neoplasm, 156–157
Liver neoplasms, 247–257
  aspiration biopsy cytology and, 247–252
  electron microscopy of, 252–253
  needle aspiration biopsy for, 298–299
Lung, complications of needle aspiration biopsy in, 293
Lung adenocarcinomas, microvilli of, 103
Lung lesions, needle aspiration biopsy for, 297–298
Lung neuroendocrine neoplasms, 167
Lymphoid hyperplasia, 229–230
Lymphoid markers
  B-cell marker as, 219
  immunoglobulins as, 219
  Ki-1 antigen as, 219
  L26 as, 219
  leu-7 as, 221
  leukocyte common antigen (LCA) as, 218–219
  leu-M1 as, 219–221
  in malignant lymphoma, 218–221
  T-cell marker as, 219
  UCHL1 as, 219
Lymphoma
  large-cell malignant, as large/polygonal neoplasm, 155–156
  malignant, 217–231
    aspiration biopsy cytology of, 217–218
    electron microscopy of, 230–231
    immunocytochemistry of, 221–230
    immunophenotyping of, 225–230
    lymphoid markers in, 218–221
    needle aspiration biopsy for, 296
    versus nonlymphoma neoplasms, immunocytochemistry of, 223
    processing of for immunocytochemistry, 221–223
    as small-cell neoplasm, 159, 177
    needle aspiration biopsy of, 217
  Non-Hodgkin's, diagnosis of, 223
Lysosomes, 118

# M

Malignant cells, in immunocytochemistry, 45–46
Malignant fibrous histiocytoma (MFH)
  as mesenchymal neoplasm, 192
  as pleomorphic-cell neoplasm, 156–157
Mediastinal lesions, needle aspiration biopsy for, 297–298
Medullary carcinoma, of thyroid gland, diagnosis of, 266
Melanoma, 122, 127
  malignant
    classification of, 159–161, 163–164
    as large/polygonal neoplasm, 155–156
    as pleomorphic-cell neoplasm, 157
    as small-cell neoplasm, 159
    as spindle-cell neoplasm, 157–159
  soft tissue, as mesenchymal neoplasm, 200
Melanoma-specific antibody (HMB-45), 53, 68–69
Melanosomes, as secretory product, 120, 122, 127
Mesenchymal neoplasms
  angiosarcoma as, 194
  cartilaginous neoplasms as, 198–199
  chordoma as, 199
  classification of, 178–192
  epithelioid sarcomas as, 200
  fibrosarcoma as, 200
  liposarcoma as, 192–193
  malignant fibrous histiocytoma as, 192
  myogenic sarcoma as, 182–183
  neurogenic neoplasms as, 188–192
  osteogenic neoplasm as, 195–198
  soft-tissue melanoma as, 200
  synovial sarcoma as, 193
Mesothelioma
  versus adenocarcinoma
    aspiration biopsy cytology and, 257–260
    electron microscopy and, 260
    primary site and, 257–265
  desmoplastic, as spindle-cell neoplasm, 157–159
  epithelial type, as large/polygonal neoplasm, 155–156
  sarcomatoid, as spindle-cell neoplasm, 157–159
Microscopy, electron, *see* Electron microscopy
Microscopy, light, *see* Light microscopy
Microvilli
  cytoplasmic filaments and, 128–129
  in electron microscopy, 101, 103–110
  intracytoplasmic lumens and, 133–135
  secretory products and, 110–127
Microvillous-like cytoplasmic extensions, 109
Molecular genetic analysis, aspiration biopsy cytology specimens of, 295–296
Monoclonal antibodies, 29–31
Mucin, as secretory product, 111, 115
Muscle-specific actin (HHF35), 53, 54
Myeloma, plasma cell, 225

Myogenic sarcoma, 182–183
Myoglobin, 183
Myosin, microvilli in, 128–129

# N

NAB, see Needle aspiration biopsy (NAB)
Needle aspiration biopsy (NAB), 1
 adrenal cortical neoplasm and, 274–281
 advantages of, 305–306
 for breast neoplasm, 296–297
 clinical uses of, 294–300
 complications of, 293–294
  in bowel, 293
  in lung, 293
  in pancreas, 293
  in renal biopsy, 293
 developing applications for, 311–314
 for intrathoracic lesions, 306–308
 for liver neoplasms, 298–299
 for lung lesions, 297–298
 for malignant lymphoma, 217, 296
 for mediastinal lesions, 297–298
 minibiopsy, nature of, 2
 potential applications for, 311–314
 for prostate neoplasms, 300
 renal neoplasm and, 274–281
 specimens obtained by, 2
 for superficial lesions, 292
 thoracic surgeons and, 305–308
 for thyroid neoplasms, 299–300
 tissue fragments in, 90–91
 ultrasonic techniques and, for diagnosis, 313–314
Needle rinse, processing of, for light microscopy, 15, 16
Negative antibody control, in immunocytochemistry, 51–52
Neolumens, microvilli in, 134
Neoplasm(s)
 adrenal cortical
  needle aspiration biopsy and, 274–281
  versus renal-cell carcinoma, primary site and, 274–281
 anaplastic, as large/polygonal neoplasm, 155–156
 antibodies, and, 43–45
 breast, needle aspiration biopsy for, 296–297
 cartilaginous, 198–199
 classification of, 2, 3, 153–205
 diagnosis of, 2
  by aspiration biopsy cytology, 3
 differential, 153–155
 follicular, diagnosis of, 266
 germ-cell, classification of, 200–205
 immunophenotyping of, 43–45
 large cell, 155–156
 liver, 247–257
  aspiration biopsy cytology and, 247–252
  electron microscopy of, 252–253
  needle aspiration biopsy for, 298–299
 look-alike, classification of, 153–159
 malignant melanoma and, classification of, 159–161, 163–164
 mesenchymal
  angiosarcoma as, 194
  cartilaginous neoplasms as, 198–199
  chordoma as, 199
  classification of, 178–192
  epithelioid sarcomas as, 200
  fibrosarcoma as, 200
  liposarcoma as, 192–193
  malignant fibrous histiocytoma as, 192
  myogenic sarcoma as, 182–183
  neurogenic neoplasms as, 188–192
  osteogenic neoplasm as, 195–198
  soft-tissue melanoma as, 200
  synovial sarcoma as, 193
 metastatic
  aspiration biopsy cytology and, 235–236
  determining primary site of, 235–236
 neuroendocrine
  classification of, 164, 166–173
  as small-cell neoplasm, 159
 neurogenic, 188–192
 nonlymphoma, versus malignant lymphoma, immunocytochemistry of, 223
 osteogenic, 195–198
 pleomorphic-cell, 156–157
 polygonal cell, as large/polygonal neoplasm, 155–156
 poorly differentiated, classification of, 153–159
 small-cell, 159
  in children, 174
  classification of, 173–174, 176–178
  Ewing's sarcoma as, 177–178
  malignant lymphoma as, 177
  neuroblastoma as, 176–177
  primitive neuroectodermal tumor (PNET) as, 179
 spindle-cell, 157–159
 thyroid
  needle aspiration biopsy for, 299–300
  primary site and, 266–269
 undifferentiated, classification of, 153–159
*Neu* Oncogene, 312
Neuroblastoma, 159, 176–177
Neuroendocrine cells, 115–117
Neuroendocrine neoplasms
 classification of, 164, 166–173
 electron microscopy in diagnosis of, 171, 173
 as small-cell neoplasm, 159
Neurofibromas, 188, 192
Neurofilament proteins (NFPs), as intermediate filaments, 67
Neurogenic neoplasms, 188–192
Neuron-specific enolase (NSE), 69
Neurosecretory granules, as secretory product, 115–118, 120
NFP, see Neurofilament proteins
NKI/C3, 69
NKK/Bteb, 69
Non-Hodgkin's lymphoma, 218, 223–229
Nonneoplastic cells, in immunocytochemistry, 45–46
NSE, see Neuron-specific enolase

## O

Oat cell carcinoma, *see* Neuroendocrine neoplasms
Oncogenes, amplification of, 312
Oncologist's perspective
  of electron microscopy, 291–292
  of immunocytochemistry, 291–292
Osteogenic neoplasm, 195–198

## P

Pancreas, complications of needle aspiration biopsy in, 293
PAP, *see* Peroxidase-antiperoxidase (PAP)
Papanicolaou method, rapid, 15
Papillary carcinomas of thyroid, 269, 300
Peroxidase-antiperoxidase (PAP), 29
  as detection system, for immunocytochemistry, 31–33
Peroxidase-labeled avidin-biotin (LAB), as detection system, for immunocytochemistry, 31–33
Pheochromocytoma, 274
Placental alkaline phosphatase (PLAP), 69–70
PLAP, *see* Placental alkaline phosphatase
Plasma-cell tumor, 225
Pleomorphic-cell neoplasms, 156–157
Polyclonal antibodies, 29–31
Polygonal cell neoplasms, 155–156
Polymerase chain reaction (PCR), 312
Polypeptides, in neuroendocrine neoplasms, 167
PrAP, *see* Prostatic acid phosphatase
Primary site, determination of, 235–281
  breast carcinoma and, 269–274
  gastrointestinal adenocarcinoma and, 243–246
  hepatocellular carcinoma versus adenocarcinoma and, 247–257
  mesothelioma versus adenocarcinoma and, 257–265
  prostatic adenocarcinoma and, 240–243
  renal-cell carcinoma versus adrenal cortical neoplasm, 274–281
  squamous-cell carcinoma versus adenocarcinoma and, 236–240
  thyroid neoplasm and, 266–269
Primitive neuroectodermal tumor (PNET), as small-cell neoplasm, 178
Progesterone receptor (PR) *see also* Breast carcinoma
  prognosis of breast carcinoma with, 271–272
Proliferating cell nuclear antigen, 312–313
Prostate neoplasms, needle aspiration biopsy for, 300
Prostate-specific antigen (PSA), 70–71, 240–243
Prostatic acid phosphatase (PrAP), 70, 240–243
Prostatic adenocarcinoma, primary site, and, 240–243
PR (progesterone receptor), *see also* Breast carcinoma
  prognosis of breast carcinoma with, 271–272
PSA, *see* Prostate-specific antigen
Pulmonary adenocarcinomas, microvilli of, 105–106, 108

## Q

Quality assurance, in immunocytochemistry, 37–38

## R

Radiologist's perspective
  of electron microscopy, 301–305
  of immunocytochemistry, 301–305
Rapid Papanicolaou method, 15
Renal biopsy, complications of needle aspiration biopsy in, 293
Renal-cell carcinoma, versus adrenal cortical neoplasm, primary site and, 274–281
Renal neoplasm, needle aspiration biopsy and, 274–281
Rhabdomyosarcoma
  embryonal, as myogenic sarcoma, 182–183
  as pleomorphic-cell neoplasm, 156–157
  as small-cell neoplasm, 159, 177

## S

Sarcoma
  epithelioid, 179, 200
  Ewing's, as small-cell neoplasm, 159, 177–178
  as large/polygonal neoplasm, 155–156
  myogenic, 182–183
  as pleomorphic-cell neoplasm, 156–157
  soft-tissue, as mesenchymal neoplasm, 178–182
  as spindle-cell neoplasm, 157–159
  synovial, 193
Sarcomatoid carcinomas, as spindle-cell neoplasm, 157–159
Sarcomatoid mesotheliomas, as spindle-cell neoplasm, 157–159
Schwannoma, 188, 192
Secretory component, 274–275
Secretory products, microvilli and, 110–127
Seminoma, 200–201, 203
Site, primary, *see* Primary site
Small-cell neoplasms, 159
  in children, 174
  classification of, 173–174, 176–178
  Ewing's sarcoma as, 177–178
  malignant lymphoma as, 177
  neuroblastoma as, 176–177
  primitive neuroectodermal tumor (PNET) as, 178
  rhabdomyosarcoma as, 177
Smears
  as cytological preparations for immunocytochemistry, 38–39
  interpretation of, 43
  multiple, preparation of from single smear, 39
  processing of
    for light microscopy, 14–15
    rapid Papanicolaou method for, 15
Soft-tissue melanoma, 200
Soft-tissue sarcomas, 178–182
Soft-tissue tumors, intercellular junctions and, 101

Specimens
  processing of
    for electron microscopy, 18–26
    for immunocytochemistry, 17–18
  processing techniques for, 11–26
    aspiration biopsy specimens and, 13–14
    fixation and, 11–13
Spindle-cell neoplasms, 157–159
S-100 protein, 71–74
Squamous-cell carcinoma
  versus adenocarcinoma, primary site and, 236–240
  electron microscopy of, 237–240
Staining
  background, in immunocytochemistry, 33–34, 46
  patterns of, in immunocytochemistry, 45
Synaptophysin (SYN), 74
  for neuroendocrine neoplasms, 166–167
Synovial sarcoma, 193

# T
T-cell markers, as lymphoid marker, 219
Terminal bar, intercellular junctions and, 101
Thoracic surgeons, needle aspiration biopsy and, 305–308
Thyroglobulin (TG), 74, 76
  as marker for specifying primary site, 236, 266, 269
Thyroid neoplasm
  needle aspiration biopsy for, 299–300
  primary site and, 266–269
Tissue control, in immunocytochemistry, 34–35, 51–52
Tissue fragments, in needle aspiration biopsy, 90–91
Titers, of antibodies, 31
Tonofilaments, as cytoplasmic filament, 129
Tumor cells, in electron microscopy, 89

# U
UCHL1, as lymphoid marker, 219
Ulex europaens agglutinin-1, (UEA-1), 195, 248–250
Ultrasonic techniques, needle aspiration biopsy and, for diagnosis, 313–314
Undifferentiated neoplasms, 153–159

# V
Vascular antigens
  factor VIII-related antigen, 195
  Ulex europaens agglutinin-1 (UEA-1), 195, 248–250
Vimentin, 65, 131

# W
Weibel-Palade bodies, 195
Wilms' tumor, 174, 176

# Y
Yolk sac tumor, 200–201

# Z
Z-bands, 131–133